Ill-Posed Problems for Integrodifferential Equations in Mechanics and Electromagnetic Theory

SIAM Studies in Applied Mathematics

WILLIAM F. AMES, Managing Editor

This series of monographs focuses on mathematics and its applications to problems of current concern to industry, government, and society. These monographs will be of interest to applied mathematicians, numerical analysts, statisticians, engineers, and scientists who have an active need to learn useful methodology for problem solving.

The first two titles in this series are *Lie-Bäcklund Transformations in Applications*, by Robert L. Anderson and Nail H. Ibragimov, and *Methods and Applications of Interval Analysis*, by Ramon E. Moore.

Frederick Bloom

Ill-Posed Problems for Integrodifferential Equations in Mechanics and Electromagnetic Theory

siam *Philadelphia/1981*

Library of Congress Catalog Card Number: 80-53713
ISBN: 0-89871-171-1

for
Leah and Amir
for just being there

Contents

Acknowledgment

The author would like to avail himself of this opportunity to thank the editors of the series SIAM Studies in Applied Mathematics for the invitation to prepare this volume. Much of the work presented here, which relates to recent work of this author on ill-posed problems for integrodifferential equations in general and to ill-posed problems for material dielectrics with memory in particular, was carried out under the auspices of AFOSR grant 77-3396; the author wishes to thank the Air Force for its continued support of my work in this area. This is also the proper place to thank Pamela Loomis for the excellent job she did in preparing the typewritten version of the text. Finally, a debt of gratitude is due to the many colleagues who have encouraged me during the years when most of the research presented here was being pursued and to my family, who put up with my varying moods during the course of the writing of this work and whose support has been invaluable.

Introduction

In the last decade, considerable interest has been generated in the study of classes of ill-posed problems for partial differential equations and integro-differential equations, with particular emphasis on the respective improperly posed initial-boundary value and initial-history boundary value problems; many of the problems that have been studied are intimately connected with some real world problem in mathematical physics. According to the classical definition of Hadamard, problems of the type indicated above are said to be well-posed if there exists a unique solution, in some function space, which depends continuously on the given data. Two well-known problems which are not well-posed are the Dirichlet problem for the wave equation and the Cauchy problem for the Laplace equation; in fact, for the latter problem, Hadamard showed that a global smooth solution cannot exist unless the initial data satisfy certain compatibility conditions.

A fundamental concept connected with establishing uniqueness, stability, and continuous data dependence for solutions to ill-posed Cauchy problems for partial differential equations evolved in the 1950's in the work of F. John [69], Pucci [131], and Lavrentiev [84], [85]; this idea, extensions of which will be employed in this treatise for ill-posed initial-history boundary value problems associated with systems of partial-integrodifferential equations, involves restricting the set of admissable solutions to an appropriate class of uniformly bounded functions. A differential inequality argument (typically, logarithmic convexity) then yields stabilization of solutions for the ill-posed problem under consideration; in this connection we note, in particular, the work in the 1960's of S. Agmon [2] and Agmon and Nirenberg [3] on the stabilization of solutions to ill-posed initial-value problems for abstract equations in Hilbert space. Besides the logarithmic convexity argument, which depends for its success on the a priori restriction of solutions to classes of uniformly bounded functions, various other techniques for treating ill-posed Cauchy problems for linear and nonlinear PDE were developed in the 1960's; among the most important of these were the weighted energy methods of Murray and Protter [117], [118],

Lagrange identity arguments [30], quasireversibility methods [83], [51], [57], [116], [135], and eigenfunction methods. An introduction to all of the various techniques indicated above, including logarithmic convexity, may be found in the recent monograph by Payne [126] and in the survey articles by Knops [72] and Payne [124], [125]; these articles contain extensive bibliographies as well as a survey of some important applications to ill-posed Cauchy problems for partial differential equations arising in various physical disciplines, e.g., linear elastodynamics, fluid dynamics, and heat conduction.

Besides the work alluded to above on problems of uniqueness, stability, and continuous data dependence, for ill-posed initial boundary value problems associated with partial differential equations, much effort has been expended in the last decade on the problem of proving nonexistence of global smooth solutions to classes of initial-boundary value problems associated with a wide variety of nonlinear partial differential equations; among the earlier work in this area we may cite the papers of Kaplan [71], Fujita [54], Tsutsumi [144], Glassey [55], Friedman [53], and F. John [70], there being very little uniformity of technique among the approaches taken in these various expositions. However, in the mid to late 1960's Levine and Payne introduced the concept of the so-called concavity argument, a differential inequality argument which is, at least conceptually, closely related to logarithmic convexity, but does not require the a priori restriction of solutions to uniformly bounded classes of functions which is inherent in the latter technique; employing concavity arguments, Levine [94]–[97], Levine and Payne [99]–[102], and Knops, Levine, and Payne [73] have been able to establish global nonexistence of smooth solutions to a variety of nonlinear initial-boundary value problems associated with partial differential equations. The more delicate problem of establishing global nonexistence of solutions via finite-time blow-up, and its relationship to the global nonexistence results alluded to above, has recently been considered by J. Ball [4], [5]. With the exception of the work of Brun [30], which employs a Lagrange-identity argument, and the work of Beevers [9], [10], which makes use of weighted energy arguments, all of the techniques we have referenced, including those based on differential inequality arguments, have been employed in connection with either ill-posed initial-boundary value problems for partial differential equations or ill-posed initial-value problems for operator equations in Hilbert space.

We consider, in this treatise, the problem of establishing results on uniqueness, stability, and continuous data dependence for solutions to ill-posed initial-history boundary value problems associated with systems of partial-integrodifferential equations and for solutions of the related ill-posed initial-history value problems associated with integrodifferential equations in Hilbert space;[1] for such problems we will also establish certain growth estimates as well

[1] For the most part, our attention will be confined to problems which give rise to hyperbolic situations, i.e., to integrodifferential equations (or systems of such equations) which involve second

as lower bounds for solutions as the time parameter $t \to +\infty$. In most cases, the problem under consideration will arise from a specific initial-history boundary value problem in either linear or nonlinear viscoelasticity or in a theory of rigid nonconducting dielectric materials with memory. The specific initial-history boundary value problems we will consider in this work will, in general, be ill-posed due to a lack of sufficiently strong definiteness assumptions on the operators involved; in most of the applications, such definiteness assumptions as would be required to secure the existence of a unique, stable solution would impose overly harsh restrictions on the constitutive relations which define the material under consideration. One key feature of the present work is that, with certain noted exceptions, we will employ just two basic techniques, namely, variants of the logarithmic convexity and concavity arguments which have been referenced above;[2] as with the application of such differential inequality arguments to ill-posed problems for PDE we will always assume the existence of a solution in some appropriate function space. We now offer a brief survey of the contents of the Monograph.

In the first chapter we explain the use of the logarithmic convexity and concavity arguments through applications to problems in linear elastodynamics and nonlinear heat conduction; no attempt is made here to give a comprehensive survey of the application of these two differential inequality arguments to ill-posed initial-boundary value problems for PDE, but it has been deemed desirable to introduce the basic concepts within the context of applications to problems for partial differential rather than integrodifferential equations. As indicated above, a comprehensive survey of techniques which have been used successfully to deal with ill-posed problems for PDE may be found in the text by Payne [126] which could be considered as companion reading for the present work.

In Chapter II we discuss ill-posed initial-history boundary value problems associated with the partial-integrodifferential equations of linear and nonlinear viscoelasticity. We begin by presenting the equations of motion for isothermal linear viscoelasticity and proceed with a discussion of both suggested and proven restrictions on the associated viscoelastic relaxation tensors. The initial-history boundary value problem is then reformulated as an initial-history value problem for an integrodifferential equation in Hilbert space and we proceed to survey the basic results in the literature connected with the

---

order time derivatives of the unknown function (or functions). Some recent work, however, on first order integrodifferential equations, particularly those of the type that arise in connection with the problem of heat flow in a nonlinear material with memory, will be considered in Chapter IV.

[2] Exceptions to this rule include, among others, the use of energy estimates in Chapter II in connection with the discussion of the problem of proving global existence, uniqueness, and asymptotic stability of solutions in nonlinear viscoelasticity and the extensive use of Riemann Invariant arguments in Chapter IV, in connection with the discussion of the breakdown of smooth solutions to initial-history boundary value problems for nonlinear viscoelastic fluids, as well as other applications of the same general Riemann Invariant argument.

existence, uniqueness, stability, and asymptotic stability of solutions to such problems when sufficient definiteness assumptions are imposed on the relevant operators in the equation so as to insure well-posedness. Two different logarithmic convexity arguments are then presented and are used, respectively, to establish growth estimates and continuous dependence theorems for initial-history boundary value problems in linear viscoelasticity under minimal assumptions about the viscoelastic relaxation tensor. We also present a more recent application of the concavity argument to the derivation of growth estimates for solutions to initial-history boundary value problems associated with a one-dimensional problem in nonlinear viscoelasticity.

In the third chapter we consider ill-posed initial-history boundary value problems for some partial-integrodifferential equations associated with theories of rigid nonconducting material dielectrics; the two materials singled out for analysis are the Maxwell-Hopkinson dielectric and a special case of the so-called isotropic holohedral dielectric. Logarithmic convexity and concavity arguments are used to establish stability and growth estimates for solutions to initial-history boundary value problems associated with each of these two respective theories of dielectric response; we also establish, by means of a mixed logarithmic convexity-concavity argument, asymptotic lower bounds for electric displacement fields in a holohedral isotropic dielectric when the initial energy is negative and sufficiently large in magnitude. Some of our growth estimates are used to obtain bounds for constitutive constants which appear in the memory functions of Maxwell-Hopkinson materials.

The final chapter surveys some recent research results and directions for ill-posed initial-history boundary value problems for integrodifferential equations with an emphasis on nonlinear first-order equations.

Throughout the text we will employ the standard summation convention and will use, interchangeably, the $\partial/\partial t\,(\cdot)$ and $(\cdot)'_t$ notation for differentiation. In some places we have duplication of notation; e.g., $\mathscr{K}(\cdot)$ is the indefinite integral of the strong operator norm of a certain linear operator in Chapter III, but it is also used frequently in the text to denote the kinetic energy associated with the solution of an initial-history boundary value problem; the context will always be clear from the particular way in which the symbol is introduced into a theorem or a computation. Finally, we assume throughout an elementary knowledge of Hilbert space theory (e.g., the Schwarz inequality), of basic formulae for linear PDE theory (e.g., the Green's formulas) and of the elementary properties of convex and concave real-valued functions of a single variable. Appropriate background material on Sobolev spaces and their duals is introduced in Chapter II; only the most elementary facts about Sobolev spaces and linear operators from a Sobolev space into its dual space are used in the text.

Chapter I

Ill-Posed Initial-Boundary Value Problems in Mathematical Physics:

Examples of the Basic Logarithmic Convexity and Concavity Arguments for PDE

As indicated in the introduction, the results to be presented in Chapters II and III will depend on the implementation of just two basic differential inequality type arguments—logarithmic convexity and concavity; these arguments have been used with great success during the past two decades to derive results relating to uniqueness, stability, continuous data dependence, and instability of solutions to initial-boundary value problems for both linear and nonlinear partial differential equations. The presentation in Chapters II and III will indicate the usefulness of logarithmic convexity and concavity arguments for deriving analogous qualitative results for solutions to initial-history boundary value problems for integrodifferential equations of the types which arise in linear and nonlinear theories of viscoelastic response as well as in various theories of rigid nonconducting material dielectrics with memory. Our aim in this chapter is to illustrate the application of these two basic differential inequality type arguments by employing two examples that have appeared in the recent literature on ill-posed problems and which involve two elementary but important problems in mathematical physics whose governing evolution equations are well-known linear partial differential equations; the first problem has associated with it linear boundary data and is amenable to a logarithmic convexity argument while the second problem has associated nonlinear boundary data and will be treated via a concavity argument. We emphasize that our purpose in presenting these two problems, and the generalizations thereof which will be indicated, is merely to illustrate, in a simple manner, the nature of the arguments to be presented in the applications to initial-history boundary value problems in the next two chapters; detailed expositions on the application of both logarithmic convexity and concavity arguments to ill-posed problems for partial differential equations are already available and, in particular, we mention again [72], [124], [125], and [126].

1. Stability and uniqueness theorems in linear elasticity: A first example of logarithmic convexity.

As a first illustration of the application of a logarithmic convexity type argument we will present an argument of Knops and Payne [76] which shows that equilibrium solutions of the equations of linear elasticity are always stable provided that the perturbations lie in a suitable class of uniformly bounded functions. No restrictions are imposed on the elasticity tensor of the material other than the usual symmetry condition; it is well known that without certain definiteness assumptions on the elasticity tensor the associated initial-boundary value problems will be ill-posed; in fact, a nonpositive definite strain-energy alone will be sufficient to produce asymptotic instability. Restriction of the perturbations to lie in a special class of uniformly bounded functions thus stabilizes solutions of the initial-boundary value problem and this idea of so restricting the admissible class of solutions is central to the logarithmic convexity argument and goes back to F. John [69]. For a positive definite strain energy function the uniqueness, stability (and existence) results are well known in linear elasticity and the reader may consult [59] or [74]. In the next section we present a slightly different logarithmic convexity argument in connection with an abstract linear equation in Hilbert space and indicate how the result can be applied to obtain a variety of continuous data dependence theorems in linear elastodynamics.

We consider a linear elastic solid occupying a regular bounded region $\Omega \subseteq R^3$ with $\partial\Omega$ smooth enough to admit applications of the divergence theorem. As the evolution equations involved are linear it is clear that we need only consider the stability of the null solution of the homogeneous equations of linear elastodynamics, i.e.,

$$\frac{\partial}{\partial x_j}\left(C_{ijkl}(\mathbf{x})\frac{\partial u_k}{\partial x_l}\right)=\rho(\mathbf{x})\frac{\partial^2 u_i}{\partial t^2}, \qquad \Omega\times[0,T), \tag{I.1}$$

subject to homogeneous boundary data

$$\begin{aligned} u_i&=0, &&\partial\bar{\Omega}_1\times[0,T),\\ C_{ijkl}\frac{\partial u_k}{\partial x_l}n_j&=0, &&\partial\Omega_2\times[0,T)\end{aligned} \tag{I.2}$$

and initial data of the form

$$u_i(\mathbf{x},0)=f_i(\mathbf{x}), \qquad \frac{\partial u_i}{\partial t}(\mathbf{x},0)=g_i(\mathbf{x}) \quad \text{on } \bar{\Omega}. \tag{I.3}$$

In (I.1)–(I.3) the elasticity tensor $C_{ijkl}(\mathbf{x})$, $\mathbf{x}\in\Omega$, is restricted only to satisfy the usual symmetry condition, i.e., $C_{ijkl}=C_{klij}$ on Ω; $\rho(\mathbf{x})>0$ on Ω (positive density), $T>0$, $\partial\Omega_1\cap\partial\Omega_2=\varnothing$ with $\partial\Omega_1\cup\partial\Omega_2=\partial\Omega$, and the data $f_i(\mathbf{x})$, $g_i(\mathbf{x})$, $i=1,2,3$ are assumed to be smooth. The u_i, $i=1,2,3$ are, of course, the components of the elastic displacement vector. As we do not make the assumption of positive definite elasticities, the above initial-boundary value problem is, in general, ill-posed. In fact (Knops and Payne [76]), if $\exists c_0 > 0$ such that

$$-\int_\Omega C_{ijkl}\xi_{ij}\xi_{kl}\,d\mathbf{x} \geqq c_0 \int_\Omega \xi_{ij}\xi_{ij}\,d\mathbf{x},$$

for all $\boldsymbol{\xi} \in \mathcal{L}(\mathbb{R}^3, R^3)$, it can be shown that

$$\int_{\Omega(t)} \rho u_i u_i\,d\mathbf{x} \geqq \frac{2}{\kappa}\sinh \kappa t \int_{\Omega(0)} \rho u_i \frac{\partial u_i}{\partial t}\,d\mathbf{x} + \cosh \kappa t \int_{\Omega(0)} \rho u_i u_i\,d\mathbf{x},$$

where $\kappa = c_0\lambda_1/\rho$ max, λ_1 being the smallest eigenvalue problem of the corresponding membrane problem. Thus the weighted L_2-norm $\int_\Omega \rho u_i u_i\,d\mathbf{x}$ of the solution (assuming that a classical solution exists) would grow exponentially in t if $\int_{\Omega(0)} \rho u_i(\partial u_i/\partial t)\,d\mathbf{x} > 0$, clearly implying instability of the null solution in the indicated norm; i.e., if

$$\int_{\Omega(0)} \rho u_i u_i\,d\mathbf{x} + \int_{\Omega(0)} \rho u_i \frac{\partial u_i}{\partial t} d\mathbf{x} < \delta$$

for any small $\delta > 0$ it will always be possible to choose t so large that $\int_{\Omega(t)} \rho u_i u_i\,d\mathbf{x} > \varepsilon$ for any $\varepsilon > 0$. Even if the strain energy is indefinite, it is possible to show [76] that for data satisfying $u_i(\mathbf{x}, 0) = 0$ and $\int_{\Omega(0)} C_{ijkl}(\partial^2 u_i/\partial x_j\,\partial t)(\partial^2 u_k/\partial x_l\,\partial t)\,d\mathbf{x} \leqq 0$ the weighted L_2-norm of $\mathbf{u}$ grows parabolically in time, i.e.,

$$\int_{\Omega(t)} \rho u_i u_i\,d\mathbf{x} \geqq t^2 \int_{\Omega(0)} \rho \frac{\partial u_i}{\partial t}\frac{\partial u_i}{\partial t}\,d\mathbf{x}.$$

In view of the instability results cited above, it is clear that some device is needed to stabilize solutions to the above initial-boundary value problem which, without the assumption of positive-definite elasticities, would be expected to be unstable. We begin with the following

DEFINITION I.1. The null solution is stable under perturbations u_i satisfying (I.1)–(I.3) if for any $\varepsilon > 0$, $\exists \delta_\varepsilon > 0$ such that

$$\left[\int_{\Omega(0)} \rho u_i u_i\,d\mathbf{x} + \mathcal{D}\right] < \delta_\varepsilon \Rightarrow \sup_{0 \leqq t < T}\left[\int_{\Omega(t)} u_i u_i\,d\mathbf{x} + \mathcal{D}\right] < \varepsilon,$$

where $\mathcal{D} = 2\max[\mathcal{E}(0), 0]$, $\mathcal{E}(0)$ being the initial energy, i.e.,

$$\mathcal{E}(0) = \frac{1}{2}\int_{\Omega(0)} \left(\rho \frac{\partial u_i}{\partial t}\frac{\partial u_i}{\partial t} + C_{ijkl}\frac{\partial u_i}{\partial x_j}\frac{\partial u_k}{\partial x_1}\right) d\mathbf{x}.$$

We now restrict the perturbations $\mathbf{u}$ to lie in the class of bounded functions defined by

$$M_1=\left\{\mathbf{u}(\cdot\,,t)\in C^2(\bar{\Omega})\,\middle|\,\int_{\Omega(t)}\rho u_i u_i\,d\mathbf{x}\leqq M^2\right\}$$

for some $M\neq 0$; we claim that solutions u_i of (I.1)–(I.3) which lie in M_1 are stable in the sense of the above definition, and in order to demonstrate this fact we employ the following version of the logarithmic convexity argument: Let $F(t)=\int_{\Omega(t)}\rho u_i u_i\,d\mathbf{x}$ and set $G(t)=\ln\,[F(t)+\mathscr{D}]+t^2$; we will show that $G(t)$ is a *convex function* on $[0,\,T]$. A series of straightforward computations using I.1, I.2 and the divergence theorem yields the following:

$$\frac{dF}{dt}=2\int_{\Omega(t)}\rho u_i\frac{\partial u_i}{\partial t}\,d\mathbf{x} \tag{I.4}$$

$$\begin{aligned}\frac{d^2F}{dt^2}&=2\int_{\Omega(t)}\left(\rho\frac{\partial u_i}{\partial t}\frac{\partial u_i}{\partial t}+\rho u_i\frac{\partial^2 u_i}{\partial t^2}\right)d\mathbf{x}\\
&=2\int_{\Omega(t)}\left(\rho\frac{\partial u_i}{\partial t}\frac{\partial u_i}{\partial t}+u_i\frac{\partial}{\partial x_j}\left(C_{ijkl}\frac{\partial u_k}{\partial x_l}\right)\right)d\mathbf{x}\\
&=2\int_{\Omega(t)}\rho\frac{\partial u_i}{\partial t}\frac{\partial u_i}{\partial t}\,d\mathbf{x}\\
&\quad+2\int_{\Omega(t)}\frac{\partial}{\partial x_j}\left(u_iC_{ijkl}\frac{\partial u_k}{\partial x_l}\right)d\mathbf{x}\\
&\quad-2\int_{\Omega(t)}C_{ijkl}\frac{\partial u_i}{\partial x_j}\frac{\partial u_k}{\partial x_l}\,d\mathbf{x}\\
&=2\int_{\Omega(t)}\left(\rho\frac{\partial u_i}{\partial t}\frac{\partial u_i}{\partial t}-C_{ijkl}\frac{\partial u_i}{\partial x_j}\frac{\partial u_k}{\partial x_l}\right)d\mathbf{x}.\end{aligned} \tag{I.5}$$

If we now define the total energy $\mathscr{E}(t)$ by

$$\mathscr{E}(t)=\frac{1}{2}\int_{\Omega(t)}\left(\rho\frac{\partial u_i}{\partial t}\frac{\partial u_i}{\partial t}+C_{ijkl}\frac{\partial u_i}{\partial x_j}\frac{\partial u_k}{\partial x_l}\right)d\mathbf{x},$$

then a simple computation using the symmetry of the elasticity tensor and the boundary conditions (I.2) shows that $\dot{\mathscr{E}}(t)=0$ so that $\mathscr{E}(t)=\mathscr{E}(0)$ (conservation of energy); note that this result is independent of any definiteness assumptions on the elasticities. We now consider the expression

$$\begin{aligned}(F+\mathscr{D})\frac{d^2F}{dt^2}-\left(\frac{dF}{dt}\right)^2&=2\left(\int_{\Omega(t)}\rho u_iu_i\,d\mathbf{x}+\mathscr{D}\right)\left(\int_{\Omega}\left(\rho\frac{\partial u_i}{\partial t}\frac{\partial u_i}{\partial t}-C_{ijkl}\frac{\partial u_i}{\partial x_j}\frac{\partial u_k}{\partial x_l}\right)d\mathbf{x}\right)\\
&\quad-4\left(\int_{\Omega(t)}\rho u_i\frac{\partial u_i}{\partial t}\,d\mathbf{x}\right)^2\end{aligned} \tag{I.6}$$

by (I.4), (I.5), and the definition of $F(t)$. Rewriting this expression we have the following series of equations:

$$
\begin{aligned}
(F+\mathcal{D})\frac{d^2F}{dt^2}-\left(\frac{dF}{dt}\right)^2 &= 2\left(\int_{\Omega(t)}\rho u_iu_i\,d\mathbf{x}+\mathcal{D}\right)\left(\int_{\Omega(t)}\rho\frac{\partial u_i}{\partial t}\frac{\partial u_i}{\partial t}\,d\mathbf{x}\right)\\
&\quad -2(F(t)+\mathcal{D})\int_{\Omega}C_{ijkl}\frac{\partial u_i}{\partial x_j}\frac{\partial u_k}{\partial x_l}\,d\mathbf{x}-4\left(\int_{\Omega(t)}u_i\frac{\partial u_i}{\partial t}\,d\mathbf{x}\right)^2\\
&= 2\left(\int_{\Omega(t)}\rho u_iu_i\,d\mathbf{x}\right)\left(\int_{\Omega(t)}\rho\frac{\partial u_i}{\partial t}\frac{\partial u_i}{\partial t}\,d\mathbf{x}\right)\\
&\quad +2\mathcal{D}\left(\int_{\Omega(t)}\rho\frac{\partial u_i}{\partial t}\frac{\partial u_i}{\partial t}\right)d\mathbf{x}-4\left(\int_{\Omega(t)}\rho u_i\frac{\partial u_i}{\partial t}\,d\mathbf{x}\right)^2\\
&\quad -2(F+\mathcal{D})\int_{\Omega(t)}C_{ijkl}\frac{\partial u_i}{\partial x_j}\frac{\partial u_k}{\partial x_l}\,d\mathbf{x}\\
&= 4\left[\left(\int_{\Omega(t)}\rho u_iu_i\,d\mathbf{x}\right)\left(\int_{\Omega(t)}\rho\frac{\partial u_i}{\partial t}\frac{\partial u_i}{\partial t}\,d\mathbf{x}\right)\right.\\
&\qquad\left.-\left(\int_{\Omega(t)}\rho u_i\frac{\partial u_i}{\partial t}\,d\mathbf{x}\right)^2\right]\\
&\quad -2\left(\int_{\Omega(t)}\rho u_iu_i\,d\mathbf{x}\right)\left(\int_{\Omega(t)}\rho\frac{\partial u_i}{\partial t}\frac{\partial u_i}{\partial t}\,d\mathbf{x}\right)\\
&\quad +4\mathcal{D}\int_{\Omega(t)}\rho\frac{\partial u_i}{\partial t}\frac{\partial u_i}{\partial t}\,d\mathbf{x}-2\mathcal{D}\int_{\Omega(t)}\rho\frac{\partial u_i}{\partial t}\frac{\partial u_i}{\partial t}\,d\mathbf{x}\\
&\quad -2(F+\mathcal{D})\int_{\Omega(t)}C_{ijkl}\frac{\partial u_i}{\partial x_j}\frac{\partial u_k}{\partial x_l}\,d\mathbf{x}.
\end{aligned}
\tag{I.7}
$$

Now the expression in the square bracket above is nonnegative (Cauchy-Schwarz inequality) and we may further decrease the value of the remaining expression by subtracting the additional nonnegative term $-2F\int_{\Omega(t)}\rho(\partial u_i/\partial t)(\partial u_i/\partial t)\,d\mathbf{x}$. Thus,

$$
\begin{aligned}
(F+\mathcal{D})\frac{d^2F}{dt^2}-\left(\frac{dF}{dt}\right)^2 &\geqq 4\mathcal{D}\int_{\Omega(t)}\rho\frac{\partial u_i}{\partial t}\frac{\partial u_i}{\partial t}\,d\mathbf{x}\\
&\quad -2(F+\mathcal{D})\int_{\Omega(t)}\frac{\partial u_i}{\partial t}\frac{\partial u_i}{\partial t}\,d\mathbf{x}\\
&\quad -2(F+\mathcal{D})\int_{\Omega(t)}C_{ijkl}\frac{\partial u_i}{\partial x_j}\frac{\partial u_k}{\partial x_l}\,d\mathbf{x}\\
&= 4\mathcal{D}\int_{\Omega(t)}\frac{\partial u_i}{\partial t}\frac{\partial u_i}{\partial t}\,d\mathbf{x}-4(F+\mathcal{D})\mathcal{E}(t)\\
&\geqq -4(F+\mathcal{D})\mathcal{E}(0)
\end{aligned}
\tag{I.8}
$$

as $\mathcal{D} \geqq 0$ (by definition) and $\mathcal{E}(t) = \mathcal{E}(0)$ for all t. Therefore,

$$
\begin{aligned}
(F+\mathcal{D})\frac{d^2F}{dt^2} - \left(\frac{dF}{dt}\right)^2 &\geqq -4(F+\mathcal{D})\mathcal{E}(0) \\
&\geqq -2\mathcal{D}(F+\mathcal{D}) \qquad \text{(I.9)} \\
&\geqq -2(F+\mathcal{D})^2
\end{aligned}
$$

as $\mathcal{D} \geqq 2\mathcal{E}(0)$ (by definition). Thus

$$(F+\mathcal{D})\frac{d^2F}{dt^2} - \left(\frac{dF}{dt}\right)^2 \geqq -2(F+\mathcal{D})^2. \tag{I.10}$$

However, by the definition of $G(t)$ it is clear that

$$(F+\mathcal{D})^2\frac{d^2G}{dt^2} = (F+\mathcal{D})\frac{d^2F}{dt^2} - \left(\frac{dF}{dt}\right)^2 + 2(F+\mathcal{D})^2,$$

so for any $T>0$,

$$\frac{d^2G}{dt^2} \geqq 0, \qquad 0 \leqq t \leqq T, \tag{I.11}$$

i.e., $G(\cdot)$ is a convex function; it then follows immediately that

$$G(t) \leqq \left(\frac{t}{T}\right)G(T) + \left(1-\frac{t}{T}\right)G(0), \qquad 0 \leqq t \leqq T$$

or, as $G(t) = \ln[F(t)+\mathcal{D}] + t^2$, that

$$F(t)+\mathcal{D} \leqq e^{t(T-t)}[F(T)+\mathcal{D}]^{t/T}[F(0)+\mathcal{D}]^{1-t/T} \quad \text{for } 0 \leqq t \leqq T. \tag{I.12}$$

Clearly (I.12) is only meaningful for $t<T$; as we are assuming the existence of a classical solution $\mathbf{u}(\cdot, t) \in \mathcal{M}_1$, $F(T)$ is bounded as $\int_{\Omega(0)} \rho u_i u_i \, d\mathbf{x} + \mathcal{D} \to 0$. Thus the estimate (I.12) immediately implies that $\sup_{0 \leqq t < T}[F(t)+\mathcal{D}] \to 0$ as $F(0)+\mathcal{D} \to 0$, providing the desired result on stability of the null solution. It should be noted that not only is (I.12) not meaningful at $t=T$ but (Knops and Payne [76]) one may produce examples where stability fails on the closed interval $[0, T]$ even for perturbations $\mathbf{u}(\cdot, t) \in \mathcal{M}_1$. It is to be noted that a key feature of the method is the need to assume the existence of a suitable classical (smooth) solution; the argument can be repeated for suitably defined weak solutions (Knops and Payne [75]) but again the existence of a solution of the ill-posed problem in an appropriate class of bounded functions must be assumed. Arguments of the type presented above may also be used to obtain stability theorems for a class of ill-posed problems in linear thermoelasticity [78], [89]. While we have not explicitly stated a uniqueness theorem it is clear that (I.12) implies that any solution $\mathbf{u}(\cdot, t) \in \mathcal{M}_1$ of (I.1)–(I.3), corresponding to zero initial data, must satisfy $\int_{\Omega(t)} \rho u_i u_i \, d\mathbf{x} = 0$, $0 \leqq t < T$ and, hence, by the positivity of ρ on Ω, must vanish identically on $\Omega \times [0, T)$.

2. Continuous data dependence for an abstract equation in Hilbert space: Applications to linear elastodynamics.

As a second illustration of the application of logarithmic convexity techniques we consider the problem of deriving continuous data dependence results for an abstract differential equation in Hilbert space; our approach is based on the work in [25] and is a generalization of the exposition given by Knops and Payne [77] for the problem of continuous data dependence for ill-posed initial-boundary value problems in linear elastodynamics. The logarithmic convexity argument presented here differs somewhat from that employed in § 1 and, more importantly, will serve as a model for the types of arguments to be used in the next two chapters where initial-history value problems for integrodifferential equations in Hilbert space will arise, in a natural way, from associated initial-history boundary value problems in viscoelasticity and the theory of nonconducting material dielectrics.

Thus, let $W \subseteq H$ be a dense subset of a real Hilbert space H with inner product $\langle \cdot, \cdot \rangle$; we assume that W is equipped with the $\langle \cdot, \cdot \rangle$ inner product as well. Let $\mathbf{M}: W \to H$ be a positive-definite and self-adjoint linear operator and $\mathbf{N}(t): W \to H$ be a symmetric linear operator for all $t < T$, $T > 0$. Suppose that $\mathbf{u}(\cdot, t) \in C^1([0, T); H)$ with $\mathbf{u}(\cdot, t)$, $\partial\mathbf{u}/\partial t(\cdot, t)$ taking values in W. We assume that $\partial^2\mathbf{u}/\partial t^2$ exists a.e. on $[0, T)$ and that $\mathbf{M}\mathbf{u}$, $\partial\mathbf{u}/\partial t$, $\mathbf{N}(t)\mathbf{u}$, and $\mathbf{N}(t)(\partial\mathbf{u}/\partial t)$ are in $C([0, T); H)$. Let $\mathcal{F}: [0, T) \to H$. For fixed $t \in [0, T)$ the domain J of $\mathbf{u}(\cdot, t)$ is a topological space with a positive measure μ_x. Let $\mathbf{f}, \mathbf{g}: J \to H$ be continuous functions on J. We are interested in stability properties of weak solutions of the initial value problem

$$\text{(I.13)} \qquad \begin{gathered} \mathbf{M}\left(\frac{\partial^2\mathbf{u}}{\partial t^2}\right) - \mathbf{N}(t)\mathbf{u}(t) = \mathcal{F}(t), \qquad 0 \leqq t < T, \\ \mathbf{u}(\cdot, 0) = \mathbf{f}(\cdot), \qquad \frac{\partial\mathbf{u}}{\partial t}(\cdot, 0) = \mathbf{g}(\cdot) \quad \text{a.e. on } J. \end{gathered}$$

By a weak solution to (I.13) we understand a function $\mathbf{u} \in C^1([0, T); H)$ such that

$$\text{(I.14)} \qquad \begin{aligned} \left\langle \boldsymbol{\phi}(t), \mathbf{M}\frac{\partial\mathbf{u}}{\partial t}\right\rangle &= \int_0^t \left\langle \frac{\partial\boldsymbol{\phi}}{\partial\eta}, \mathbf{M}\frac{\partial\mathbf{u}}{\partial t}\right\rangle d\eta \\ &\quad + \int_0^t \langle \boldsymbol{\phi}(\eta), \mathbf{N}(\eta)\mathbf{u}(\eta)\rangle\, d\eta + \int_0^t \langle \boldsymbol{\phi}(\eta), \mathcal{F}(\eta)\rangle\, d\eta \\ &\quad + \langle \boldsymbol{\phi}(0), \mathbf{M}\mathbf{u}_t(0)\rangle, \end{aligned}$$

for each $\boldsymbol{\phi}(\cdot, t) \in C^\infty([0, T); W)$ and each $t \in [0, T)$ and

$$\text{(I.15)} \qquad \begin{gathered} \mathbf{u}(\cdot, 0) = \mathbf{f}(\cdot), \qquad \mathbf{u}_t(\cdot, 0) = \mathbf{g}(\cdot), \\ \mathbf{N}(0)\mathbf{u}(\cdot, 0) = \mathbf{N}(0)f(\cdot), \qquad \mathbf{M}\mathbf{u}_t(\cdot, 0) = \mathbf{M}\mathbf{g}(\cdot) \end{gathered}$$

a.e. on J. We now define

$$\mathcal{M}_2 = \{\mathbf{v} \in C^1([0, T); W) \mid \|\mathbf{v}\|_T^2 \leqq M_1, \|\mathbf{v}_t\|_T^2 \leqq M_2\} \tag{I.16}$$

for some $M_1, M_2 > 0$ where

$$\|\mathbf{v}\|_t^2 = \int_0^t \langle \mathbf{v}(\eta), \mathbf{M}\mathbf{v}(\eta)\rangle \, d\eta, \qquad 0 \leqq t < T. \tag{I.17}$$

We will show that given appropriate conditions on $\mathbf{N}(t)$ (but not definiteness conditions, i.e., (I.13) is an ill-posed problem) $\exists K, L > 0$ and $\delta : [0, T) \to [0, 1)$ such that

$$\|\mathbf{u}\|_t^2 \leqq K L^{2\delta} \|\mathbf{M}^{-1}\mathcal{F}\|_T^{2(1-\delta)}, \qquad 0 \leqq t < T, \tag{I.18}$$

whenever $\mathbf{M}^{-1}$ exists and $\mathbf{u}(\cdot, t)$ is a weak solution of (I.13), with $\mathbf{f} = \mathbf{g} = \mathbf{0}$, which lies in the class $\mathcal{M}_1$; from the stability estimate (I.18) we will deduce continuous dependence of $\mathbf{u}$, in the $\|(\cdot)\|_t$-norm, on perturbations of the initial data, and on the initial geometry. Our results may then be applied to a class of initial-boundary value problems for linear elastic materials with time-dependent elasticities.

Remark. Other stability arguments for ill-posed initial-value problems of the type (I.13), which employ logarithmic convexity, may be found in Knops and Payne [79] ($\mathcal{F} \equiv \mathbf{0}$) and Levine [93].

We now state and prove the following basic theorem.

THEOREM (Bloom [25]). *Let* $\mathbf{u}(\cdot, t) \in \mathcal{M}_2$ *be a weak solution of* (I.13) *with* $\mathbf{f} = \mathbf{g} = \mathbf{0}$ *such that* $\mathbf{M}^{-1}$ *exists and* $\dot{\mathbf{N}}(t)$ *exists a.e. on* $[0, T)$ *with* $\langle \mathbf{w}, \dot{\mathbf{N}}(t)\mathbf{w}\rangle \geqq 0$, $\forall \mathbf{w} \in W$. *Then for* $0 \leqq t < T$, $\mathbf{u}(\cdot, t)$ *satisfies the estimate* (I.18).

Remark. $\dot{\mathbf{N}}(t)\mathbf{w} = \lim_{h\to 0} 1/h[\mathbf{N}(t+h)\mathbf{w} - \mathbf{N}(t)\mathbf{w}]$, $\mathbf{w} \in W$; i.e., $\dot{\mathbf{N}}(t)$ exists if

$$\lim_{h\to 0} \left\| \dot{\mathbf{N}}(t)\mathbf{w} - \frac{1}{h}[\mathbf{N}(t+h)\mathbf{w} - \mathbf{N}(t)\mathbf{w}] \right\| = 0$$

for all $\mathbf{w} \in W$, $t \in [0, T)$.

Proof. Define the real-valued function

$$G(t) \equiv \int_0^t \langle \mathbf{u}(\eta), \mathbf{M}\mathbf{u}(\eta)\rangle \, d\eta + T^4 \|\mathbf{M}^{-1}\mathcal{F}\|_T^2. \tag{I.19}$$

We will show that

$$H(t) = \ln [e^{t^2/T^2} G(t)] \tag{I.20}$$

is a convex function on $[0, T)$. First of all,

$$\frac{dG}{dt} = 2 \int_0^t \langle \mathbf{u}(\eta), \mathbf{M}\mathbf{u}_\eta\rangle \, d\eta, \tag{I.21}$$

as $\mathbf{f}=\mathbf{0}$ and $\mathbf{M}$ is symmetric. Also

$$\begin{aligned}\frac{d^2G}{dt^2}&=2\langle \mathbf{u}(t), \mathbf{M}\mathbf{u}_t(t)\rangle\\ &=2\int_0^t \langle \mathbf{u}_\eta, \mathbf{M}\mathbf{u}_\eta\rangle\, d\eta+2\int_0^t \langle \mathbf{u}(\eta), \mathbf{N}(\eta)\mathbf{u}(\eta)\rangle\, d\eta\\ &\quad+2\int_0^t \langle \mathbf{u}(\eta), \mathcal{F}(\eta)\rangle\, d\eta,\end{aligned} \tag{I.22}$$

as (I.14) holds in the limit for $\{\boldsymbol{\phi}_n(\cdot, t)\}\subset C^\infty([0, T); W)$ such that $\boldsymbol{\phi}_n(\cdot, t)\to \mathbf{u}(\cdot, t)$ with $(\boldsymbol{\phi}_n)_t(\cdot, t)\to \mathbf{u}_t(\cdot, t)$, and $\mathbf{u}_t(\cdot, 0)=\mathbf{0}$. Now if $\mathbf{u}_{tt}(\cdot, t)$ exists a.e. on J we may apply the same argument to (I.14) with a sequence $\{\boldsymbol{\psi}_n(\cdot, t)\}$ such that $\boldsymbol{\psi}_n(\cdot, t)\to \mathbf{u}_t(\cdot, t)$. Thus

$$\begin{aligned}\langle \mathbf{u}_\eta, \mathbf{M}\mathbf{u}_\eta\rangle&=\int_0^\eta \langle \mathbf{u}_{\gamma\gamma}, \mathbf{M}\mathbf{u}_\gamma\rangle\, d\gamma\\ &\quad+\int_0^\eta \langle \mathbf{u}_\gamma, \mathbf{N}(\gamma)\mathbf{u}(\gamma)\rangle\, d\gamma+\int_0^\eta \langle \mathbf{u}_\gamma, \mathcal{F}(\gamma)\rangle\, d\gamma.\end{aligned} \tag{I.23}$$

However,

$$\begin{aligned}\int_0^\eta \langle \mathbf{u}_\gamma, \mathbf{N}(\gamma)\mathbf{u}(\gamma)\rangle\, d\gamma&=\frac{1}{2}\langle \mathbf{u}(\eta), \mathbf{N}(\eta)\mathbf{u}(\eta)\rangle\\ &\quad-\frac{1}{2}\int_0^\eta \langle \mathbf{u}(\gamma), \dot{\mathbf{N}}(\gamma)\mathbf{u}(\gamma)\rangle\, d\gamma,\end{aligned}$$

so by integrating (I.23) we obtain

$$\begin{aligned}\frac{1}{2}\|\mathbf{u}_\eta\|_t^2&-\frac{1}{2}\int_0^t \langle \mathbf{u}(\eta), \mathbf{N}(\eta)\mathbf{u}(\eta)\rangle\, d\eta\\ &=-\frac{1}{2}\int_0^t (t-\eta)\langle \mathbf{u}(\eta), \dot{\mathbf{N}}(\eta)\mathbf{u}(\eta)\rangle\, d\eta\\ &\quad+\int_0^t (t-\eta)\langle \mathbf{u}_\eta, \mathcal{F}(\eta)\rangle\, d\eta.\end{aligned} \tag{I.24}$$

Now, as $\mathbf{M}$ is self-adjoint it has a unique self-adjoint square root $\mathbf{M}^{1/2}$. Thus, by the Cauchy-Schwarz inequality and (I.21),

$$\left(\frac{dG}{dt}\right)^2=4\left(\int_0^t \langle \mathbf{u}, \mathbf{M}\mathbf{u}_\eta\rangle\, d\eta\right)^2\leqq 4\|\mathbf{u}\|_t^2\|\mathbf{u}_\eta\|_t^2, \tag{I.25}$$

for $0\leqq t<T$. However, in view of the definition of $G(t)$, $\|\mathbf{u}\|_t^2=G(t)-T^4\|\mathbf{M}^{-1}\mathcal{F}\|_T^2$ so

$$\left(\frac{dG}{dt}\right)^2\leqq 4\|\mathbf{u}_\eta\|_t^2(G(t)-T^4\|\mathbf{M}^{-1}\mathcal{F}\|_T^2). \tag{I.26}$$

Combining this last estimate with (I.22) then yields the differential inequality

$$G\frac{d^2G}{dt^2}-\left(\frac{dG}{dt}\right)^2 \geqq -2G\left\{\|\mathbf{u}_\eta\|_t^2-\int_0^t \langle \mathbf{u}, \mathbf{N}(\eta)\mathbf{u}\rangle\, d\eta\right\} \tag{I.27}$$
$$+2G\int_0^t \langle \mathbf{u}, \mathscr{F}\rangle\, d\eta+4T^4\|\mathbf{u}_\eta\|_t^2\|\mathbf{M}^{-1}\mathscr{F}\|_T^2,$$

and substitution from (I.24) yields

$$G\frac{d^2G}{dt^2}-\left(\frac{dG}{dt}\right)^2 \geqq 2G\int_0^t \langle \mathbf{u}, \mathscr{F}\rangle\, d\eta$$
$$-4G\int_0^t (t-\eta)\langle \mathbf{u}_\eta, \mathscr{F}\rangle\, d\eta+4T^4\|\mathbf{u}_\eta\|_t^2\|\mathbf{M}^{-1}\mathscr{F}\|_T^2 \tag{I.28}$$
$$+2G\int_0^t (t-\eta)\langle \mathbf{u}(\eta), \dot{\mathbf{N}}(\eta)\mathbf{u}(\eta)\rangle\, d\eta.$$

By using the definition of $G(t)$ in conjunction with the Cauchy-Schwarz and arithmetic-geometric mean inequalities it is a straightforward matter to show that

$$2G\int_0^t \langle \mathbf{u}(\eta), \mathscr{F}(\eta)\rangle\, d\eta \geqq -T^{-2}G^2(t)-4G\int_0^t (t-\eta)\langle \mathbf{u}_\eta, \mathscr{F}(\eta)\rangle\, d\eta \tag{I.29}$$
$$\geqq -T^{-2}G^2(t)-4T^4\|\mathbf{u}_\eta\|_t^2\|\mathbf{M}^{-1}\mathscr{F}\|_T^2.$$

We now use the hypothesis of the theorem to drop the last expression in (I.28) and then combine the remaining expressions with (I.29) so as to obtain the differential inequality

$$G\frac{d^2G}{dt^2}-\left(\frac{dG}{dt}\right)^2 \geqq -T^{-2}G^2(t), \qquad 0\leqq t<T, \tag{I.30}$$

which (if $G(t)\not\equiv 0$) is equivalent to

$$\frac{d^2}{dt^2}(\ln[e^{t^2/T^2}G(t)])\geqq 0, \qquad 0\leqq t<T; \tag{I.31}$$

i.e., $\ln[e^{t^2/T^2}G(t)]$ is a convex function on the interval $[0, T)$. If we set $H(t)=\ln[e^{t^2/T^2}G(t)]$ then the convexity of $H(t)$, $0\leqq t<T$ implies that

$$H(t)\leqq\frac{H(T)-H(0)}{T}\cdot t+H(0), \qquad 0\leqq t<T,$$

or

$$\ln F(t) \leqq \frac{\log F(T) - \ln F(0)}{T} \cdot t + \ln F(0), \qquad 0 \leqq t < T, \tag{I.32}$$
$$F(t) = e^{t^2/T^2} G(t).$$

Thus,

$$F(t) \leqq F(0)\left[\frac{F(T)}{F(0)}\right]^{t/T}, \qquad 0 \leqq t < T. \tag{I.33}$$

But, $F(0) = G(0)$, $F(T) = eG(T)$, so $e^{t^2/T^2}G(t) \leqq G(0)[eG(T)/G(0)]^{t/T}$, $0 \leqq t < T$. From the definition of $G(t)$, $\|\mathbf{u}\|_t^2 \leqq G(t)$, $G(0) = T^4\|\mathbf{M}^{-1}\mathcal{F}\|_T^2$, and $G(T) = \|\mathbf{u}\|_T^2 + T^4\|\mathbf{M}^{-1}\mathcal{F}\|_T^2$ is bounded, in view of our hypothesis that $\mathbf{u}(\cdot, t) \in \mathcal{M}_2$. Thus

$$\begin{aligned}\|\mathbf{u}\|_t^2 &\leqq e^{-t^2/T^2} G(0)^{1-t/T}[eG(T)]^{t/T} \\ &= e^{-t^2/T^2}(T^4)^{1-t/T}\|\mathbf{M}^{-1}\mathcal{F}\|_T^{2(1-t/T)}[eG(T)]^{t/T} \\ &\leqq KL^{\delta}\|\mathbf{M}^{-1}\mathcal{F}\|_T^{2(1-\delta)}, \qquad 0 \leqq t < T,\end{aligned} \tag{I.34}$$

where $\delta(t) = t/T$. Q.E.D.

We now indicate how the stability estimate (I.34) can be used to obtain continuous dependence on initial data for solutions $\mathbf{u} \in \mathcal{M}_2$ of (I.13) with $\mathcal{F} = \mathbf{0}$. To this end we define $\mathbf{v}(\cdot, t) = \mathbf{u}(\cdot, t) - t\mathbf{g}(\cdot) - \mathbf{f}(\cdot)$. Then, clearly $\mathbf{v}(\cdot, 0) = \mathbf{0}$, $\mathbf{v}_t(\cdot, 0) = \mathbf{0}$ a.e. on J. Also

$$\mathbf{M}\mathbf{v}_{tt} - \mathbf{N}(t)\mathbf{v} = \mathbf{N}(t)[t\mathbf{g} + \mathbf{f}] \equiv \hat{\mathcal{F}}(t).$$

If we set $\boldsymbol{\lambda}(t) = t\mathbf{g} + \mathbf{f}$ then the estimate (I.34) implies that

$$\|\mathbf{v}\|_t^2 \leqq KL^{\delta}\left\{\int_0^t \langle \mathbf{N}(\eta)\boldsymbol{\lambda}(\eta), \mathbf{M}^{-1}\mathbf{N}(\eta)\boldsymbol{\lambda}(\eta)\rangle \, d\eta\right\}^{1-\delta}, \qquad 0 \leqq t < T, \tag{I.35}$$

provided $\mathcal{R}(\mathbf{N}(t)) \subset \mathcal{D}(\mathbf{M}^{-1})$, $0 \leqq t < T$. By using the relationship between $\mathbf{u}$, $\mathbf{v}$ it is easy to show that

$$\|\mathbf{u}\|_t \leqq \|\mathbf{v}\|_t + (\kappa(T))^{1/2}(\max\{\|\mathbf{f}\|_*^2, \|\mathbf{g}\|_*^2\})^{1/2}, \tag{I.36}$$

where $\kappa(T) = T^3/3 + T^2 + T$, $\|\mathbf{h}\|_*^2 = \langle \mathbf{h}, \mathbf{M}\mathbf{h}\rangle$. Thus

$$\begin{aligned}\|\mathbf{u}\|_t \leqq\; & K^{1/2}L^{\delta/2}\|\mathbf{M}^{-1}\mathbf{N}(\eta)\boldsymbol{\lambda}(\eta)\|_T^2 \\ & + \kappa^{1/2}(\max\{\|\mathbf{f}\|_*^2, \|\mathbf{g}\|_*^2\})^{1/2}.\end{aligned} \tag{I.37}$$

If $\mathbf{M}^{-1}\mathbf{N}(t)$ is bounded in the strong operator topology generated by $\|(\cdot)\|_*$, for each t, $0 \leqq t < T$, then (I.37) implies that

$$\|\mathbf{u}\|_t \leqq (K^{1/2}L^{\delta/2}\alpha^{1-\delta} + \kappa^{1/2})\psi(\mathbf{f}, \mathbf{g})(\psi^{-\delta}(\mathbf{f}, \mathbf{g}) + 1), \tag{I.38}$$

where $\psi^2(\mathbf{f}, \mathbf{g}) \equiv \max\{\|\mathbf{f}\|_*^2, \|\mathbf{g}\|_*^2\}$ and $\alpha = \sup_{[0,T)} \|\|\mathbf{M}^{-1}\mathbf{N}(t)\|\|_*$. The estimate

(I.38) shows, in fact, that $\|\mathbf{u}\|_t \to 0$, $0 \leqq t < T$, as $\max\{\|\mathbf{f}\|_*^2, \|\mathbf{g}\|_*^2\} \to 0$. Also, if $\mathbf{u}^{(i)}$, $i = 1, 2$, are two solutions of (I.13) for which $\mathbf{u} = \mathbf{u}^{(1)} - \mathbf{u}^{(2)}$ lies in $\mathcal{M}_1$, then $\mathbf{u}$ satisfies (I.13) with $\mathcal{F} = \mathbf{f} = \mathbf{g} = 0$ so that $\boldsymbol{\lambda} \equiv 0$ and (I.36) then implies that $\|\mathbf{u}\|_t = 0$, $0 \leqq t < T$, even if $\mathbf{M}^{-1}\mathbf{N}(t)$ is not bounded in the stronger operator topology generated by $\|(\cdot)\|_*$ (actually, this last assumption on $\mathbf{M}^{-1}\mathbf{N}(t)$ is not needed in the application to linear elastodynamics even for the discussion of continuous dependence on initial data; the conclusion that $\mathbf{u} = 0$, $0 \leqq t < T$, then follows directly from the definition of $\|(\cdot)\|_t$ and yields uniqueness of solutions to the ill-posed initial value problem (I.13)).

One interesting result which may be obtained as a consequence of the stability estimate (I.34) concerns continuous dependence of solutions of (I.13) on perturbations of the initial geometry: Let $\chi: J \to R^+$ be a continuous function, $\chi \geqq 0$ on J with $\sup_{x \in J} |\chi(\mathbf{x})| < \varepsilon$, for some $\varepsilon > 0$. Let $\mathbf{u}^\chi(\cdot, t)$ be any solution of (I.13$_1$) which satisfies initial data of the form

$$\mathbf{u}^\chi(\cdot, -\chi(\cdot)) = \mathbf{f}(\cdot), \qquad \mathbf{u}_t^\chi(\cdot, -\chi(\cdot)) = \mathbf{g}(\cdot); \tag{I.39}$$

i.e., $\mathbf{u}^\chi(\cdot, t)$ has initial data prescribed on the surface $t = -\chi(\mathbf{x})$, $\mathbf{x} \in J$ instead of on $t = 0$. Let $\mathbf{u}^\varepsilon(\cdot, t) = \mathbf{u}^\chi(\cdot, t) - \mathbf{u}(\cdot, t)$ where $\mathbf{u}(\cdot, t)$ is the solution of (I.13). We want to show that provided the initial data are restricted in an appropriate fashion, $\|\boldsymbol{u}^\varepsilon\|_t \to 0$, as $\varepsilon \to 0$, for each t, $0 \leqq t < T$. To this end we compute

$$\begin{aligned} \mathbf{u}_t^\varepsilon(\cdot, 0) &= \mathbf{u}_t^\chi(\cdot, 0) - \mathbf{u}_t(\cdot, 0) \\ &= \mathbf{u}_t^\chi(\cdot, 0) - \mathbf{u}_t^\chi(\cdot, -\chi(\cdot)) \\ &= \int_{-\chi(\cdot)}^{0} \frac{\partial^2 \mathbf{u}^\chi}{\partial t^2}\, dt \end{aligned} \tag{I.40}$$

and in a similar manner

$$\mathbf{u}^\varepsilon(\cdot, 0) = \chi(\cdot)\mathbf{g}(\cdot) + \int_0^{-\chi(\cdot)} t \frac{\partial^2 \mathbf{u}^\chi}{\partial t^2}\, dt. \tag{I.41}$$

Thus, $\mathbf{u}(\cdot, t)$ is a solution of the Cauchy problem

$$\begin{aligned} &\mathbf{M}\mathbf{u}_{tt}^\varepsilon - \mathbf{N}(t)\mathbf{u}^\varepsilon = \mathbf{0}, \qquad 0 \leqq t < T, \\ &\mathbf{u}^\varepsilon(\cdot, 0) = \chi(\cdot)\mathbf{g}(\cdot) + \int_0^{-\chi(\cdot)} t \frac{\partial^2 \mathbf{u}^\chi}{\partial t^2}\, dt, \\ &\mathbf{u}_t^\varepsilon(\cdot, 0) = \int_{-\chi(\cdot)}^{0} \frac{\partial^2 \mathbf{u}^\chi}{\partial t^2}\, dt. \end{aligned} \tag{I.42}$$

By our previous results on continuous dependence on initial data, if $\dot{\mathbf{N}}(t)$ exists and is positive definite, $0 \leqq t < T$, and $\mathbf{M}^{-1}\mathbf{N}(t)$ is bounded in the strong operator topology generated by $\|(\cdot)\|_*$, $0 \leqq t < T$, (as already indicated, this assumption may be avoided in the application to linear elastodynamics we present below) then

$$\|\mathbf{u}^\varepsilon\|_t \to 0 \quad \text{as } \psi(\mathbf{u}^\varepsilon(0), \mathbf{u}_t^\varepsilon(0)) \to 0, \tag{I.43}$$

provided $\mathbf{u}^\varepsilon \in \mathcal{M}_2$ (for ε sufficiently small). It may be shown (Bloom [23], [25]) that the following result obtains: If $\|\chi \mathbf{g}\| \geqq (1+\varepsilon)\|\mathbf{u}_t^\varepsilon(0)\|$, for ε sufficiently small and $\lim_{\varepsilon\to 0} \|\mathbf{M}\mathbf{u}^\varepsilon(0)\| < \infty$, $\lim_{\varepsilon\to 0} \|\mathbf{M}\mathbf{u}_t^\varepsilon(0)\| < \infty$, then $\psi(\mathbf{u}^\varepsilon(0), \mathbf{u}_t^\varepsilon(0)) \to 0$ as $\varepsilon \to 0$; for the details we refer the reader to the references cited above.

As in § 1 we let $\Omega \subseteq R^3$ be a bounded regular region with smooth $\partial\Omega$. We assume that the elasticities $C_{ijkl}(\mathbf{x}, t)$ (time-dependent) are bounded measurable functions for each $t \in [0, T)$. For our basic Hilbert spaces we take

$$H = (L^2(\Omega))^3,$$

$$W = \{\mathbf{w} \in H \mid \mathbf{w} \in C^1(\Omega),\ \mathbf{w} = \mathbf{0} \text{ on } \partial\Omega\}.$$

Also

$$\langle \mathbf{u}, \mathbf{v} \rangle_H = \int_\Omega u_i(\mathbf{x}) v_i(\mathbf{x})\, d\mathbf{x}, \qquad \mathbf{u}, \mathbf{v} \in H,$$

and for any $\mathbf{w} \in W$ we define $\mathbf{M}$, $\mathbf{N}(t)$ via

$$\begin{aligned} (\mathbf{M}\mathbf{w})_i(\mathbf{x}) &= \rho(\mathbf{x}) w_i(\mathbf{x}), \\ (\mathbf{N}(t)\mathbf{w})_i(\mathbf{x}) &= \frac{\partial}{\partial x_j}\left(C_{ijkl}(\mathbf{x}, t) \frac{\partial w_k}{\partial x_l} \right), \end{aligned} \tag{I.44}$$

for any $\mathbf{w} \in W$. The derivatives in (I.44) may be interpreted in the sense of distributions but this idea is, in fact, already implicit in the idea of weak solution of the elasticity equations and we shall, in any case, soon impose strong smoothness requirements on the elasticities so as to avoid, at least until the next chapter, the problem of dealing with the elasticity operator $\mathbf{N}$ as a map from W into H as defined above; in this context $\mathbf{N}$ is, of course, an unbounded linear operator (this problem is circumvented in Chapters II and III via the introduction of certain elementary Sobolev spaces and their duals). We want to consider the nonhomogeneous initial-boundary value problem in $\Omega \times [0, T)$

$$\begin{aligned} &\rho(\mathbf{x}) \frac{\partial^2 u_i}{\partial t^2} - \frac{\partial}{\partial x_j}\left(C_{ijkl}(\mathbf{x}, t) \frac{\partial u_k}{\partial x_l} \right) = \rho(\mathbf{x}) \mathcal{F}_i, \\ &u_i(\mathbf{x}, 0) = f_i(\mathbf{x}), \qquad \frac{\partial u_i}{\partial t}(\mathbf{x}, 0) = g_i(\mathbf{x}) \text{ in } \Omega, \\ &u_i(\mathbf{x}, t) = 0, \qquad \partial\Omega \times [0, T), \end{aligned} \tag{I.45}$$

where $\rho(\mathbf{x}) > 0$ on Ω, $\mathcal{F} \in C(\Omega)$, and $\mathbf{f}, \mathbf{g} \in C^1(\Omega)$. For $\mathbf{u}(\cdot, t): \Omega \to (L^2(\Omega))^3$ we have

$$\|\mathbf{u}\|_t^2 = \int_0^t \int_\Omega \rho(\mathbf{x}) u_i(\mathbf{x}, \eta) u_i(\mathbf{x}, \eta)\, d\mathbf{x}\, d\eta.$$

By a weak solution of (I.45) we understand a function $\mathbf{u}(\cdot, t) \in C^1([0, T); W)$ such that $\mathbf{u}_{tt}$ exists a.e. on $[0, T)$ and

$$\text{(I.46)} \quad \begin{aligned} &\int_{\Omega(t)} \rho\phi_i \frac{\partial u_i}{\partial t}\, d\mathbf{x} - \int_0^t \int_{\Omega(\eta)} \left[\rho \frac{\partial \phi_i}{\partial \eta}\frac{\partial u_i}{\partial \eta} - C_{ijkl}(\mathbf{x}, \eta)\frac{\partial \phi_i}{\partial x_j}\frac{\partial u_k}{\partial x_l}\right] d\mathbf{x}\, d\eta \\ &\qquad = \int_0^t \int_{\Omega(t)} \rho \mathscr{F}_i \phi_i \, d\mathbf{x}\, d\eta + \int_{\Omega(0)} \rho\phi_i g_i \, d\mathbf{x}, \end{aligned}$$

for each $\boldsymbol{\phi}(\cdot, t) \in C^\infty([0, T); W)$, and

$$\text{(I.47)} \quad \begin{aligned} &\mathbf{u}(\cdot, 0) = \mathbf{f}(\cdot), \qquad \mathbf{u}_t(\cdot, 0) = \mathbf{g}(\cdot) \quad \text{a.e. on } \Omega, \\ &u_{i,j}(\cdot, 0) = f_{i,j}(\cdot) \quad \text{a.e. on } \Omega \end{aligned}$$

($u_{i,j} = \partial u_i / \partial x_j$, understood in the sense of distributions). In order to satisfy our hypotheses in the abstract setting we require that $\mathbf{u}(\cdot, t) \in \bar{\mathcal{M}}_2$ where

$$\bar{\mathcal{M}}_2 = \Big\{ \mathbf{v}(\cdot, t) \in C^1([0, T); W) \Big| \int_0^T \int_{\Omega(\eta)} \rho(\mathbf{x}) v_i(\mathbf{x}, \eta) v_i(\mathbf{x}, \eta)\, d\mathbf{x}\, d\eta \leqq M_1, \int_0^T \int_{\Omega(\eta)} \rho(\mathbf{x}) \frac{\partial v_i}{\partial \eta}(\mathbf{x}, \eta) \frac{\partial v_i}{\partial \eta}(\mathbf{x}, \eta)\, d\mathbf{x}\, d\eta \leqq M_2 \Big\}$$

for some $M_1, M_2 > 0$. We also require that the elasticities $C_{ijkl}(\mathbf{x}, t)$ satisfy

(i) $\left(\dfrac{\partial}{\partial t}\right) C_{ijkl}(\mathbf{x}, t) = \dot{C}_{ijkl}(\mathbf{x}, t)$ exist a.e. on Ω,

(ii) $\displaystyle\int_{\Omega(t)} C_{ijkl}(\mathbf{x}, t) \frac{\partial w_i}{\partial x_j} \frac{\partial w_k}{\partial x_l} d\mathbf{x} \leqq 0$ for each $t \in [0, T)$ and all $\mathbf{w} \in W$,

(iii) $C_{ijkl}(\mathbf{x}, t) = C_{klij}(\mathbf{x}, t), \qquad (\mathbf{x}, t) \in \Omega \times [0, T)$.

Clearly, the operator $\mathbf{M}^{-1}\mathbf{N}(t)$ is not bounded as a map from W into H, but continuous dependence on initial data for solutions $\mathbf{u} \in \bar{\mathcal{M}}_2$ can still be deduced if we further restrict the class of elasticities and the class of perturbations; i.e., we require that

(iv) $C_{ijkl}(\cdot, t) \in C^1(\bar{\Omega}), \qquad t \in [0, T)$,

(v) $u_i, \dfrac{\partial u_i}{\partial t} \in C^2(\bar{\Omega})$.

The estimates obtained for the abstract problem then yield

$$\left[\int_0^t \int_{\Omega(\eta)} \rho(\mathbf{x}) u_i(\mathbf{x}, \eta) u_i(\mathbf{x}, \eta)\, d\mathbf{x}\, d\eta\right]^{1/2}$$

(I.48)
$$\leqq \left[\int_0^t \int_{\Omega(\eta)} \lambda_i(\mathbf{x}, \eta)\lambda_i(\mathbf{x}, \eta)\, d\mathbf{x}\, d\eta\right]^{1/2} + K^{1/2} L^{\delta}\left[\int_0^T \int_{\Omega(\eta)} \rho^{-1}(\mathbf{x}) \times [\mathbf{N}(\eta)\boldsymbol{\lambda}(\mathbf{x}, \eta)]_i [\mathbf{N}(\eta)\boldsymbol{\lambda}(\mathbf{x}, \eta)]_i\, d\mathbf{x}\, d\eta\right]^{1/2-\delta/2},$$

where

$$\lambda_k(\mathbf{x}, \eta) = \eta\left(\frac{\partial u_k}{\partial \eta}\right)(\mathbf{x}, 0) + u_k(\mathbf{x}, 0),$$

(I.49)
$$[\mathbf{N}(\eta)\boldsymbol{\lambda}(\mathbf{x}, \eta)]_i = \frac{\partial}{\partial x_j} C_{ijkl}(\mathbf{x}, \eta)\frac{\partial \lambda_k}{\partial x_l} + C_{ijkl}(\mathbf{x}, \eta)\frac{\partial^2 \lambda_k}{\partial x_j \partial x_l}.$$

A tedious but straightforward computation then produces the estimate

$$\left[\int_0^t \int_{\Omega(\eta)} \rho(\mathbf{x}) u_i(\mathbf{x}, \eta) u_i(\mathbf{x}, \eta)\, d\mathbf{x}\, d\eta\right]^{1/2}$$

(I.50)
$$\leqq (\bar{\gamma} T \sup_{[0,T)} [\mu(\Omega(t))])^{1/2} + K^{1/2} L^{\delta} (\bar{\gamma} T \sup_{[0,T)} [\mu(\Omega(t))])^{1/2-\delta/2},$$

where

$$\bar{\lambda} = \sup_{[0,T)} \max_{\bar{\Omega}(t)} [\lambda_i(\mathbf{x}, t)\lambda_i(\mathbf{x}, t)],$$

$\mu(\Omega(t))$ is the Lebesgue volume measure of $\Omega(t)$, and $\bar{\gamma}$ is a positive constant satisfying

$$\mathcal{Q}(T) \equiv \int_0^T \int_{B(\eta)} \rho^{-1}(\mathbf{x}) [\mathbf{N}(\eta)\boldsymbol{\lambda}(\mathbf{x}, \eta)]_i [\mathbf{N}(\eta)\boldsymbol{\lambda}(\mathbf{x}, \eta)]_i\, d\mathbf{x}\, d\eta \leqq \bar{\gamma} T \sup_{[0,T)} [\mu(\Omega(t))].$$

$\mathcal{Q}(T)$ is bounded by virtue of our smoothness hypothesis relative to the elasticities $C_{ijkl}(\cdot, t)$. If we set $\Gamma = \max(\bar{\lambda}, \bar{\gamma})$ then clearly (I.50) implies that

$$\int_0^t \int_{\Omega(\eta)} \rho(\mathbf{x}) u_i(\mathbf{x}, \eta) u_i(\mathbf{x}, \eta)\, d\mathbf{x}\, d\eta \to 0, \qquad 0 \leqq t < T$$

as $\Gamma \to 0$. It is a straightforward matter to show that $\bar{\lambda}$ satisfies

$$\text{(I.51)} \qquad \bar{\lambda} \leqq \max_{x \in \bar{\Omega}} \left[T\left(\frac{\partial u_i}{\partial t}(\mathbf{x}, 0) \frac{\partial u_i}{\partial t}(\mathbf{x}, 0)\right)^{1/2} + (u_i(\mathbf{x}, 0) u_i(\mathbf{x}, 0))^{1/2} \right]^2 .$$

Therefore, if $\bar{\lambda} \geqq \bar{\gamma}$ then

$$\int_0^t \int_{\Omega(\eta)} \rho(\mathbf{x}) u_i(\mathbf{x}, \eta) u_i(\mathbf{x}, \eta)\, d\mathbf{x}\, d\eta \to 0, \qquad 0 \leqq t < T$$

as

$$\max\left(\max_{\bar{\Omega}} \left(\frac{\partial u_i}{\partial t}(\mathbf{x}, 0) \frac{\partial u_i}{\partial t}(\mathbf{x}, 0)\right), \max_{\bar{\Omega}} (u_i(\mathbf{x}, 0) u_i(\mathbf{x}, 0))\right) \to 0.$$

If $\bar{\lambda} \leqq \bar{\gamma}$, a more complicated result is then obtained which involves second order spatial derivatives of u_i and $\partial u_i/\partial t$ at $t = 0$.

Remark. The logarithmic convexity arguments presented in §§ 1, 2 have been used to derive stability, uniqueness, and continuous data dependence results with particular emphasis on ill-posed initial-boundary value problems in linear elastodynamics; the basic logarithmic convexity estimates in these two sections can also be used to derive growth estimates for solutions of these problems. However, we will delay our discussion of growth estimates until the next chapter when we will treat ill-posed initial-history boundary value problems for linear, isothermal, viscoelastic materials.

3. A nonexistence theorem for the heat equation with nonlinear boundary data: An application of the basic concavity argument.

In order to illustrate the second type of differential inequality argument that we will use later in our discussion of initial-history boundary value problems (in particular, in the applications to Maxwell-Hopkinson and isotropic holohedral dielectrics in Chapter III) we will present an application that has been made by Levine and Payne [101] to prove global nonexistence of smooth solutions for the linear heat equation with nonlinear boundary data. While our use of the concavity argument in Chapter III will involve the derivation of certain growth estimates for electric induction fields, the derivation of those growth estimates will hinge on certain results concerning nonexistence of global solutions for the associated initial-history value problem; the application presented below should, therefore, serve as a good introduction to the type of argument that we will employ in our application to dielectric behavior. As indicated in the introduction, the type of concavity argument we present below has been used with great success in recent years to prove theorems concerning global nonexistence (and finite-time blow-up) of solutions to initial-boundary value problems for a variety of nonlinear partial differential equations that arise in applications to physical problems; we mention, in particular, the recent work of Knops, Levine, and Payne [73], Levine and Payne [99], Levine [94], [95], [96], [97], and [98]. However, one must be quite careful as concerns some of the arguments related to finite-time blow-up of solutions in the papers cited above: as has been recently shown by J. Ball [4] and [5], while the concavity argument is perfectly all right as a means of proving global nonexistence of smooth solutions it often leads to specious conclusions as regards finite-time blow-up of solutions. More specifically, Ball shows the following: suppose that $\mathbf{u}:[t_0, t_{\max})\to \mathcal{B}$, $\mathcal{B}$ a Banach space, is the solution of a nonlinear equation for which $\mathbf{u}(t_0)=\mathbf{u}_0\in\mathcal{B}$. By means of the concavity argument (to be illustrated below) it is often possible to show that for some norm π on $\mathcal{B}$ and some time $t_1\in(t_0,\infty)$, $\lim_{t\to t_1}\pi(\mathbf{u}(t))=\infty$. This clearly implies that $t_{\max}<\infty$ and that $t_{\max}\leqq t_1$. It does not, however, establish that $\lim_{t\to t_{\max}}\pi(\mathbf{u}(t))=\infty$ since it may happen that $t_{\max}<t_1$. In fact, Ball gives as an example [4] the problem

$$v_t=-\Delta v+\left\{\int_\Omega v^2\,d\mathbf{x}\right\}v,\qquad \mathbf{x}\in\Omega,\quad t>0,$$

$$v=0,\qquad \mathbf{x}\in\partial\Omega,\quad t>0,$$

$$v(\mathbf{x},0)=\psi(\mathbf{x}),\qquad \mathbf{x}\in\Omega,$$

$\Omega\subseteq R^n$ a bounded open set with smooth $\partial\Omega$, $\psi\in L^2(\Omega)$, for which he proves global nonexistence with $t_{\max}=1$, in fact, for an appropriate choice of ψ, and yet $\lim_{t\to 1}\|v\|^2_{L_2(\Omega)}$ exists. For the problems cited above it is often possible to show that if $t_{\max}<\infty$ then $\lim_{t\to t_{\max}}\tilde{\pi}(u(t))=\infty$ for some norm $\tilde{\pi}$ on $\mathcal{B}$ which is usually weaker than the π norm employed in the logarithmic concavity arguments; this is usually accomplished by employing certain continuation theorems for ordinary differential equations in Banach space and for the details we refer the reader to the two papers of Ball which are referenced above.

We now proceed with illustrating the use of the concavity argument to prove global nonexistence of smooth solutions for the heat equation with a particular class of associated nonlinear boundary data. Let $\Omega\subseteq R^n$ be a bounded open domain with smooth $\partial\Omega$, let $\mathbf{n}=(n_1,\cdots,n_n)$ be the unit outward normal to $\partial\Omega$, and $f\in C^1(R^1)$. Let u be a real-valued classical solution of

$$\frac{\partial u}{\partial t}=\Delta u\quad\text{in }\Omega\times[0,T),$$

(I.52)
$$\frac{\partial u}{\partial n}=f(u)\quad\text{on }\partial\Omega\times[0,T),$$

$$u(\mathbf{x},0)=u_0(\mathbf{x})\quad\text{on }\bar{\Omega},$$

where $u_0 \in C^2(\bar{\Omega})$ and $\partial u/\partial n = \sum_{i=1}^n (\partial u/\partial x_i)n_i$ is the normal derivative of u on $\partial\Omega$. We will exhibit a class of nonlinearities f for which there does not exist a global solution u of (I.52), i.e., a solution on $\bar{\Omega}\times[0,\infty)$. Clearly, (I.52) is a heat conduction problem with a nonlinear radiation (absorption) law prescribed on $\partial\Omega$. The key idea to be used in proving the global nonexistence of a smooth solution to the initial-boundary value (I.52) is the following: we want to exhibit a real-valued function $\tilde{F}(u(t)) = F(t)$, depending on $u(\mathbf{x}, t)$, which is nonnegative (as in the logarithmic convexity technique) and which satisfies on $[0, T)$ a differential inequality of the form

$$FF'' - (\alpha+1)F'^2 \geqq 0 \quad \text{for some } \alpha > 0. \tag{I.53}$$

This last inequality is, in turn, equivalent to $(F^{-\alpha})'' \leqq 0$, which implies that

$$\begin{aligned}(F^{-\alpha})(t) &\leqq (F^{-\alpha})'(0)t + F^{-\alpha}(0)\\ &= -\alpha F^{-\alpha-1}(0)F'(0)t + F^{-\alpha}(0).\end{aligned} \tag{I.54}$$

Thus, on $[0, T)$

$$F^{-\alpha}(t) \leqq F^{-\alpha-1}(0)[F(0) - \alpha t F'(0)], \tag{I.55}$$

or as $\alpha > 0$, $F(t) \geqq 0$, $0 \leqq t < T$,

$$F^{\alpha}(t) \geqq F^{\alpha+1}(0)[F(0) - \alpha t F'(0)]^{-1}, \tag{I.56}$$

which clearly implies that

$$\lim_{t\to t_\infty} F(t) = +\infty, \qquad t_\infty = \frac{F(0)}{\alpha F'(0)}. \tag{I.57}$$

If $F(0)/F'(0) > 0$ and $F(0)/\alpha F'(0) < T$ (perhaps for sufficiently large α) then (I.57) establishes global nonexistence of u (in the norm defined by $\tilde{F}$). Note that $t_{\max} \leqq t_\infty$ and if, in fact, $t_{\max} < t_\infty$ then (I.57) does not imply nonexistence by finite-time blow-up in the norm defined by $\tilde{F}$. For the initial-boundary value problem (I.52) we now exhibit, as an application of the above series of ideas, an argument due to Levine and Payne [101] which demonstrates the global nonexistence of smooth solutions to (I.52) within a prescribed class of nonlinear boundary data f:

THEOREM. *Suppose that*

(i) $f(y) = |y|^{2\alpha+1}h(y)$ *for some* $\alpha > 0$ *with* $h(y)$ *monotone increasing.*

(ii) $\displaystyle\oint_{\partial\Omega}\left(\int_0^{u_0(s)} f(y)\,dy\right)ds \geqq \frac{1}{2}\int_\Omega |\nabla u_0|^2\,d\mathbf{x}.$

Then $\nexists$ *a classical smooth solution* $u:\bar{\Omega}\times[0,\infty)\to R^1$ *of* (I.52).

Proof. We assume that $\exists u: \bar{\Omega} \times [0, T) \to R^1$, a classical solution of (I.52), for each $T > 0$, and then deduce a contradiction. Define

$$F(t) = \int_0^t \int_\Omega u^2(\mathbf{x}, \eta)\, d\mathbf{x}\, d\eta + (T - t) \int_\Omega u_0^2(\mathbf{x})\, d\mathbf{x} + \beta(t + \tau)^2, \tag{I.58}$$

where $\beta \geqq 0$, $\tau \geqq 0$ are arbitrary real numbers. We want to demonstrate that $F(t)$ satisfies (I.53) for $t \in [0, T)$. Clearly

$$\begin{aligned} F'(t) &= \int_\Omega u^2(\mathbf{x}, t)\, d\mathbf{x} - \int_\Omega u_0^2(\mathbf{x})\, d\mathbf{x} + 2\beta(t + \tau) \\ &= \int_0^t \left(\frac{\partial}{\partial \eta} \int_\Omega u^2(\mathbf{x}, \eta)\, d\mathbf{x} \right) d\eta + 2\beta(t + \tau) \\ &= 2 \int_0^t \int_\Omega u u_\eta\, d\mathbf{x}\, d\eta + 2\beta(t + \tau) \\ &= 2 \int_0^t \int_\Omega u\, \Delta u\, d\mathbf{x}\, d\eta + 2\beta(t + \tau), \end{aligned} \tag{I.59}$$

where we have used the governing (heat) equation in the last step. But

$$\begin{aligned} \int_\Omega u \Delta u\, d\mathbf{x} &= \oint_{\partial\Omega} u \frac{\partial u}{\partial \eta}\, ds - \int_\Omega |\nabla u|^2\, d\mathbf{x} \\ &= \oint_{\partial\Omega} u f(u)\, ds - \int_\Omega |\nabla u|^2\, d\mathbf{x} \end{aligned} \tag{I.60}$$

by Green's formula and the (nonlinear) boundary condition. Thus

$$\begin{aligned} F'(t) = -2 \int_0^t \int_\Omega |\nabla u|^2\, d\mathbf{x}\, d\eta + 2 \int_0^t \oint_{\partial\Omega} u f(u)\, ds\, d\eta \\ + 2\beta(t + \tau), \end{aligned} \tag{I.61}$$

so

$$F''(t) = -2 \int_\Omega |\nabla u|^2\, d\mathbf{x} + 2 \oint_{\partial\Omega} u f(u)\, ds + 2\beta. \tag{I.62}$$

As

$$\begin{aligned} \int_0^t \int_\Omega \nabla u_\eta \cdot \nabla u\, d\mathbf{x}\, d\eta &= \frac{1}{2} \int_0^t \frac{\partial}{\partial \eta} \int_\Omega \nabla u \cdot \nabla u\, d\mathbf{x}\, d\eta \\ &= \frac{1}{2} \int_\Omega |\nabla u|^2\, d\mathbf{x} - \frac{1}{2} \int_\Omega |\nabla u_0|^2\, d\mathbf{x}, \end{aligned}$$

$$\begin{aligned} F''(t) = -4 \int_0^t \int_\Omega \nabla u_\eta \cdot \nabla u\, d\mathbf{x}\, d\eta - 2 \int_\Omega |\nabla u_0|^2\, d\mathbf{x} \\ + 2 \oint_{\partial\Omega} u f(u)\, ds + 2\beta. \end{aligned} \tag{I.63}$$

But,

$$\int_0^t \int_\Omega \nabla u_\eta \cdot \nabla u \, d\mathbf{x} \, d\eta = \int_0^t \int_\Omega \frac{\partial u_\eta}{\partial x_i} \frac{\partial u}{\partial x_i} \, d\mathbf{x} \, d\eta$$

$$\text{(I.64)} \qquad = \int_0^t \int_\Omega \frac{\partial}{\partial x_i}\left(u_\eta \frac{\partial u}{\partial x_i}\right) d\mathbf{x} \, d\eta - \int_0^t \int_\Omega u_\eta \, \Delta u \, d\mathbf{x} \, d\eta$$

$$= \int_0^t \oint_{\partial\Omega} u_\eta f(u) \, ds \, d\eta - \int_0^t \int_\Omega u_\eta^2 \, d\mathbf{x} \, d\eta,$$

where we have again used the heat equation and the boundary condition in conjunction with integration by parts. Therefore, (I.63) may be put in the form

$$\text{(I.65)} \qquad \begin{aligned} F''(t) = 4 \int_0^t \int_\Omega u_\eta^2 \, d\mathbf{x} \, d\eta - 4 \int_0^t \oint_{\partial\Omega} u_\eta f(u) \, ds \, d\eta \\ - 2 \int_\Omega |\nabla u_0|^2 \, d\mathbf{x} + 2 \oint_{\partial\Omega} u f(u) \, ds + 2\beta. \end{aligned}$$

Splitting $2\beta = 4(\alpha+1)\beta - 2(2\alpha+1)\beta$, and using (I.64) again, we may put (I.65) in the form

$$\text{(I.66)} \qquad \begin{aligned} F''(t) &= 4(\alpha+1)\left[\int_0^t \int_\Omega u_\eta^2 \, d\mathbf{x} \, d\eta + \beta\right] \\ &\quad + 2\left[-2 \int_0^t \oint_{\partial\Omega} u_\eta f(u) \, ds \, d\eta - \int_\Omega |\nabla u_0|^2 \, d\mathbf{x} \right. \\ &\quad \left. - 2\alpha \int_0^t \int_\Omega u_\eta \, \Delta u \, d\mathbf{x} \, d\eta + \oint_{\partial\Omega} u f(u) \, ds - (2\alpha+1)\beta\right] \\ &= 4(\alpha+1)\left[\int_0^t \int_\Omega u_\eta^2 \, d\mathbf{x} \, d\eta + \beta\right] \\ &\quad + 2\left\{\alpha \int_\Omega |\nabla u|^2 \, d\mathbf{x} - (\alpha+1) \int_\Omega |\nabla u_0|^2 \, d\mathbf{x} \right. \\ &\quad + \oint_{\partial\Omega} \left[u f(u) - 2(2\alpha+1) \int_0^t \frac{\partial}{\partial \eta}\left(\int_{u_0(s)}^{u(s,\eta)} f(y) \, dy\right) d\eta\right] ds \\ &\quad \left. - (2\alpha+1)\beta\right\}, \end{aligned}$$

where we have substituted for the expression $\int_0^t \int_\Omega u_\eta \Delta \, d\mathbf{x} \, d\eta$ and, in addition, used the fact that

$$\text{(I.67)} \qquad \oint_{\partial\Omega} \int_0^t \frac{\partial}{\partial \eta}\left(\int_{u_0(s)}^{u(s,\eta)} f(y) \, dy\right) d\eta \, ds = \oint_{\partial\Omega} \int_0^t u_\eta f(u) \, d\eta \, ds.$$

More precisely, in the second bracket in (I.66$_1$)

$$
\begin{aligned}
-2\int_0^t \oint_{\partial\Omega} & u_\eta f(u)\, ds\, d\eta - 2\alpha \int_0^t \int_\Omega u_\eta\, \Delta u\, d\mathbf{x}\, d\eta \\
&= -2\int_0^t \oint_{\partial\Omega} u_\eta f(u)\, ds\, d\eta \\
&\quad -2\alpha\left[\int_0^t \int_{\partial\Omega} u_\eta f(u)\, ds\, d\eta - \int_0^t \int_\Omega \nabla u_\eta \cdot \nabla u\, d\mathbf{x}\, d\eta\right] \quad \text{(by Green's formula)} \\
&= -2(a+1)\int_0^t \oint_{\partial\Omega} u_\eta f(u)\, ds\, d\eta + 2\alpha \int_0^t \int_\Omega \nabla u_\eta \cdot \nabla u\, d\mathbf{x}\, d\eta \\
&= -2(\alpha+1)\int_0^t \oint_{\partial\Omega} u_\eta f(u)\, ds\, d\eta + \alpha\left[\int_\Omega |\nabla u|^2\, d\mathbf{x} - \int_\Omega |\nabla u_0|^2\, d\mathbf{x}\right].
\end{aligned}
$$

To recapitulate, we have

$$
\begin{aligned}
F(t) &= \int_0^t \int_\Omega u^2\, d\mathbf{x}\, d\eta + (T-t)\int_\Omega u_0^2\, d\mathbf{x} + \beta(t+\tau)^2, \\
F'(t) &= 2\int_0^t \int_\Omega u u_\eta\, d\mathbf{x}\, d\eta + 2\beta(t+\tau), \\
F''(t) &= 4(\alpha+1)\left[\int_0^t \int_\Omega u_\eta^2\, d\mathbf{x}\, d\eta + \beta\right] \\
&\quad + 2\Big\{\alpha \int_\Omega |\nabla u|^2\, d\mathbf{x} - (\alpha+1)\int_\Omega |\nabla u_0|^2\, d\mathbf{x} \\
&\quad + \oint_{\partial\Omega}\left[u f(u) - 2(\alpha+1)\int_0^t \left(\frac{\partial}{\partial\eta}\int_{u_0(s)}^{u(s,\eta)} f(y)\, dy\right) d\eta\right] ds - (2\alpha+1)\beta\Big\}.
\end{aligned}
$$

We now split

$$F''(t) = F''_{(1)}(t) + F''_{(2)}(t),$$

where

(I.68)
$$F''_{(1)}(t) = 4(\alpha+1)\left[\int_0^t \int_\Omega u_\eta^2\, d\mathbf{x}\, d\eta + \beta\right].$$

Then,

(I.69)
$$
\begin{aligned}
FF'' - (\alpha+1)F'^2 &= FF''_{(1)} + FF''_{(2)} - (\alpha+1)F'^2 \\
&= (FF''_{(1)} - (\alpha+1)F'^2) + FF''_{(2)} \\
&\geqq (\hat{F}F''_{(1)} - (\alpha+1)F'^2) + FF''_{(2)},
\end{aligned}
$$

where

$$\hat{F}(t)=\int_0^t\int_\Omega u^2\,d\mathbf{x}\,d\eta+\beta(t+\tau)^2<F(t),\qquad t<T. \tag{I.70}$$

Now,

$$\hat{F}F''_{(1)}-(\alpha+1)F'^2=4(\alpha+1)S^2, \tag{I.71}$$

where

$$\begin{aligned}S^2\equiv&\left(\int_0^t\int_\Omega u^2\,d\mathbf{x}\,d\eta+\beta(t+\tau)^2\right)\left(\int_0^t\int_\Omega u_\eta^2\,d\mathbf{x}\,d\eta+\beta\right)\\&-\left[\int_0^t\int_\Omega uu_\eta\,d\mathbf{x}\,d\eta+\beta(t+\tau)\right]^2\\&\geqq 0,\end{aligned}$$

by virtue of the Cauchy-Schwarz inequality. Therefore, by (I.69) and our earlier computations,

$$\begin{aligned}FF''-(\alpha+1)F'^2&\geqq FF_{(2)}\\&=2F\left\{\alpha\int_\Omega|\nabla u|^2\,d\mathbf{x}-(\alpha+1)\int_\Omega|\nabla u_0|^2\,d\mathbf{x}-(2\alpha+1)\beta\right.\\&\quad\left.+\oint_{\partial\Omega}\left[uf(u)-2(\alpha+1)\int_0^t\left(\frac{\partial}{\partial\eta}\int_{u_0(s)}^{u(s,\eta)}f(y)\,dy\right)d\eta\right]ds\right\}.\end{aligned} \tag{I.72}$$

We now note that

$$\begin{aligned}\oint_{\partial\Omega}uf(u)\,ds&=\oint_{\partial\Omega}\int_0^u\frac{d}{dy}(yf(y))\,dy\,ds\\&=\oint_{\partial\Omega}\int_0^u(yf'(y)+f(y))\,dy\,ds\end{aligned} \tag{I.73}$$

and that

$$\begin{aligned}-2(\alpha+1)&\oint_{\partial\Omega}\int_0^t\left(\frac{\partial}{\partial\eta}\int_{u_0(s)}^{u(s,\eta)}f(y)\,dy\right)d\eta\,ds\\&=-2(\alpha+1)\oint_{\partial\Omega}\int_{u_0(s)}^{u(s,t)}f(y)\,dy\,ds\\&=-2(\alpha+1)\oint_{\partial\Omega}\int_{u_0(s)}^{0}f(y)\,dy\,ds-2(\alpha+1)\oint_{\partial\Omega}\int_0^{u(s,t)}f(y)\,dy\,ds\\&=2(\alpha+1)\oint_{\partial\Omega}\int_0^{u_0(s)}f(y)\,dy\,ds-2(\alpha+1)\oint_{\partial\Omega}\int_0^{u(s,t)}f(y)\,dy\,ds.\end{aligned} \tag{I.74}$$

Therefore, combining (I.73), (I.74) yields

$$\oint_{\partial\Omega}\left[uf(u)-2(\alpha+1)\int_0^t\left(\frac{\partial}{\partial\eta}\int_{u_0(s)}^{u(s,\eta)} f(y)\,dy\right)d\eta\right]ds$$

$$= 2(\alpha+1)\oint_{\partial\Omega}\int_0^{u_0(s)} f(y)\,dy\,ds \tag{I.75}$$

$$+\oint_{\partial\Omega}\int_0^{u(s,t)} (yf'(y)-(2\alpha+1)f(y))\,dy\,ds$$

and (I.72) becomes

$$FF''-(\alpha+1)F'^2$$

$$\geqq 2F\left\{\alpha\int_\Omega |\nabla u|^2\,d\mathbf{x}-(2\alpha+1)\beta\right.$$

$$+2(\alpha+1)\left[\oint_{\partial\Omega}\int_0^{u_0(s)} f(y)\,dy\,ds-\frac{1}{2}\int_\Omega|\nabla u_0|^2\,d\mathbf{x}\right] \tag{I.76}$$

$$\left.+\oint_{\partial\Omega}\int_0^{u_0(s,t)}[yf'(y)-(2\alpha+1)f(y)]\,dy\,ds\right\}.$$

In view of our assumption that

$$f(y)=|y|^{2\alpha+1}h(y), \qquad h'(y)>0,$$

we have

$$\oint_{\partial\Omega}\int_0^u [yf'(y)-(2\alpha+1)f(y)]\,dy\,ds>0.$$

Dropping this latter term in (I.76), as well as the lead term $2\alpha F\int_\Omega |\nabla u|^2\,d\mathbf{x}\geqq 0$, leaves us with

$$FF''-(\alpha+1)F'^2\geqq 2F\left\{-(2\alpha+1)\beta\right. \tag{I.77}$$

$$\left.+2(\alpha+1)\left[\oint_{\partial\Omega}\int_0^{u_0(s)} f(y)\,dy\,ds-\frac{1}{2}\int_\Omega|\nabla u_0|^2\,d\mathbf{x}\right]\right\}.$$

To complete the proof we need only choose

$$\beta=\frac{2(\alpha+1)}{(2\alpha+1)}\left[\oint_{\partial\Omega}\left(\int_0^{u_0(s)} f(y)\,dy\right)ds-\frac{1}{2}\int_\Omega|\nabla u_0|^2\,d\mathbf{x}\right]\geqq 0 \tag{I.78}$$

in view of hypothesis (ii) of the theorem. Thus, for $0\leqq t<T$,

$$FF''-(\alpha+1)F'^2\geqq 0 \Leftrightarrow (F^{-\alpha})''\leqq 0, \tag{I.79}$$

which, as already indicated, implies that

$$F^\alpha(t)\geqq F^{\alpha+1}(0)[F(0)-\alpha tF'(0)]^{-1}.$$

However, $F'(0)=2\beta\tau$ and $F(0)=T\int_\Omega u_0^2(\mathbf{x})\,d\mathbf{x}+\beta\tau^2$ so $\lim_{t\to t_\infty}F(t)=+\infty$, where

$$t_\infty=\frac{T\int_\Omega u_0^2\,d\mathbf{x}+\beta\tau^2}{2\alpha\beta\tau}<T \tag{I.80}$$

if and only if

$$\beta\tau^2<T\left[2\alpha\beta\tau-\int_\Omega u_0^2(\mathbf{x})\,d\mathbf{x}\right]. \tag{I.81}$$

Clearly, (I.81) is satisfied for sufficiently large T provided we choose the constant

$$\tau>\int_\Omega \frac{u_0^2(\mathbf{x})\,d\mathbf{x}}{2\alpha\beta}. \tag{I.82}$$

Q.E.D.

To summarize, we have the following result: If we choose β, τ so as to satisfy (I.78) and (I.82), respectively, then $F(t)$ as given by (I.58) satisfies $\lim_{t\to t_\infty}F(t)=+\infty$ for some $t_\infty<T$ when T is sufficiently large; this contradicts the assumed existence of a classical solution $u(\mathbf{x},t)$ on $\Omega\times[0,T)$ for every $T>0$. The key to the whole argument has been the establishing of the basic concavity estimate (I.79). In the present example α has been fixed essentially by hypothesis (i) of the theorem; for the linear initial-history value problems associated with the integrodifferential equations arising in Chapter III the parameter α which appears in the basic concavity argument will be arbitrary (but nonnegative) and may be used in certain instances to insure that the point $t_\infty<T$ for given $T>0$ provided α is chosen large. The global nonexistence result proven above may be extended in many different directions (to more general heat equations, i.e.,

$$\frac{\partial u}{\partial t}=\sum_{i,j}\frac{\partial}{\partial x_i}\left(a_{ij}(\mathbf{x})\frac{\partial u}{\partial x_j}\right),\qquad a_{ij}(\cdot)\in C^1(\Omega),$$

with $a_{ij}=a_{ji}$ on Ω, and boundary data of the form $\partial u/\partial\nu=\sum_{i,j}a_{ij}(\mathbf{x})\,(\partial u/\partial x_j)n_i=f(u)$ on $\partial\Omega\times[0,T)$, to initial-boundary value problems for the wave equation with associated nonlinear boundary data, and to systems of parabolic equations with coupled nonlinear boundary data but, as the main thrust of the present work concerns itself with integrodifferential rather than partial differential equations, we will not concern ourselves with such extensions and generalizations here; rather, we will simply indicate that the interested reader may consult [101] and [102] for an in depth discussion of global nonexistence results for a variety of nonlinear initial-boundary value problems for linear partial differential equations and systems of such equations.

Chapter II

Ill-Posed Problems for the Partial-Integrodifferential Equations of Linear and Nonlinear Viscoelasticity

In Chapter I we presented two concrete examples of the application of differential inequality arguments to problems in mathematical physics: logarithmic convexity arguments to derive stability, uniqueness, and continuous data dependence theorems for ill-posed initial-boundary value problems in linear elastodynamics and a concavity argument to prove global nonexistence of solutions for a nonlinear initial-boundary value problem in the theory of heat conduction. In the present chapter we will (a) show how logarithmic convexity arguments can be used to derive stability and growth estimates for solutions to ill-posed initial-history value problems in the theory of linear isothermal viscoelasticity and (b) how a modified concavity argument can be used to derive growth estimates for a specific nonlinear viscoelastic model. In the course of our discussion we will reference some of the more recent work on the problems of existence, uniqueness, stability, and asymptotic stability of solutions to the corresponding well-posed initial-history boundary value problems in both the linear and nonlinear cases.

1. Equations of motion and restrictions on relaxation functions: Isothermal linear viscoelasticity.

Let $\Omega \subseteq R^3$ be a bounded domain with smooth boundary $\partial\Omega$; we assume that in Ω the following constitutive relation of isothermal linear viscoelasticity obtains

$$t_{ij}(\mathbf{x}, t) = \int_{-\infty}^{t} \mathcal{G}_{ijkl}(\mathbf{x}, t-\tau) \frac{\partial^2 u_k}{\partial x_l \, \partial \tau} \, d\tau, \qquad t > 0, \tag{II.1}$$

where the $t_{ij}(\mathbf{x}, t)$ are the components of the (symmetric) Cauchy stress tensor at the point $\mathbf{x}$ and the time t, the $\mathcal{G}_{ijkl}(\mathbf{x}, t)$ are the components of the viscoelastic relaxation tensor and the u_i are the components of the displacement vector. Taken in conjunction with Cauchy's balance equation, i.e., $\operatorname{div} \mathbf{t} = \rho \ddot{\mathbf{u}}$, ρ the density on Ω (we assume zero external body force until the discussion of continuous data dependence) (II.1) yields the following system of equations of motion in $\Omega\times(0,T)$, $T>0$:

$$\text{(II.2)}\qquad \rho(\mathbf{x})\frac{\partial^2 u_i}{\partial t^2}-\frac{\partial}{\partial x_j}\left[\int_{-\infty}^{t} g_{ijkl}(\mathbf{x},t-\tau)\frac{\partial^2 u_k}{\partial x_l\,\partial\tau}\,d\tau\right]=0,\qquad i=1,2,3.$$

We associate with (II.2) initial and boundary data of the form

$$\text{(II.3)}\qquad \begin{aligned} &\mathbf{u}(\mathbf{x},t)=\mathbf{0}, && \partial\Omega\times(-\infty,T)\\ &\mathbf{u}(\mathbf{x},0)=\mathbf{f}(\mathbf{x}), && \frac{\partial\mathbf{u}}{\partial t}(\mathbf{x},0)=\mathbf{g}(\mathbf{x})\quad\text{on }\Omega,\end{aligned}$$

and assume that the displacement vector $\mathbf{u}$ is prescribed on Ω for all $t<0$, i.e.,

$$\text{(II.4)}\qquad u_i(\mathbf{x},\tau)=U_i(\mathbf{x},\tau),\qquad \Omega\times(-\infty,0).$$

In (II.2)–(II.3) the density $\rho(\mathbf{x})$ is assumed to be Lebesgue measurable with ess inf $\rho>0$. We also assume that for each t, $-\infty<t<\infty$, the components of the relaxation tensor g_{ijkl} are Lebesgue measurable and essentially bounded on Ω, with $g_{ijkl}(\cdot,t)\in C^1(\Omega)$, and that $(\partial/\partial t)g_{ijkl}(\mathbf{x},t)$ exists for each $t\in(-\infty,\infty)$ a.e. on Ω. The initial values $\mathbf{f}$ and $\mathbf{g}$ are taken to be continuous on Ω. If we make the usual assumption that $\lim_{\tau\to-\infty} g_{ijkl}(\mathbf{x},t-\tau)=0$, uniformly on Ω, then we may recast (II.2) in the form

$$\text{(II.2}'\text{)}\qquad \begin{aligned}&\rho(\mathbf{x})\frac{\partial^2 u_i}{\partial t^2}(\mathbf{x},t)-\frac{\partial}{\partial x_j}\left[g_{ijkl}(\mathbf{x},0)\frac{\partial u_k}{\partial x_l}(\mathbf{x},t)\right]\\ &\quad+\int_{-\infty}^{t}\frac{\partial}{\partial\tau}g_{ijkl}(\mathbf{x},t-\tau)\frac{\partial u_k}{\partial x_l}(\mathbf{x},\tau)\,d\tau=0,\qquad i=1,2,3,\end{aligned}$$

on $\Omega\times[0,T)$, $T>0$. The corresponding one-dimensional equations of motion are of the form

$$\text{(II.5)}\qquad \rho(x)\frac{\partial^2 u}{\partial t^2}-g(x,0)\frac{\partial^2 u}{\partial x^2}+\int_{-\infty}^{t}\frac{\partial}{\partial\tau}g(x,t-\tau)\frac{\partial^2 u}{\partial x^2}\,d\tau=0$$

on $[a,b]\times[0,T)$, $T>0$, where $b>a>0$ and $g(t)$ is called the relaxation function of the material.

A great deal of work has been done in the last decade on the problem of deriving physically reasonable restrictions on the relaxation tensors of linear viscoelastic materials; the most notable efforts in this direction have been made by Coleman [33], Gurtin and Herrera [60], and Day [46]–[48] with the basic results as follows: if the constitutive relations (II.1) of linear isothermal are compatible with thermodynamics (i.e., with the Clausius-Duhem inequality) then:

(i) $g_{ijkl}(\mathbf{x},0)$, the instantaneous elastic modulus is both positive semi-definite and symmetric.

(ii) $\dot{g}_{ijkl}(\mathbf{x}, 0)$ is negative semi-definite.

This latter result is actually a consequence of the relaxation tensor g_{ijkl} being dissipative in the sense of König and Meixner [81], i.e., that the work done on a strain-path, which starts from a state of zero strain, be nonnegative; the relaxation tensor is compatible with thermodynamics if and only if the work done on every closed strain-path is nonnegative. Precise connections between the concepts of dissipativity and compatibility with thermodynamics are derived in [48]; property (ii) of the viscoelastic relaxation tensor was actually shown by Coleman [33] to be a consequence of the Clausius-Duhem inequality for a more general class of materials with fading memory. The two properties of the relaxation function delineated in (i) and (ii) above can be shown to ensure the decay of acceleration waves in linear isothermal viscoelastic materials of the type governed by the constitutive relations (II.1). We may note here that stronger conditions than those given above on $g_{ijkl}(\mathbf{x}, 0)$, $\dot{g}_{ijkl}(\mathbf{x}, 0)$ will be needed to ensure that the initial-history boundary value problem (II.2'), (II.3), (II.4) is well-posed; some of these conditions are described in § 2. One condition which figures prominently in the existence and asymptotic stability results of Dafermos [37] and [38] is that the relaxation tensor $g_{ijkl}(\mathbf{x}, t)$ be symmetric not only at $t = 0$ but for all t, i.e., that $g_{ijkl}(\mathbf{x}, t) = g_{klij}(\mathbf{x}, t)$ on $\Omega \times [0, T)$, $T > 0$. However, Shu and Onat [136] have given an example of a dissipative relaxation function for which this symmetry does not hold and in conjunction with the results of Day [48] this implies that the symmetry of $g_{ijkl}(\mathbf{x}, t)$ on $\Omega \times [0, T)$ is implied neither by compatibility with thermodynamics nor by dissipativity; the symmetry of $g_{ijkl}(\mathbf{x}, t)$ on $\Omega \times [0, T)$ is also not implied by the fact that the relaxation tensor admits a free energy functional. In fact, Day [48] has constructed a relaxation tensor which is compatible with thermodynamics, is dissipative, admits a free-energy functional and yet is not symmetric for any $t > 0$; such results cast serious doubts on some of the mathematical assumptions which have been made in the recent literature to show that solutions of the initial-history boundary value problems of isothermal linear viscoelasticity exist and have the desirable properties of stability and asymptotic stability and validate the study of the corresponding ill-posed problems which result by dropping some of those mathematical assumptions (while we will retain the symmetry assumption on $g_{ijkl}(\mathbf{x}, t)$, we will drop, in our considerations in § 3, the definiteness assumptions on $g_{ijkl}(\mathbf{x}, t)$ and its time derivatives which have been used in the literature to prove well-posedness of the initial-history boundary value problem).

In the one-dimensional case it seems (at least from experimental evidence) that the corresponding one-dimensional relaxation functions $g(x, t)$ are always monotonically decreasing functions of t; yet Gurtin and Herrera [60] have constructed a relaxation function which is not monotonically decreasing in time but is dissipative. From the results of Day [48] it again follows that this property, the monotonicity of one-dimensional relaxation functions, is implied neither by compatibility with thermodynamics, nor by dissipativity nor by the fact that the relaxation function admits a free energy functional. We do note, however, the following results of Day ([47] and [46], respectively):

(iii) The relaxation function $g_{ijkl}(\mathbf{x}, t)$ is symmetric for every t, $0 \leqq t < \infty$, if and only if the work done on every closed path starting from a state of zero strain is invariant under time-reversal.

(iv) The relaxation function $g(x, t)$ of a one-dimensional viscoelastic material is monotonically decreasing in time if and only if the work done is always increased by delay on retraced paths; i.e., suppose we take any path in strain space which starts from equilibrium and arrives at a given strain $\mathcal{S}$ at time t; the material can be taken around a closed path either by retracing the given path immediately or by holding the strain fixed at the value $\mathcal{S}$ for a time τ and then retracing the given path. In general $\mathcal{W}(\tau) \neq \mathcal{W}$ where $\mathcal{W}$ is the work done on the first closed path and $\mathcal{W}(\tau)$ is the work done on the second closed path. If $\mathcal{W}(\tau) \geqq \mathcal{W}$, $\forall \tau \geqq 0$, and all initial strain paths which start from equilibrium, the work done is said to be increased by delay on retraced paths. For further details and a proof of the above characterization we refer the reader to [46]; with the above characterization, Day actually proves that the one-dimensional relaxation functions g must be completely monotone, i.e., that

$$(-1)^n g^{(n)}(x, t) \geqq 0, \quad \text{on } [a, b] \times [0, T), \qquad T > 0,$$

for all positive integers n.

We note in closing that in the application of our growth estimates to ill-posed initial-history value problems in one-dimensional isothermal viscoelasticity we will not need to make use of the full monotonicity of the relaxation function g but shall only require that $\dot{g}(x, 0) < -k$ for $k > 0$ sufficiently large (i.e., that g be decreasing sufficiently fast at $t = 0$).

2. Structure of the initial-history value problem in Hilbert space: Basic results on existence, uniqueness, stability, and asymptotic stability.

In order to expedite the discussion of the qualitative behavior of solutions to ill-posed initial-history boundary value problems of the form (II.2′), (II.3), (II.4) we will first recast the problem as an initial-history value problem in Hilbert space; this necessitates the introduction of certain elementary Sobolev spaces.

Let $(C_0^\infty(\Omega))^3$ denote the set of three-dimensional vector fields with compact support in Ω whose components belong to $C^\infty(\Omega)$; we define the space H to be the completion of $(C_0^\infty(\Omega))^3$ under the norm induced by the inner-product

$$\langle \mathbf{v}, \mathbf{w} \rangle = \int_\Omega v_i(\mathbf{x}) w_i(\mathbf{x}) \, d\mathbf{x} \tag{II.6}$$

and H_+ to be the completion of $(C_0(\Omega))^3$ under the norm induced by the inner-product

$$\langle \mathbf{v}, \mathbf{w} \rangle_+ \equiv \int_\Omega \frac{\partial v_i}{\partial x_j}(\mathbf{x}) \frac{\partial w_i}{\partial x_j}(\mathbf{x})\, d\mathbf{x}. \tag{II.7}$$

The natural norms associated with $\langle \cdot , \cdot \rangle$ and $\langle \cdot , \cdot \rangle_+$ will be denoted by $\|(\cdot)\|$ and $\|(\cdot)\|_+$, respectively. Finally, we define H_- to be the completion of $(C_0^\infty(\Omega))^3$ under the norm

$$\|\mathbf{v}\|_- = \sup_{\mathbf{w} \in H_+} \left[\frac{\left| \int_\Omega v_i w_i \, d\mathbf{x} \right|}{\left(\int_\Omega \frac{\partial w_i}{\partial x_j} \frac{\partial w_i}{\partial x_j} d\mathbf{x} \right)^{1/2}} \right]. \tag{II.8}$$

The reader familiar with the basic elements of Sobolev space theory (for a comprehensive text see [1]) will recognize that $H = (L_2(\Omega))^3$, $H_+ = (H_0^1(\Omega))^3$ and $H_- = (H^{-1}(\Omega))^3$; these spaces may be characterized in an equivalent but entirely different fashion as follows: Let $\boldsymbol{\alpha} = (\alpha_1, \cdots, \alpha_n)$ where $\alpha_i = \text{integer} \geqq 0$. Set $|\boldsymbol{\alpha}| = \alpha_1 + \cdots + \alpha_n$ and $\mathbf{x}^{\boldsymbol{\alpha}} = x_1^{\alpha_1} x_2^{\alpha_2} \cdots x_n^{\alpha_n}$, where $\mathbf{x} = (x_1, \cdots, x_n) \in R^n$. Then, if $\mathbf{D} = (\partial/\partial x_1, \cdots, \partial/\partial x_n)$ denotes the gradient operator in R^n, $\mathbf{D}^{\boldsymbol{\alpha}} = \partial^{|\boldsymbol{\alpha}|}/(\partial x_1^{\alpha_1} \cdots \partial x_n^{\alpha_n})$; this multi-index notation is now standard in the modern theory of partial differential equations. The Sobolev space of order (m, p), $1 \leqq p \leqq m$ (integers), which we denote as $W_p^m(\Omega)$, is defined as the space of real-valued functions on Ω whose *generalized* (*distributional*) derivatives of order $\leqq m$ are in $L_p(\Omega)$, i.e.,

$$W_p^m(\Omega) = \{\mathbf{u} : \mathbf{D}^{\boldsymbol{\alpha}}\mathbf{u} \in L_p(\Omega) \forall \boldsymbol{\alpha}, |\boldsymbol{\alpha}| \leqq m\}. \tag{II.9}$$

Note. We say that $\mathbf{v} = \mathbf{D}^{\boldsymbol{\alpha}}\mathbf{u}$ is the $\boldsymbol{\alpha}$th distributional (or generalized) derivative of $\mathbf{u}$, $|\boldsymbol{\alpha}| = m$ if $\forall \phi \in C_0^\infty(\Omega)$

$$\int_\Omega \mathbf{u} \mathbf{D}^{\boldsymbol{\alpha}} \boldsymbol{\phi}\, d\mathbf{x} = (-1)^{|\boldsymbol{\alpha}|} \int_\Omega \mathbf{v} \boldsymbol{\phi}\, d\mathbf{x}. \tag{II.10}$$

The space $W_p^m(\Omega)$ is assigned the (Sobolev) norm

$$\begin{aligned} \|\mathbf{u}\|_{m,p} &= \left\{ \sum_{|\boldsymbol{\alpha}| \leqq m} \int_\Omega |\mathbf{D}^{\boldsymbol{\alpha}}\mathbf{u}|^p \, d\mathbf{x} \right\}^{1/p} \\ &= \left\{ \sum_{|\boldsymbol{\alpha}| \leqq m} \|\mathbf{D}^{\boldsymbol{\alpha}}\mathbf{u}\|_{L_p(\Omega)}^p \right\}^{1/p}. \end{aligned} \tag{II.11}$$

($W_p^m(\Omega)$ can be shown to coincide with the completion of $C^m(\Omega)$ with respect to the $\|(\cdot)\|_{m,p}$-norm.) For $m \geqq k, p \geqq r$ it follows that $W_p^m(\Omega) \subseteq W_r^k(\Omega)$. An important subspace of $W_p^m(\Omega)$ is

$$\mathring{W}_p^m(\Omega) \equiv \{\mathbf{u} \in W_p^m(\Omega) | \mathbf{D}^{\boldsymbol{\alpha}}\mathbf{u} = 0 \text{ on } \partial\Omega, |\boldsymbol{\alpha}| \leqq m-1\}. \tag{II.12}$$

(If one defines $\mathring{W}_p^m(\Omega)$ to be the completion of $C_0^m(\Omega)$, the space of m-times continuously differentiable functions with compact support in Ω, then (II.12) follows as a consequence of a standard trace theorem). For $p=2$ the Sobolev spaces become Hilbert spaces with inner-products

$$\langle \mathbf{u}, \mathbf{v}\rangle_{m,2} \equiv \langle \mathbf{u}, \mathbf{v}\rangle_m = \sum_{|\alpha|\leq m} \int_\Omega \mathbf{D}^\alpha \mathbf{u} \cdot \mathbf{D}^\alpha \mathbf{v}\, dx \tag{II.13}$$

and the notation

$$H^m(\Omega) = W_2^m(\Omega), \qquad H_0^m(\Omega) = \mathring{W}_2^m(\Omega) \tag{II.14}$$

is then fairly standard. ($\|\mathbf{u}\|_{m,2}$ will then be denoted as just $\|\mathbf{u}\|_m$ for $\mathbf{u} \in H^m(\Omega)$). In the applications presented in this work we will only have to concern ourselves with the special case of $m=1$, $p=2$. We note that "negative" and "fractional" Sobolev spaces can be defined for negative and fractional values of m, respectively; we will have no need to employ "fractional" Sobolev spaces in this book and, thus, do not enter into a discussion of them here. The "negative" Sobolev spaces $W_q^{-m}(\Omega)$ may be defined as the dual spaces of $\mathring{W}_p^m(\Omega)$, $m>0$, where $(1/p)+(1/q)=1$, i.e.,

$$\begin{aligned} &W_q^{-m}(\Omega) \equiv (\mathring{W}_p^m(\Omega))', \\ &H^{-m}(\Omega) \equiv (H_0^m(\Omega))', \qquad (p=q=2). \end{aligned} \tag{II.15}$$

In particular, $H^{-1}(\Omega) \equiv (H_0^1(\Omega))'$. The norm on $W_q^{-m}(\Omega)$ is then given by

$$\|\mathbf{u}\|_{-m,q} = \sup_{v \in \mathring{W}_p^m(\Omega)} \frac{|\langle \mathbf{u}, \mathbf{v}\rangle|}{\|\mathbf{v}\|_{W_p^m(\Omega)}} \tag{II.16}$$

where $\langle \cdot, \cdot\rangle$ denotes the duality pairing on $(\mathring{W}_p^m(\Omega))' \times \mathring{W}_p^m(\Omega)$. It can be shown that the definition of $H^{-1}(\Omega) \equiv (H_0^1(\Omega))'$ is in agreement with the definition of $H^{-1}(\Omega)$ as the completion of $C_0^\infty(\Omega)$ in the $\|(\cdot)\|_{-1,2}$ norm; note that for $p=q=2$ the duality pairing $\langle \cdot, \cdot\rangle$ may be replaced by the standard $L_2(\Omega)$ inner-product as in (II.8). Finally, we note the well-known fact that $H_0^1(\Omega) \subseteq L_2(\Omega) \subseteq H^{-1}(\Omega)$, both topologically as well as algebraically so that, in particular, $\exists \tilde{\gamma} > 0$ (depending only on Ω) such that $\|\mathbf{u}\|_1 \leqq \tilde{\gamma}\|\mathbf{u}\|_0$, $\forall \mathbf{u} \in H_0^1(\Omega)$, where $\|(\cdot)\|_0$ denotes the $L_2(\Omega)$-norm of $\mathbf{u}$; this also implies that $H_+ \equiv (H_0^1(\Omega))^3 \subseteq (L_2(\Omega))^3 \equiv H$, both algebraically as well as topologically, so that there exists $\gamma > 0$ such that $\|\mathbf{u}\| \leqq \gamma \|\mathbf{u}\|_+$, $\forall \mathbf{u} \in H$; we call γ the embedding constant for the inclusion map $i: H_+ \to H$. No additional material related to the so-called Sobolev embedding theorems will be needed in this treatise.

Now, let $\mathscr{L}_S(H_+, H_-)$ denote the space of all symmetric linear operators which are bounded as maps of $H_+(\equiv (H_0^1(\Omega))^3)$ into $H_-(\equiv (H^{-1}(\Omega))^3)$. We define, as in [22], an operator $\mathbf{G}(\cdot) \in L^2((-\infty, \infty); \mathscr{L}_S(H_+, H_-))$ as follows: for any $\mathbf{w} \in H_+$ and any $t \in (-\infty, \infty)$ let

$$[\mathbf{G}(t)\mathbf{w}]_i(\mathbf{x}) \equiv \frac{1}{\rho(\mathbf{x})} \frac{\partial}{\partial x_j}\left[g_{ijkl}(\mathbf{x}, t) \frac{\partial w_k}{\partial x_l}\right], \tag{II.17}$$

where the g_{ijkl} are the components of the viscoelastic relaxation tensor and the spatial derivatives are taken in the distributional sense. Thus

$$G_{ik}(\mathbf{x}, t) = \frac{1}{\rho(\mathbf{x})} \frac{\partial}{\partial x_j} \left[g_{ijkl}(\mathbf{x}, t) \frac{\partial}{\partial x_l} \right] \tag{II.18}$$

for $(\mathbf{x}, t) \in \Omega \times (-\infty, \infty)$. Clearly, if we make the previously discussed symmetry assumption relative to the relaxation tensor, i.e., that $g_{ijkl}(\mathbf{x}, t) = g_{klij}(\mathbf{x}, t)$ on $\Omega \times (-\infty, \infty)$ then $\mathbf{G}(t)$ is a symmetric operator for each $t \in (-\infty, \infty)$. To show that $\mathbf{G}(t)$ is bounded as a map from H_+ into H_- we compute as follows: Let $\mathbf{v} \in H_+$; then

$$\|\mathbf{G}(t)\mathbf{v}\|_- = \sup_{\mathbf{w} \in H_+} \frac{|\langle \mathbf{G}(t)\mathbf{v}, \mathbf{w} \rangle|}{\|\mathbf{w}\|_+}$$

$$= \sup_{\mathbf{w} \in H_+} \frac{\left| \int_\Omega \frac{1}{\rho(\mathbf{x})} \frac{\partial}{\partial x_j} \left[g_{ijkl}(\mathbf{x}, t) \frac{\partial v_k}{\partial x_l} \right] w_i \, d\mathbf{x} \right|}{\|\mathbf{w}\|_+}$$

$$\leqq \sup_\Omega \left(\frac{1}{\rho} \right) \sup_{\mathbf{w} \in H_+} \frac{\left| \int_\Omega \frac{\partial}{\partial x_j} \left[g_{ijkl}(\mathbf{x}, t) \frac{\partial v_k}{\partial x_l} \right] w_i \, d\mathbf{x} \right|}{\|\mathbf{w}\|_+}$$

$$= \sup_\Omega \left(\frac{1}{\rho} \right) \sup_{\mathbf{w} \in H_+} \frac{1}{\|\mathbf{w}\|_+} \left| \left(\oint_{\partial\Omega} g_{ijkl}(\mathbf{x}, t) \frac{\partial v_k}{\partial x_l} w_i n_j \, dS_\mathbf{x} - \int_\Omega g_{ijkl}(\mathbf{x}, t) \frac{\partial w_i}{\partial x_j} \frac{\partial v_k}{\partial x_l} \, d\mathbf{x} \right) \right|$$

$$= \sup_\Omega \left(\frac{1}{\rho} \right) \sup_{\mathbf{w} \in H_+} \frac{1}{\|\mathbf{w}\|_+} \left| \int_\Omega g_{ijkl}(\mathbf{x}, t) \frac{\partial w_i}{\partial x_j} \frac{\partial v_k}{\partial x_l} \, d\mathbf{x} \right| \tag{II.19}$$

$$= \sup_\Omega \left(\frac{1}{\rho} \right) \sup_{\mathbf{w} \in H_+} \frac{\left| \int_\Omega g_{ijkl}(\mathbf{x}, t) \frac{\partial w_i}{\partial x_j} \frac{\partial v_k}{\partial x_l} \, d\mathbf{x} \right|}{\left[\int_\Omega \frac{\partial w_i}{\partial x_j} \frac{\partial w_i}{\partial x_j} \, d\mathbf{x} \right]^{1/2}}$$

$$\leqq \sup_\Omega \left(\frac{1}{\rho} \right) \left(\max_{i,j,k,l} \sup_\Omega |g_{ijkl}(\mathbf{x}, t)| \right) \times \sup_{\mathbf{w} \in H_+} \left(\frac{\left| \int_\Omega \frac{\partial w_i}{\partial x_j} \frac{\partial v_k}{\partial x_l} \, d\mathbf{x} \right|}{\left[\int_\Omega \frac{\partial w_i}{\partial x_j} \frac{\partial w_i}{\partial x_j} \, d\mathbf{x} \right]^{1/2}} \right)$$

$$\leqq \sup_\Omega \left(\frac{1}{\rho} \right) \left(\max_{i,j,k,l} \sup_\Omega |g_{ijkl}(\mathbf{x}, t)| \right) \times \left[\int_\Omega \frac{\partial v_k}{\partial x_l} \frac{\partial v_k}{\partial x_l} \, d\mathbf{x} \right]^{1/2}$$

$$= \sup_\Omega \left(\frac{1}{\rho} \right) \left(\max_{i,j,k,l} \sup_\Omega |g_{ijkl}(\mathbf{x}, t)| \right) \|\mathbf{v}\|_+,$$

where we have used integration by parts, the fact that $w_i = 0$ on $\partial\Omega$ (as $\mathbf{w} \in H_+ \equiv (H_0^1(\Omega))^3$), and the Cauchy-Schwarz inequality. Therefore, for any $\mathbf{v} \in H_+$, $t \in (-\infty, \infty)$,

$$\text{(II.20)} \qquad \frac{\|\mathbf{G}(t)\mathbf{v}\|_-}{\|\mathbf{v}\|_+} \leqq \sup_\Omega \left(\frac{1}{\rho}\right)(\max_{i,j,k,l} \sup_\Omega |g_{ijkl}(\mathbf{x}, t)|)$$

and so

$$\text{(II.21)} \qquad \begin{aligned} \|\mathbf{G}(t)\|_{\mathcal{L}_S(H_+,H_-)} &\equiv \sup_{\mathbf{v}\in H_+} \frac{\|\mathbf{G}(t)\mathbf{v}\|_-}{\|\mathbf{v}\|_+} \\ &\leqq \sup_\Omega \left(\frac{1}{\rho}\right)\left(\max_{i,j,k,l} \sup_\Omega |g_{ijkl}(\mathbf{x}, t)|\right). \end{aligned}$$

Thus, $\mathbf{G}(t)$ is bounded, as a (linear) map of H_+ into H_-, for each $t \in (-\infty, \infty)$.

We now define

$$\text{(II.22)} \qquad \mathbf{N} = \mathbf{G}(0) \quad \text{and} \quad \mathbf{K}(t-\tau) = \frac{\partial}{\partial\tau}\mathbf{G}(t-\tau).$$

It should be clear from the discussion above that $\mathbf{N} \in \mathcal{L}_S(H_+, H_-)$, $\mathbf{K}(\cdot) \in L^2((-\infty, \infty); \mathcal{L}_S(H_+, H_-))$ and that with these definitions of $\mathbf{N}$, $\mathbf{K}(t)$, the initial-history boundary value problem (II.2′), (II.3), (II.4) may be recast as the following initial-history value problem in Hilbert space: for $0 \leqq t < T$, $T > 0$, find $\mathbf{u} \in C^2([0, T); H_+)$ for which $\mathbf{u}_t \in C^1([0, T); H_+)$, $\mathbf{u}_{tt} \in C([0, T); H_-)$ and

$$\text{(II.23)} \qquad \mathbf{u}_{tt} - \mathbf{N}\mathbf{u} + \int_{-\infty}^{t} \mathbf{K}(t-\tau)\mathbf{u}(\tau)\,d\tau = \mathbf{0},$$

$$\text{(II.24)} \qquad \mathbf{u}(0) = \mathbf{f}, \qquad \mathbf{u}_t(0) = \mathbf{g}, \qquad \mathbf{f}, \mathbf{g} \in H_+,$$

$$\text{(II.25)} \qquad \mathbf{u}(\tau) = \mathbf{U}(\tau), \qquad -\infty < \tau < 0,$$

where we assume that $\mathbf{U}(\cdot) \in \mathrm{C}^1((-\infty, \infty); \mathrm{H}_+)$ with

(i) $\lim_{t\to 0^-}\|\mathbf{U}(t) - \mathbf{f}\| = 0, \qquad \lim_{t\to 0^-}\|\mathbf{U}_t(t) - \mathbf{g}\| = 0;$

(ii) $\lim_{t\to -\infty}\|\mathbf{U}(t)\|_+ = 0;$

(iii) $\int_{\infty}^{0} \|\mathbf{U}(\tau)\|_+ \, d\tau < +\infty.$

Our goal, in the next several sections, will be to derive a series of stability and growth estimates for solutions $\mathbf{u} \in C^2([0, T); H_+)$ of the initial-history value problem (II.23)–(II.25) under very mild conditions on the operators $\mathbf{N}$ and $\mathbf{K}(t)$; in particular, no definiteness assumptions will be made on $\mathbf{K}(t)$, $t \in (-\infty, \infty)$, and as the results in the recent literature suggest, without such definiteness assumptions, the aforementioned initial-history value problem will, in general, be ill-posed. The stability and growth estimates which we will derive will be based on logarithmic convexity and concavity arguments of the type employed in Chapter I. However, before proceeding to the derivation of our basic estimates it is worthwhile, at this point, to pause and indicate the kind of conditions on the relaxation tensor which must be imposed if the initial-history boundary value problem (II.2′), (II.3), (II.4) is to be proven well-posed.

Dafermos has considered the problem of existence, uniqueness, stability, and asymptotic stability of smooth solutions to the initial-history boundary value problem (II.2′), (II.3), (II.4) in a series of papers [37], [38], [39], and [40] and by employing a variety of methods; in [37] he recasts the problem within the dynamical systems framework introduced by Hale [65] and shows, via a modification of the invariance principle for dynamical systems, that there exist unique smooth solutions of (II.23)–(II.25) if the operators $\mathbf{N}$, $\mathbf{K}(t)$ satisfy the definiteness hypotheses

(II.26) $$\langle \mathbf{N}\mathbf{v}, \mathbf{v}\rangle \geqq c_0\|\mathbf{v}\|^2, \qquad c_0>0, \quad \forall \mathbf{v}\in H_+,$$

(II.27) $$\langle \mathbf{K}(t)\mathbf{v}, \mathbf{v}\rangle \leqq 0, \quad \forall \mathbf{v}\in H_+, \quad t\in[0,\infty).$$

In order to prove asymptotic stability of solutions within this framework the following (stronger) conditions in [37] must be appended to (II.26) and (II.27):

(II.28) $$\langle \mathbf{K}_t(t)\mathbf{v}, \mathbf{v}\rangle \geqq 0 \quad \forall \mathbf{v}\in H_+, \qquad t\in[0,\infty),$$

(II.29) $$\int_0^\infty \|\mathbf{K}(t)\|_{\mathscr{L}_S(H_+,H_-)}\,dt \equiv M_\infty < 1,$$

(II.30) $$\langle \mathbf{A}\mathbf{v}, \mathbf{v}\rangle \geqq a_0\|\mathbf{v}\|_+^2, \qquad a_0>0 \quad \forall \mathbf{v}\in H_+,$$

where $\mathbf{A}=\mathbf{N}+\int_0^\infty \mathbf{K}(t)\,dt$. If we set

(II.31) $$\mathscr{G}_{ijkl}(\mathbf{x}, t-\tau) = \frac{\partial}{\partial \tau}\, g_{ijkl}(\mathbf{x}, t-\tau), \qquad \mathbf{x}\in\Omega,$$

then in terms of the relaxation tensor, the hypotheses in [37] which guarantee uniqueness, stability, and asymptotic stability of solutions, assume the following forms:

(II.32) $$\int_\Omega g_{ijkl}(\mathbf{x},0)\frac{\partial v_i}{\partial x_j}\frac{\partial v_k}{\partial x_l}\,d\mathbf{x} \geqq c_1\int_\Omega \frac{\partial v_i}{\partial x_j}\frac{\partial v_i}{\partial x_j}\,d\mathbf{x}, \qquad c_1>0, \quad \forall \mathbf{v}\in (C_0^\infty(\Omega))^3,$$

(II.33) $$\int_\Omega \mathscr{G}_{ijkl}(\mathbf{x},t)\frac{\partial v_i}{\partial x_j}\frac{\partial v_k}{\partial x_l}\,d\mathbf{x} \leqq 0 \quad \forall \mathbf{v}\in (C_0^\infty(\Omega))^3, \quad t\in[0,\infty),$$

(II.34) $$\int_\Omega \dot{\mathscr{G}}_{ijkl}(\mathbf{x},t)\frac{\partial v_i}{\partial x_j}\frac{\partial v_k}{\partial x_l}\,d\mathbf{x} \leqq 0 \quad \forall \mathbf{v}\in (C_0(\Omega))^3, \quad t\in[0,\infty).$$

In addition, the conditions which correspond to (II.29) and (II.30) must be satisfied as well. For the one-dimensional problem, that is, the initial-history boundary value problem associated with the integrodifferential equation (II.5), the conditions are somewhat simpler to state: in order for existence, uniqueness, and asymptotic stability of smooth solutions to obtain, the hypotheses of [37] require that the function $\mathcal{G}(t-\tau)=(\partial/\partial t)g(t-\tau)$ (we assume, for the sake of simplicity that the one-dimensional viscoelastic material is homogeneous) be nonnegative, monotonically nonincreasing, and satisfy

$$g(0)-\int_0^\infty \mathcal{G}(\tau)\,d\tau>0. \tag{II.35}$$

(We note that the proof of asymptotic stability given in [38] requires that $\mathcal{G}(t)$ also be a convex function.) In [39] and [40] Dafermos again establishes the existence, uniqueness, and asymptotic stability of solutions to the initial-history boundary value problem (II.2′), (II.3), (II.4) by using the theory of contraction semi-groups on Hilbert space; more specifically, for the one-dimensional problem with governing evolution equation (II.5) it is shown by Dafermos that one may recast the associated initial-history boundary value problem in the form

$$\begin{aligned}\dot{\mathbf{\Psi}}(t)&=\mathbf{A}\mathbf{\Psi}(t),\\ \mathbf{\Psi}(0)&=\boldsymbol{\chi},\end{aligned} \tag{II.36}$$

where $\mathbf{\Psi}:[0,\infty)\to\mathcal{H}$ (a Hilbert space), $\boldsymbol{\chi}\in\mathcal{H}$, and $\mathbf{A}$ is an unbounded operator on $\mathcal{H}$ which can be extended to be the generator of a contraction semigroup $\mathbf{S}(t)$ on $\mathcal{H}$; the unique generalized solution of (II.36) is then given by $\mathbf{\Psi}(t)-\mathbf{S}(t)\boldsymbol{\chi}$. The fact that $\mathbf{S}(t)$ is a contraction semigroup ($\|\mathbf{S}(t)\mathbf{\Phi}\|_{\mathcal{H}}\leqq\|\mathbf{\Phi}\|_{\mathcal{H}}\forall\mathbf{\Phi}\in\mathcal{H}$, $t\in[0,\infty)$) immediately implies stability of solutions; asymptotic stability of solutions then follows from results of Dafermos and Slemrod [42] on the asymptotic behavior of motions of contraction semigroups in Hilbert space. What is basically involved in [39] and [40] is the idea of writing the equation of motion (II.5) in the form of a system

$$\begin{aligned}\dot{u}&=\frac{1}{\rho}v,\\ \dot{v}&=g(0)\frac{\partial^2 u}{\partial x^2}-\int_0^\infty \mathcal{G}(\tau)\frac{\partial^2 u_h}{\partial x^2}\,d\tau,\\ \dot{u}_h&=-\frac{\partial u_h}{\partial \tau},\end{aligned} \tag{II.37}$$

where

$$\begin{aligned}u_h(x,t,\tau)&=u(x,t-\tau),\qquad \tau\in[0,\infty),\\ v(x,t)&=\rho\dot{u}(x,t).\end{aligned} \tag{II.38}$$

The state of the material is then characterized by the triple $\Psi(t)=(u(x,t), v(x,t), u_h(x,t))$; the evolution equation (II.5) and the prescription of the past history of the displacement are then formally equivalent to the operator equation in (II.36) where $\mathcal{H}$, the underlying Hilbert space is chosen to be the set of all triples $(\tilde{u}(x), \tilde{v}(x), \tilde{u}_h(x))$ such that $\tilde{u}(\cdot)\in H_0^1(\Omega)$, ($\Omega=[a,b]$, an interval), $\tilde{v}(\cdot)\in L^2(\Omega)$, $\tilde{u}_h(\cdot,\tau)\in H_0^1(\Omega)$, a.e. on $[0,\infty)$ and

$$\text{(II.39)}\qquad \int_0^\infty \int_\Omega \mathcal{G}(\tau)[\tilde{u}_x(x)-(\tilde{u}_h)_x(x,\tau)]^2\,dx\,d\tau<\infty.$$

For the inner-product on $\mathcal{H}$ we have

$$\text{(II.40)}\qquad \begin{aligned}\langle(\tilde{u},\tilde{v},\tilde{u}_h),(\tilde{u}^*,\tilde{v}^*,\tilde{u}_h^*)\rangle &= \int_\Omega\left(\frac{1}{\rho}v\tilde{v}^*+g(0)\tilde{u}_x\tilde{u}_x^*\right)dx\\ &\quad+\int_0^\infty\int_\Omega \mathcal{G}(\tau)(\tilde{u}_x-(\tilde{u}_h)_x)(\tilde{u}_x^*-(\tilde{u}_h^*)_x)\,dx\,d\tau.\end{aligned}$$

This inner-product is suggested by an identity associated with solutions of (II.5), i.e.,

$$\text{(II.41)}\qquad \begin{aligned}&\frac{d}{dt}\int_\Omega\left(\rho\dot{u}^2(x,t)+g(0)u_x^2(x,t)+\int_{-\infty}^t \mathcal{G}(t-\tau)[u_x(x,t)-u_x(x,\tau)]^2\,d\tau\right)dx\\ &=\int_0^\infty\int_\Omega \dot{\mathcal{G}}(t-\tau)[u_x(x,t)-u_x(x,\tau)]^2\,dx\,d\tau\\ &\leqq 0,\end{aligned}$$

which is valid under the hypothesis of monotonically nonincreasing $\mathcal{G}(\cdot)$. By employing the inner-product (II.40) in conjunction with the identity (II.41) Dafermos ([39], [40]) is able to show that if **A** stands for the operator defined by the right-hand side of the system (II.37) then for all triples $(\tilde{u},\tilde{v},\tilde{u}_h)$ in the domain of **A**

$$\text{(II.42)}\qquad \langle \mathbf{A}(\tilde{u},\tilde{v},\tilde{u}_h),(\tilde{u},\tilde{v},\tilde{u}_h)\rangle=\frac{1}{2}\int_0^\infty\int_\Omega \dot{\mathcal{G}}(\tau)(\tilde{u}_x-(\tilde{u}_h)_x)^2\,dx\,d\tau\leqq 0,$$

i.e., **A** is a dissipative operator on $\mathcal{H}$; if one now demonstrates, as in [40], that the range of $A-I$ is $\mathcal{H}$ then the dissipativity of **A** implies, by virtue of the Lumer-Phillips theorem (see Pazy [127]), that **A** generates a contraction semigroup $S(t)$ on $\mathcal{H}$. The essential conditions, again, are that $\mathcal{G}(t)$ be nonnegative, monotonically nonincreasing, and satisfy (II.35). To obtain well-posedness by this latter method, for the three-dimensional problem (II.2′), (II.3), (II.4), one would again have to impose on the relaxation tensor the conditions implied by the rather strong definiteness hypotheses (II.26)–(II.31); our basic aim in the next two sections will be to examine the type of information that one can obtain concerning the stability and growth behavior of solutions to initial-history boundary value problems in linear viscoelasticity when definiteness hypotheses of this type are not imposed, a priori, on the viscoelastic relaxation tensor.

Remarks. It seems worthwhile at this point to indicate that some of the methods delineated above, e.g. in [39] and [40] in particular, have been used with success to prove results on existence, uniqueness, stability and asymptotic stability for problems which arise in a variety of situations in mathematical physics and whose governing evolution equations are integrodifferential equations. For example, in [137] and [138] Slemrod considers the problem of proving the existence, stability and asymptotic stability of solutions to a linearized initial-history boundary value problem for simple fluids which obey a Boltzmann type constitutve equation with fading memory as first formulated by Coleman and Noll [34]; the Coleman-Noll constitutive equation is of the form

$$\mathbf{t}+p\boldsymbol{\delta}=2\int_0^t m(s)(\mathbf{E}(t-s)-\mathbf{E}(t))\,ds, \tag{II.43}$$

where $\mathbf{t}$ is the stress, $\boldsymbol{\delta}$ the Kronecker delta, p the indeterminate mean normal stress, $\mathbf{E}$ the infinitesimal strain tensor, and $m(s)$ a material function. The shear relaxation modulus G is defined by $G=\int_\infty^s m(\tau)\,d\tau$ and Slemrod [137] shows by using a contraction semigroup argument, akin to those in [39] and [40], that if G satisfies the hypotheses

$$\begin{gathered} G\in C^2[0,\infty), \qquad G(s)\to 0 \quad \text{as } s\to\infty, \\ (-1)^k\frac{d^kG(x)}{ds^k}>0, \qquad k=0,1, \\ G''(s)\geqq 0, \end{gathered} \tag{II.44}$$

then the rest state of the fluid is stable in the fading memory norm associated with the inner-product

$$\begin{aligned} \langle(\mathbf{v},\mathbf{y}),(\mathbf{v}^*,\mathbf{y}^*)\rangle_{\#} &= \int_\Omega \nabla_x\left\{\int_0^\infty G(s)\mathbf{y}(\mathbf{x},s)\,ds\right\}:\nabla_x\left\{\int_0^\infty G(s)\mathbf{y}^*(\mathbf{x},s)\,ds\right\}d\mathbf{x} \\ &\quad -\int_\Omega\int_0^\infty G'(s)(\mathbf{y}(\mathbf{x},s)-\mathbf{v}(\mathbf{x}))\cdot(\mathbf{y}^*(\mathbf{x},s)-\mathbf{v}^*(\mathbf{x}))\,ds\,d\mathbf{x}, \end{aligned} \tag{II.45}$$

where, as previously, $\Omega\subseteq R^3$ is an open bounded domain, $\mathbf{v}(\mathbf{x},t)$ is the velocity of the fluid at $\mathbf{x}\in\Omega$ at time t, and $\mathbf{y}(\mathbf{x},s,t)=\mathbf{v}(\mathbf{x},t-s)$ ($\mathbf{y}(\mathbf{x},s,t)=\mathbf{v}_h(\mathbf{x},t,s)$ in the notation of [39], [40]). The additional assumption that

$$-\int_0^\infty G'(s)s^2\,ds<\infty \tag{II.46}$$

yields asymptotic stability in the $\|(\cdot)\|_{\#}$-norm associated with $\langle\cdot,\cdot\rangle_{\#}$, i.e.,

$\lim_{t\to\infty}\|(\mathbf{v}(t), \mathbf{y}(t))\|_{\#} = 0$, where by the constitutive hypothesis (II.43), Cauchy's equations of motion and the incompressibility of the fluid, $\mathbf{v}$ satisfies the initial-history boundary value problem

$$\text{(II.47)}\qquad \begin{aligned} &\dot{\mathbf{v}}(\mathbf{x}, t) = -\nabla p + \Delta_x \int_0^\infty G(s)\mathbf{v}(\mathbf{x}, t-s)\, ds \quad \text{in } \Omega\times[0,\infty),\\ &\nabla\cdot\mathbf{v} = 0,\\ &\mathbf{v}(\mathbf{x}, t) = \mathbf{0} \quad \text{on } \partial\Omega\times[0,\infty),\\ &\mathbf{v}(\mathbf{x}, \tau) = \mathbf{v}^0(\mathbf{x}, \tau), \qquad \Omega\times(-\infty, 0], \end{aligned}$$

with $\mathbf{v}^0$ the prescribed velocity history. Taking $\tilde{\mathcal{H}}$ to be the Hilbert space obtained by completing the solenoidal vectors $\mathbf{v}\in(C_0^\infty(\Omega))^3$, $\mathbf{y}\in (C_0^\infty([0,\infty)\times\Omega))^3$, $\mathbf{y}(\mathbf{x}, 0) = \mathbf{v}(\mathbf{x})$ in the $\langle\cdot,\cdot\rangle_{\#}$ inner-product, Slemrod [137] is able to show that the operator $\mathbf{B}$, defined by

$$\text{(II.48)}\qquad \mathbf{B}\binom{\mathbf{v}}{\mathbf{y}} = \begin{pmatrix} \Delta_x \displaystyle\int_0^\infty G(s)\mathbf{y}(\mathbf{x}, s)\, ds \\ -\dfrac{d\mathbf{y}}{ds}(\mathbf{x}, s) \end{pmatrix},$$

with domain

$$\text{(II.49)}\qquad \mathcal{D}(\mathbf{B}) = \{(\mathbf{v}, \mathbf{y}) \in \tilde{\mathcal{H}} : \mathbf{B}\tbinom{\mathbf{v}}{\mathbf{y}} \in \tilde{\mathcal{H}},\ \mathbf{y}(\mathbf{x}, 0) = \mathbf{v}(\mathbf{x}),\ \mathbf{x}\in\Omega\},$$

generates a contraction semigroup on $\tilde{\mathcal{H}}$; this ensures the existence, uniqueness, and stability of generalized solutions to the initial value problem

$$\text{(II.50)}\qquad \begin{aligned} &\frac{d}{dt}\begin{pmatrix}\mathbf{v}(t)\\ \mathbf{y}(t)\end{pmatrix} = \mathbf{B}\begin{pmatrix}\mathbf{v}(t)\\ \mathbf{y}(t)\end{pmatrix},\\ &\begin{pmatrix}\mathbf{v}(0)\\ \mathbf{y}(0)\end{pmatrix} = \begin{pmatrix}\mathbf{v}^0\\ \mathbf{y}^0\end{pmatrix}, \end{aligned}$$

which is the appropriate abstract form of the system (II.47). Slemrod's results [137] may be extended to problems where the constitutive equation is of the form

$$\text{(II.51)}\qquad \begin{aligned} &\mathbf{t} + p\boldsymbol{\delta} = 2\int_0^\infty \mathbf{m}(\mathbf{x}, s)[\mathbf{E}(t-s) - \mathbf{E}(t)]\, ds,\\ &\mathbf{G}(\mathbf{x}, s) = \int_\infty^s \mathbf{m}(\mathbf{x}, \tau)\, d\tau, \end{aligned}$$

where $\mathbf{G}$ is required to be a sufficiently smooth fourth order tensor which satisfies the same kind of symmetry conditions as the relaxation tensor in the (isothermal) linear viscoelasticity theories, as well as certain definiteness conditions which are the analogues of the inequalities delineated in (II.44) and (II.46). Using the results in [137], Slemrod is able to prove [138], on the basis of some energy estimates, that with the additional assumptions

$$\text{(II.52)} \quad G'(s)+\lambda G(s)\leqq 0, \quad G''(s)+\lambda G'(s)\geqq 0 \quad \text{on } [0,\infty) \text{ for some } \lambda>0,$$

solutions of (II.47) decay exponentially in the norm $\|(\cdot)\|_{\#}$. Many of the results we will derive in the next two sections for ill-posed initial-history boundary value problems in isothermal linear viscoelasticity may be modified so as to apply to the initial-history boundary value problems associated with the constitutive equation (II.43) (or its generalization (II.51)) when the hypotheses which guarantee well-posedness; i.e., (II.44), (II.46), and their analogues are either weakened or deleted.

3. Growth estimates for ill-posed initial-history boundary value problems in linear viscoelasticity.

We consider in this section the problem of obtaining growth estimates for solutions $\mathbf{u}\in C^2([0,T);H_+)$ of the initial-history value problem (II.23)–(II.25) when no definiteness assumptions are imposed on the operator $\mathbf{K}(t)$ and its strong operator derivative $\mathbf{K}_t(t)$ for $t\in(-\infty,\infty)$. Our basic results will be obtained via logarithmic convexity arguments not unlike those which were employed to obtain the stability estimates in Chapter I for the first two problems considered there, i.e., the initial-boundary value problems for the equations (I.1), ($\mathrm{I}.45_1$) of linear elastodynamics and the initial-value problems for the abstract equation ($\mathrm{I}.13_1$) which may be used to model the elastodynamics problem with a nonzero body force. As with the logarithmic convexity arguments presented in Chapter I, a key ingredient in terms of stabilizing solutions to the ill-posed problems at hand is that of restricting attention to solutions which lie in some uniformly bounded class; for the abstract integrodifferential (II.23) it turns out that the appropriate classes of functions to look at are of the form

$$\text{(II.53)} \qquad \mathcal{N}=\{\mathbf{w}\in C^2([0,T);H_+)\,|\,\sup_{[0,T)}\|\mathbf{w}(t)\|_+\leqq N^2\}$$

for some real number $N\neq 0$. We now set $\mathcal{K}(t)=\frac{1}{2}\langle\mathbf{u}_t,\mathbf{u}_t\rangle$ (the kinetic energy), $\mathcal{P}(t)=-\frac{1}{2}\langle\mathbf{u}(t),\mathbf{N}\mathbf{u}(t)\rangle$ (the potential energy) and define $\mathcal{E}(t)=\mathcal{K}(t)+\mathcal{P}(t)$ (the total energy). Our first basic result is then contained in the following:

THEOREM (Bloom [22]). *Let* $\mathbf{u}\in\mathcal{N}$ *be any solution of* (II.23)–(II.25) *and define*

$$\text{(II.54)} \qquad F(t;\beta,t_0)=\|\mathbf{u}(t)\|^2+\beta(t+t_0)^2, \qquad 0\leqq t<T,$$

where β, t_0 *are arbitrary nonnegative real numbers. Then provided* $\mathbf{K}(t)$ *satisfies*

$$\text{(II.55)} \qquad -\langle\mathbf{v},\mathbf{K}(0)\mathbf{v}\rangle\geqq\kappa\|\mathbf{v}\|_+^2 \quad \forall\mathbf{v}\in H_+,$$

with

$$\kappa \geqq \gamma T \sup_{[0,\infty)} \|\mathbf{K}_t(t)\|_{\mathscr{L}_S(H_+,H_-)},$$

and

$$\sup_{[0,\infty)} \|\mathbf{K}(t)\|_{\mathscr{L}_S(H_+,H_-)} < \infty,$$

$$\sup_{[0,\infty)} \|\mathbf{K}_t(t)\|_{\mathscr{L}_S(H_+,H_-)} < \infty,$$

$F(t; \beta, t_0)$ satisfies the differential inequality

(II.56) $$FF'' - F'^2 \geqq -2F(2\mathscr{F}(0)+\beta), \qquad 0 \leqq t < T,$$

where

(II.57) $$\mathscr{F}(t) = \mathscr{E}(t) + k_1 \sup_{[0,\infty)} \|\mathbf{K}(t)\|_{\mathscr{L}_S(H_+,H_-)} + k_2 \sup_{[0,\infty)} \|\mathbf{K}_t(t)\|_{\mathscr{L}_S(H_+,H_-)},$$

with k_1, k_2 computable nonnegative constants.

Proof. Let $F(t; \beta, t_0)$ be defined by (II.54); then

(II.58) $$F'(t; \beta, t_0) = 2\langle \mathbf{u}(t), \mathbf{u}_t(t)\rangle + 2\beta(t+t_0)$$

and

(II.59) $$F''(t; \beta, t_0) = 2\langle \mathbf{u}_t(t), \mathbf{u}_t(t)\rangle + 2\langle \mathbf{u}(t), \mathbf{u}_{tt}(t)\rangle + 2\beta,$$

or, in view of the governing evolution equation (II.23),

(II.60) $$F''(t; \beta, t_0) = 2\langle \mathbf{u}_t(t), \mathbf{u}_t(t)\rangle + 2\langle \mathbf{u}(t), \mathbf{N}\mathbf{u}(t)\rangle - 2\left\langle \mathbf{u}(t), \int_{-\infty}^{t} \mathbf{K}(t-\tau)\mathbf{u}(\tau)\, d\tau \right\rangle + 2\beta.$$ Q.E.D.

To proceed further we will need the following lemma concerning the total energy $\mathscr{E}(t)$:

LEMMA. *If $\mathbf{u}(t)$ is any solution of (II.23)–(II.25) then*

(II.61) $$\begin{aligned}\mathscr{E}(t) - \mathscr{E}(0) = &\left\langle \mathbf{f}, \int_{-\infty}^{0} \mathbf{K}(-\tau)\mathbf{u}(\tau)\, d\tau \right\rangle \\ &- \left\langle \mathbf{u}(t), \int_{-\infty}^{t} \mathbf{K}(t-\tau)\mathbf{u}(\tau)\, d\tau \right\rangle \\ &+ \int_0^t \langle \mathbf{u}(\tau), \mathbf{K}(0)\mathbf{u}(\tau)\rangle\, d\tau \\ &+ \int_0^t \left\langle \mathbf{u}(\tau), \int_{-\infty}^{t} \mathbf{K}_\tau(\tau-\lambda)\mathbf{u}(\lambda)\, d\lambda \right\rangle d\tau,\end{aligned}$$

for all t, $0 \leqq t < T$.

In order to prove the above lemma we first take the inner product of (II.23) with $\mathbf{u}_\tau$ and use the symmetry of $\mathbf{N}$ and the definition of $\mathscr{E}(\tau)$ so as to obtain

$$\left(\frac{d}{d\tau}\right)\mathscr{E}(\tau)=-\left\langle \mathbf{u}_\tau, \int_{-\infty}^{\tau} \mathbf{K}(\tau-\lambda)\mathbf{u}(\lambda)\, d\lambda\right\rangle, \tag{II.62}$$

which implies that

$$\mathscr{E}(t)-\mathscr{E}(0)=-\int_0^t\left\langle \mathbf{u}_\tau, \int_{-\infty}^{\tau} \mathbf{K}(t-\lambda)\mathbf{u}(\lambda)\, d\lambda\right\rangle d\tau. \tag{II.63}$$

However,

$$\begin{aligned}\left\langle \mathbf{u}_\tau, \int_{-\infty}^{\tau} \mathbf{K}(t-\lambda)\mathbf{u}(\lambda)\, d\lambda\right\rangle &= \left(\frac{d}{d\tau}\right)\left\langle \mathbf{u}(\tau), \int_{-\infty}^{\tau} \mathbf{K}(\tau-\lambda)\mathbf{u}(\lambda)\, d\lambda\right\rangle\\ &\quad -\left\langle \mathbf{u}(\tau), \int_{-\infty}^{\tau} \mathbf{K}_\tau(\tau-\lambda)\mathbf{u}(\lambda)\, d\lambda\right\rangle\\ &\quad -\langle \mathbf{u}(\tau), \mathbf{K}(0)\mathbf{u}(\tau)\rangle,\end{aligned}$$

and, therefore,

$$\begin{aligned}\mathscr{E}(t)-\mathscr{E}(0)&=\int_0^t \langle \mathbf{u}(\tau), \mathbf{K}(0)\mathbf{u}(\tau)\rangle\, d\tau\\ &\quad+\int_0^t\left\langle \mathbf{u}(\tau), \int_{-\infty}^{\tau} \mathbf{K}_\tau(\tau-\lambda)\mathbf{u}(\lambda)\, d\lambda\right\rangle d\tau\\ &\quad-\int_0^t\frac{d}{d\tau}\left\langle \mathbf{u}(\tau), \int_{-\infty}^{\tau} \mathbf{K}(\tau-\lambda)\mathbf{u}(\lambda)\, d\lambda\right\rangle d\tau,\end{aligned} \tag{II.64}$$

from which the identity (II.61) follows immediately. We now return to the proof of the Theorem and note that (II.60) may be rewritten in the form

$$\begin{aligned}F''(t;\beta,t_0)&=-2\left\langle \mathbf{u}(t), \int_{-\infty}^{t} \mathbf{K}(t-\tau)\mathbf{u}(\tau)\, d\tau\right\rangle\\ &\quad+2\beta+4\mathscr{K}(t)-4[(\mathscr{E}(0)-\mathscr{K}(t))+(\mathscr{E}(t)-\mathscr{E}(0))].\end{aligned} \tag{II.65}$$

If we substitute in (II.65) for $\mathscr{E}(t)-\mathscr{E}(0)$ from the lemma above and then simplify we obtain

$$\begin{aligned}F''(t;\beta,t_0)&=4(2\mathscr{K}(t)+\beta)-2(2\mathscr{E}(0)+\beta)\\ &\quad-4\int_0^t \langle \mathbf{u}(t), \mathbf{K}(0)\mathbf{u}(\tau)\rangle\, d\tau\\ &\quad-4\int_0^t\left\langle \mathbf{u}(\tau), \int_{-\infty}^{\tau} \mathbf{K}_\tau(\tau-\lambda)\mathbf{u}(\lambda)\, d\lambda\right\rangle d\tau\\ &\quad+2\left\langle \mathbf{u}(t), \int_{-\infty}^{t} \mathbf{K}(t-\tau)\mathbf{u}(\tau)\, d\tau\right\rangle\\ &\quad-4\left\langle \mathbf{f}, \int_{-\infty}^{0} \mathbf{K}(-\tau)\mathbf{U}(\tau)\, d\tau\right\rangle.\end{aligned} \tag{II.66}$$

We now define the real-valued function

$$H(t;\beta,t_0)\equiv F(t;\beta,t_0)(\|\mathbf{u}_t(t)\|^2+\beta)-\tfrac{1}{4}(F'(t;\beta,t_0))^2. \tag{II.67}$$

It is a relatively simple matter to check that the definition of $F(t;\beta,t_0)$, (II.58), and the Schwarz inequality together imply that $H(t;\beta,t_0)\geqq 0$. If we now combine (II.58) with (II.66) we obtain

$$\begin{aligned} FF''-F'^2 &= 4H(t;\beta,t_0)-2F(2\mathscr{E}(0)+\beta)\\ &\quad -4F\int_0^t \langle \mathbf{u}(\tau),\mathbf{K}(0)\mathbf{u}(\tau)\rangle\,d\tau\\ &\quad -4F\int_0^t\left\langle \mathbf{u}(\tau),\int_{\infty}^{\tau}\mathbf{K}_\tau(\tau-\lambda)\mathbf{u}(\lambda)\,d\lambda\right\rangle d\tau\\ &\quad +2F\left\langle \mathbf{u}(t),\int_{-\infty}^{t}\mathbf{K}(-\tau)\mathbf{U}(\tau)\,d\tau\right\rangle\\ &\quad -4F\left\langle \mathbf{f},\int_{-\infty}^{0}\mathbf{K}(-\tau)\mathbf{U}(\tau)\,d\tau\right\rangle. \end{aligned} \tag{II.68}$$

We now need to go to work on the right-hand side of (II.68). By the Schwarz inequality, the topological embedding of H_+ into H and the fact that $\mathbf{K}(t)\in\mathscr{L}_S(H_+,H_-)$, for each $t\in(-\infty,\infty)$, we have the estimates

$$\begin{aligned} \left|\left\langle \mathbf{f},\int_{-\infty}^{0}\mathbf{K}(-\tau)\mathbf{U}(\tau)\,d\tau\right\rangle\right| &\leqq \|\mathbf{f}\|\int_{-\infty}^{0}\|\mathbf{K}(-\tau)\mathbf{U}(\tau)\|\,d\tau\\ &\leqq \gamma\sup_{[0,\infty)}\|\mathbf{K}(t)\|_{\mathscr{L}_S(H_+,H_-)}\left(\|\mathbf{f}\|_+\int_{-\infty}^{0}\|\mathbf{U}(\tau)\|_+\,d\tau\right) \end{aligned} \tag{II.69}$$

in view of our assumption that the prescribed past history $\mathbf{U}$ satisfies $\|\mathbf{U}(\cdot)\|_+\in L_1(-\infty,0]$. Also

$$\begin{aligned} &\left|\left\langle \mathbf{u}(t),\int_{-\infty}^{0}\mathbf{K}(t-\tau)\mathbf{u}(\tau)\,d\tau\right\rangle\right|\\ &\leqq \gamma\sup_{[0,\infty)}\|\mathbf{K}(t)\|_{\mathscr{L}_S(H_+,H_-)}\left(\|\mathbf{u}(t)\|_+\int_{-\infty}^{0}\|\mathbf{U}(t)\|_+\,d\tau+\|\mathbf{u}(t)\|_+\int_0^t\|\mathbf{u}(\tau)\|_+\,d\tau\right)\\ &\leqq \gamma N^2\sup_{[0,\infty)}\|\mathbf{K}(t)\|_{\mathscr{L}_S(H_+,H_-)}\left(TN^2+\int_{-\infty}^{0}\|\mathbf{U}(\tau)\|_+\,d\tau\right). \end{aligned} \tag{II.70}$$

If we now combine the differential inequality (II.68) with the estimates (II.69),

(II.70) and recall that $H(t;\beta, t_0) \geqq 0$, $0 \leqq r < T$, we obtain

$$
\begin{aligned}
FF'' - F'^2 \geqq & -2F(2\mathscr{E}(0)+\beta) \\
& -2F\gamma \sup_{[0,\infty)} \|\mathbf{K}(t)\|_{\mathscr{L}_S(H_+,H_-)} \Big(TN^4 + (N^2 + 2\|\mathbf{f}\|_+) \int_{-\infty}^{0} \|\mathbf{U}(\tau)\|_+ \, d\tau \Big) \\
& -4F \int_0^t \langle \mathbf{u}(\tau), \mathbf{K}(0)\mathbf{u}(\tau)\rangle \, d\tau \\
& -4F \int_0^t \Big\langle \mathbf{u}(\tau), \int_{-\infty}^{\tau} \mathbf{K}_\tau(\tau-\lambda)\mathbf{u}(\lambda)\, d\lambda \Big\rangle \, d\tau.
\end{aligned} \tag{II.71}
$$

However, by virtue of our hypothesis (II.55)

$$
-4F \int_0^t \langle \mathbf{u}(\tau), \mathbf{K}(0)\mathbf{u}(\tau)\rangle \, d\tau \geqq 4\kappa F \int_0^t \|\mathbf{u}(\tau)\|_+^2 \, d\tau. \tag{II.72}
$$

Also,

$$
\begin{aligned}
& \Big| \Big\langle \mathbf{u}(\tau), \int_{-\infty}^{\tau} \mathbf{K}_\tau(\tau-\lambda)\mathbf{u}(\lambda)\, d\lambda \Big\rangle \Big| \\
& \qquad \leqq \gamma \sup_{[0,\infty)} \|\mathbf{K}_t(t)\|_{\mathscr{L}_S(H_+,H_-)} \Big(\|\mathbf{u}(\tau)\|_+ \int_{-\infty}^{\tau} \|\mathbf{u}(\lambda)\|_+ \, d\lambda \Big)
\end{aligned} \tag{II.73}
$$

and, therefore,

$$
\begin{aligned}
& \int_0^t \Big\langle \mathbf{u}(\tau), \int_{-\infty}^{\tau} \mathbf{K}_\tau(\tau-\lambda)\mathbf{u}(\lambda)\, d\lambda \Big\rangle \, d\tau \\
& \quad \leqq \int_0^t \Big| \Big\langle \mathbf{u}(\tau), \int_{-\infty}^{\tau} \mathbf{K}_\tau(\tau-\lambda)\mathbf{u}(\lambda)\, d\lambda \Big\rangle \Big| \, d\tau \\
& \quad \leqq \gamma (\sup_{[0,\infty)} \|\mathbf{K}_t(t)\|_{\mathscr{L}_S(H_+,H_-)} \Big[\int_0^t \|\mathbf{u}(\tau)\|_+ \int_{-\infty}^{0} \|\mathbf{U}(\lambda)\|_+ \, d\lambda \, d\tau \\
& \qquad\qquad + \int_0^t \|\mathbf{u}(\tau)\|_+ \int_0^{\tau} \|\mathbf{u}(\lambda)\|_+ \, d\lambda \, d\tau \Big] \\
& \quad \leqq \gamma \sup_{[0,\infty)} \|\mathbf{K}_t(t)\|_{\mathscr{L}_S(H_+,H_-)} \Big[TN^2 \int_{-\infty}^{0} \|\mathbf{U}(\tau)\|_+ \, d\tau + T \int_0^t \|\mathbf{u}(\tau)\|_+^2 \, d\tau \Big],
\end{aligned} \tag{II.74}
$$

from which it follows immediately that

$$
\begin{aligned}
& -\int_0^t \Big\langle \mathbf{u}(\tau), \int_{-\infty}^{\tau} \mathbf{K}_\tau(\tau-\lambda)\mathbf{u}(\lambda)\, d\lambda \Big\rangle \, d\tau \\
& \quad \geqq -\gamma T \sup_{[0,\infty)} \|\mathbf{K}_t(t)\|_{\mathscr{L}_S(H_+,H_-)} \int_0^t \|\mathbf{u}(\tau)\|_+^2 \, d\tau \\
& \qquad - \gamma TN^2 \sup_{[0,\infty)} \|\mathbf{K}_t(t)\|_{\mathscr{L}_S(H_+,H_-)} \int_{-\infty}^{0} \|\mathbf{U}(\tau)\|_+ \, d\tau.
\end{aligned} \tag{II.75}
$$

Combining the differential inequality (II.71) with the estimates (II.72), (II.75) and using the hypothesis that $\kappa \geqq \gamma T \sup_{[0,\infty)} \|\mathbf{K}_t(t)\|_{\mathscr{L}_S(H_+, H_-)}$ we find that

$$
\begin{aligned}
&FF'' - F'^2 \geqq -2F(2\mathscr{E}(0)+\beta)\\
\text{(II.76)}\qquad &-2F\gamma \sup_{[0,\infty)} \|\mathbf{K}(t)\|_{\mathscr{L}_S(H_+,H_-)} \left[TN^2 + (N^2 + 2\|\mathbf{f}\|_+) \int_{-\infty}^{0} \|\mathbf{U}(t)\|_+ \, d\tau \right]\\
&-4F\gamma \sup_{[0,\infty)} \|\mathbf{K}_t(t)\| \left(TN^2 \int_{-\infty}^{0} \|\mathbf{U}(t)\|_+ \, d\tau \right).
\end{aligned}
$$

If we set

$$
\text{(II.77)}\qquad
\begin{aligned}
k_1 &= \frac{1}{2}\gamma(TN^4 + 2\|\mathbf{f}\|_+) \int_{-\infty}^{0} \|\mathbf{U}(\tau)\|_+ \, d\tau,\\
k_2 &= \gamma TN^2 \int_{-\infty}^{0} \|\mathbf{U}(\tau)\|_+ \, d\tau,
\end{aligned}
$$

then it is clear that (II.76) assumes the form (II.56) where $\mathscr{F}(t)$ is defined by (II.57) and k_1, k_2 are given by (II.77) and are clearly nonnegative. Q.E.D.

Remark. If the conditions

$$
\sup_{[0,\infty)} \|\mathbf{K}(t)\|_{\mathscr{L}_S(H_+,H_-)} < \infty, \qquad \sup_{[0,\infty)} \|\mathbf{K}_t(t)\|_{\mathscr{L}_S(H_+,H_-)} < \infty
$$

are not (jointly) satisfied then it should be clear that we may restrict our attention to past histories $\mathbf{U}(t)$ for which $\exists t_\infty > 0$ such that $\mathbf{U}(t) = \mathbf{0}$, $t < -t_\infty$; our basic estimates then go through with the obvious changes; one could also introduce fading memory spaces to replace the basic spaces H_+, H but this would greatly complicate matters and involve us in analysis which would be disproportionately involved as compared to the basic elementary nature of the logarithmic convexity argument itself.

Remark. Since $|\langle \mathbf{v}, \mathbf{K}(0)\mathbf{v}\rangle| \leqq \|\mathbf{K}(0)\|_{\mathscr{L}_S(H_+,H_-)} \|\mathbf{v}\|_+^2$ $\forall \mathbf{v} \in H_+$, it is useful to note that the hypotheses of the Theorem imply that for $\sup_{[0,\infty)} \|\mathbf{K}_t(t)\|_{\mathscr{L}_S(H_+,H_-)} < \infty$ we must have

$$
\text{(II.78)}\qquad \sup_{[0,\infty)} \|\mathbf{K}_t(t)\|_{\mathscr{L}_S(H_+,H_-)} \leqq \left(\frac{1}{\gamma T}\right) \|\mathbf{K}(0)\|_{\mathscr{L}_S(H_+,H_-)}.
$$

Remark. In terms of the relaxation tensor $g_{ijkl}(\mathbf{x}, t)$ our basic hypotheses (II.55) assume the following form: first of all it follows directly from (I.22) that

$$
\begin{aligned}
\mathbf{K}(0) &= \frac{\partial}{\partial \tau} \mathbf{G}(t-\tau)|_{t=\tau}\\
&= -\frac{\partial}{\partial t} \mathbf{G}(t-\tau)|_{t=\tau}\\
&= -\mathbf{G}'(0),
\end{aligned}
$$

so that the first condition in (II.55) becomes

$$\langle \mathbf{v}, \mathbf{G}'(0)\mathbf{v}\rangle \geqq \kappa \|\mathbf{v}\|_+^2, \qquad \forall \mathbf{v} \in H_+. \tag{II.79}$$

If we employ the definitions of $H((L_2(\Omega))^3)$, $H_+((H_0^1(\Omega))^3)$ and $\mathbf{G}(t)$ (II.18) then an elementary computation quickly shows that (II.79) assumes the form

$$\int_\Omega \dot{g}_{ijkl}(\mathbf{x}, 0) \frac{\partial v_i}{\partial x_j} \frac{\partial v_k}{\partial x_l} \, d\mathbf{x} \leqq -\kappa \int_\Omega \frac{\partial v_i}{\partial x_j} \frac{\partial v_i}{\partial x_j} \, d\mathbf{x} \tag{II.80}$$

for all $\mathbf{v} \in (H_0^1(\Omega))^3 (\equiv H_+)$. An argument completely analogous to the one which led to the estimate (II.21) then yields

$$\|\mathbf{K}_t(t)\|_{\mathscr{L}_S(H_+,H_-)} \leqq \left(\sup_\Omega \frac{1}{\rho}\right)\left(\max_{i,j,k,l} \sup_\Omega |\ddot{g}_{ijkl}(\mathbf{x}, t)|\right),$$

so that the second condition in (II.55) will be satisfied if the nonnegative constant κ satisfies

$$k \geqq \gamma T \sup_\Omega \left(\frac{1}{\rho}\right) \sup_{[0,\infty)} \left(\max_{i,j,k,l} \sup_\Omega |\ddot{g}_{ijkl}(\mathbf{x}, t)|\right). \tag{II.81}$$

In particular, for the one-dimensional homogeneous viscoelastic material, i.e., the material governed by the evolution equation (II.5) with relaxation function $g(x, t) = g_0(t)$, the conditions (II.80), (II.81) assume the simple form

$$g_0'(0) \leqq -\kappa \textit{ with } \kappa \geqq \frac{\gamma T}{\rho_0} \sup_{[0,\infty)} |g_0''(t)|. \tag{II.82}$$

The conditions expressed by (II.82) imply that g_0 must be decreasing sufficiently fast at $t = 0$; we do not, therefore, require that $g_0(t)$ be monotonically nonincreasing for all t, in order to apply, to the material governed by the evolution equation (II.5), the growth estimates which will result from the basic logarithmic convexity type estimate expressed by (II.56).

We now present, based on the differential inequality (II.56), some growth estimates for solutions $\mathbf{u} \in \mathcal{N}$ of the initial-history value problem (II.23)–(II.25); our growth estimates will be immediately applicable to the initial-history boundary value problem (II.2′), (II.3), (II.4) provided the relaxation tensor $g_{ijkl}(\mathbf{x}, t)$ satisfies the rather mild requirements expressed by (II.80), (II.81), and the past history satisfies the smoothness hypotheses which are delineated immediately following (II.25). As in Knops and Payne [79] our considerations will be based on an examination of the values of the initial data (and, in particular, the initial energy $\mathscr{E}(0)$) in comparison with the values of $\sup_{[0,\infty)}\|\mathbf{K}(t)\|_{\mathscr{L}_S(H_+,H_-)}$ and $\sup_{[0,\infty)}\|\mathbf{K}_t(t)\|_{\mathscr{L}_S(H_+,H_-)}$. Our arguments proceed on a case by case basis but we will not present an exhaustive treatment of all the situations which are possible.

Case I. $\mathcal{F}(0) \leqq 0$ and $\langle \mathbf{f}, \mathbf{g} \rangle \geqq 0$, $\mathbf{f} \neq 0$.

In this first case we may set $\beta = 0$, and thus $F(t; \beta, t_0)$ reduces to $F(t) = \|\mathbf{u}(t)\|^2$. From (II.56) it is then immediate that $FF'' - F'^2 \geqq 0$, $0 \leqq t < T$, so that if $F(t) \neq 0$ on $[0, T)$

(II.83) $$\frac{d^2}{dt^2}(\ln F) \geqq 0, \qquad 0 \leqq t < T.$$

Now a simple argument [79] based on an application of Jensen's inequality to (II.83) can be used to show that if $\|\mathbf{u}(\bar{t})\| = 0$ for some $\bar{t}$, $0 \leqq t < T$, then $\|\mathbf{u}(t)\| = 0$, $0 \leqq t < T$; thus, we may assume, without loss of generality that $F(t) > 0$ on $[0, T)$ so that (II.83) is valid. From the definition of $\mathcal{F}(t)$, i.e., (II.57), it is clear that the condition $\mathcal{F}(0) \leqq 0$ is equivalent to

(II.84) $$\mathscr{E}(0) \leqq -\left(k_1 \sup_{[0,\infty)} \|\mathbf{K}(t)\|_{\mathscr{L}_S(H_+, H_-)} + k_2 \sup_{[0,\infty)} \|\mathbf{K}_t(t)\|_{\mathscr{L}_S(H_+, H_-)}\right),$$

where k_1, k_2 are given by (II.77). However, in view of (II.78), (II.84) is implied by the simpler condition that

(II.85) $$\mathscr{E}(0) \leqq -k \sup_{[0,\infty)} \|\mathbf{K}(t)\|_{\mathscr{L}_S(H_+, H_-)}, \qquad k \geqq k_1 + \frac{k_2}{\gamma T}.$$

For the simple case where $\rho(\mathbf{x}) = \rho_0$ (constant) it is a relatively simple matter to compute that

(II.86) $$\mathscr{E}(0) = \frac{1}{2}\int_\Omega g_i g_i \, d\mathbf{x} + \frac{1}{2\rho_0}\int_\Omega \mathscr{g}_{ijkl}(\mathbf{x}, 0)\frac{\partial f_i}{\partial x_j}\frac{\partial f_k}{\partial x_l}\, d\mathbf{x},$$

so that (II.85) will be satisfied provided

(II.87) $$\int_\Omega g_i g_i \, d\mathbf{x} + \frac{1}{\rho_0}\int_\Omega \mathscr{g}_{ijkl}(\mathbf{x}, 0)\frac{\partial f_i}{\partial x_j}\frac{\partial f_k}{\partial x_l}\, d\mathbf{x} \leqq -\frac{2k}{\rho_0}\sup_{[0,\infty)}\left(\max_{i,j,k,l}\sup_\Omega |\dot{\mathscr{g}}_{ijkl}(\mathbf{x}, t)|\right),$$

with $k \geqq k_1 + k_2/\gamma T$. If we assume that the initial-value of the relaxation tensor is positive definite then the case of negative initial energy can not arise. For those cases where (II.84), or the stronger condition (II.85), may occur we can proceed as follows: First of all, the differential inequality (II.83) can be used, as in Chapter I, to obtain theorems on uniqueness and stability of solutions $\mathbf{u} \in \mathcal{N}$; however, as this is fairly elementary and we will in any case derive in the next section a different inequality which implies more general continuous dependence results, we will consider in this section the derivation of certain growth estimates only. To this end, we set $J(t) = \ln F(t)$ and expand $J(t)$ in a finite

Taylor series about $t=0$, i.e.,

$$J(t)=J(0)+J'(0)+\frac{J''}{2}(\zeta)t^2, \qquad 0\leqq t<T, \tag{II.88}$$

where $0<\zeta<T$. By (II.83), $J''(\zeta)\geqq 0$ so

$$\ln F(t)\geqq \ln F(0)+\frac{d}{dt}(\ln F(t)|_{t=0})t, \tag{II.89}$$

which implies, in view of the definition of $F(t)$, that $\|\mathbf{u}(t)\|$ satisfies the exponential growth estimate

$$\|\mathbf{u}(t)\|^2\geqq \|\mathbf{f}\|^2 \exp\left(\frac{2\langle \mathbf{f}, \mathbf{g}\rangle t}{\|\mathbf{f}\|^2}\right), \qquad 0\leqq t<T. \tag{II.90}$$

In particular, $\|\mathbf{u}(t)\|^2\geqq\|\mathbf{f}\|^2$, $0\leqq t<T$, if $\mathbf{g}=\mathbf{0}$.

Case II. $\mathscr{F}(0)<0$ and $\langle \mathbf{f}, \mathbf{g}\rangle<0$.

In this case we may choose $\bar{\beta}>0$ such that $2\mathscr{F}(0)+\bar{\beta}\leqq 0$. Then by (II.56)

$$\frac{d^2}{dt^2}F(t;\bar{\beta}, t_0)\geqq 0, \qquad 0\leqq t<T, \quad t_0\geqq 0. \tag{II.91}$$

If we expand $J(t;\bar{\beta}, t_0)=\ln F(t;\bar{\beta}, t_0)$ in a finite Taylor series about $t=0$ we obtain

$$J(t;\bar{\beta}, t_0)\geqq J(0;\bar{\beta}; t_0)+J'(0;\bar{\beta}, t_0)t$$

for $0\leqq t<T$, or

$$F(t;\bar{\beta}, t_0)\geqq F(0;\bar{\beta}, t_0)\exp\left(\frac{F'(0;\bar{\beta}, t_0)}{F(0;\bar{\beta}, t_0)}\cdot t\right), \qquad 0\leqq t<T. \tag{II.92}$$

We now choose $t_0=\bar{t}_0\geqq(1/\beta)|\langle \mathbf{f}, \mathbf{g}\rangle|$; with this choice of t_0 it follows immediately from (II.58) that $F'(0;\bar{\beta}, \bar{t}_0)\geqq 0$ and, thus, (II.92) yields the exponential growth estimate

$$\begin{gathered}\|\mathbf{u}(t)\|^2+\bar{\beta}(t+\bar{t}_0)^2\\ (\|\mathbf{f}\|^2+\bar{\beta}\bar{t}_0^2)\exp\left\{\frac{2t(\bar{\beta}\bar{t}_0-|\langle \mathbf{f}, \mathbf{g}\rangle|)}{\|\mathbf{f}\|^2+\bar{\beta}\bar{t}_0^2}\right\}, \qquad 0\leqq t<T.\end{gathered} \tag{II.93}$$

Remark. For the case where $\mathscr{F}(0)\leqq 0$ but $\langle \mathbf{f}, \mathbf{g}\rangle<0$ we would again choose $\beta=0$; in this situation it is not difficult to see that we would obtain in place of (II.90) the estimate

$$\|\mathbf{u}(t)\|^2\geqq\|\mathbf{f}\|^2\exp\left\{\frac{-2|\langle \mathbf{f}, \mathbf{g}\rangle|t}{\|\mathbf{f}\|^2}\right\}, \qquad 0\leqq t<T, \tag{II.94}$$

which provides a lower-bound on the rate at which $\|\mathbf{u}(t)\|$ could decay on $[0, T)$.

Remark. The exponential growth estimates obtained in cases I and II above can be viewed in a slightly different manner; for instance, suppose that $\mathscr{E}(0) < -\hat{k}$ with $\hat{k} > 0$ and that $\langle \mathbf{f}, \mathbf{g} \rangle \geqq 0$, with $\mathbf{f} \neq 0$. Then $\mathscr{F}(0) \leqq 0$ provided $\mathbf{K}(t)$ satisfies

$$k_1 \sup_{[0,\infty)} \|\mathbf{K}(t)\|_{\mathscr{L}_S(H_+,H_-)} + k_2 \sup_{[0,\infty)} \|\mathbf{K}_t(t)\|_{\mathscr{L}_S(H_+,H_-)} \leqq \hat{k} \tag{II.95}$$

and, in view of (II.78), this condition is easily seen to be implied by

$$\sup_{[0,\infty)} \|\mathbf{K}(t)\|_{\mathscr{L}_S(H_+,H_-)} \leqq \frac{\hat{k}\gamma T}{\gamma k_1 T + k_2}. \tag{II.96}$$

Thus if $\mathbf{u} \in \mathscr{N}$ is a solution of (II.23)–(II.25) for which $\mathscr{E}(0) \leqq -\hat{k}$, with $\hat{k} > 0$, and $\mathbf{K}(t)$ satisfies (II.55), (II.96), then $\|\mathbf{u}(t)\|$ satisfies the growth estimate (II.90) on $[0, T)$ whenever $\langle \mathbf{f}, \mathbf{g} \rangle \geqq 0$ with $\mathbf{f} \neq 0$.

Case III. $\mathscr{F}(0) > 0$, $\langle \mathbf{f}, \mathbf{g} \rangle \geqq (2\mathscr{F}(0))^{1/2} \|\mathbf{f}\|$.

In this case the initial energy must satisfy

$$\mathscr{E}(0) > -\Big(k_1 \sup_{[0,\infty)} \|\mathbf{K}(t)\|_{\mathscr{L}_S(H_+,H_-)} + k_2 \sup_{[0,\infty)} \|\mathbf{K}_t(t)\|_{\mathscr{L}_S(H_+,H_-)}\Big). \tag{II.97}$$

However, by (II.78),

$$\begin{aligned} \sup_{[0,\infty)} \|\mathbf{K}(t)\|_{\mathscr{L}_S(H_+,H_-)} &\geqq \|\mathbf{K}(0)\|_{\mathscr{L}_S(H_+,H_-)} \\ &\geqq \gamma T \sup_{[0,\infty)} \|\mathbf{K}_t(t)\|_{\mathscr{L}_S(H_+,H_-)}, \end{aligned} \tag{II.98}$$

so that (II.97) is implied by the simpler condition

$$\mathscr{E}(0) > -(k_1 \gamma T + k_2) \sup_{[0,\infty)} \|\mathbf{K}_t(t)\|_{\mathscr{L}_S(H_+,H_-)}. \tag{II.99}$$

We now follow the analogous argument presented in [79]; we assume that $\langle \mathbf{u}(t), \mathbf{u}_t(t) \rangle \geqq 0$, $0 \leqq t < t_1 < T$ and set $\beta = 0$. Then by (II.56)

$$\frac{d}{dt}\left(\frac{dF/dt}{F(t)}\right) \geqq -\frac{4\mathscr{F}(0)}{F(t)}, \qquad 0 \leqq t < t_1, \tag{II.100}$$

which, upon integrating, implies that

$$\left(\frac{\frac{dF}{dt}}{F(t)}\right)^2 \geqq \Gamma^2 + \frac{8\mathscr{F}(0)}{F(t)}, \qquad 0 \leqq t < t_1, \tag{II.101}$$

where

$$\begin{aligned} \Gamma^2 &\equiv \left(\frac{dF/dt(0)}{F(0)}\right)^2 - \frac{8\mathscr{F}(0)}{F(t)} \\ &= \left(\frac{2\langle \mathbf{f}, \mathbf{g} \rangle}{\|\mathbf{f}\|^2}\right)^2 - \frac{8\mathscr{F}(0)}{\|\mathbf{f}\|^2} \geqq 0, \end{aligned}$$

in view of the hypothesis that $\langle \mathbf{f}, \mathbf{g}\rangle \geqq (2\mathcal{F}(0))^{1/2}\|\mathbf{f}\|$. It follows from (II.101) that $dF/dt > 0, 0 \leqq t < t_1$; thus dF/dt is nonnegative on $[0, T)$ for any $T > 0$ so taking the square root on both sides of (II.101), and integrating, we obtain the growth estimate

(II.102)
$$\|\mathbf{u}(t)\|^2 \geqq \left(\|\mathbf{f}\|^2 + \frac{4\mathcal{F}(0)}{\Gamma^2}\right)\cosh \Gamma t + \left(\frac{2\langle \mathbf{f}, \mathbf{g}\rangle}{\Gamma}\right)\sinh \Gamma t - \frac{4\mathcal{F}(0)}{\Gamma^2}, \qquad 0 \leqq t < T,$$

for $\Gamma^2 > 0$ (i.e., for $\langle \mathbf{f}, \mathbf{g}\rangle > (2\mathcal{F}(0))^{1/2}\|\mathbf{f}\|$) while for $\Gamma^2 = 0(\langle \mathbf{f}, \mathbf{g}\rangle = (2\mathcal{F}(0))^{1/2}\|\mathbf{f}\|)$ integration of (II.101) yields the polynomial growth estimate on $[0, T)$

$$\|\mathbf{u}(t)\|^2 \geqq \|\mathbf{f}\|^2 + 2^{3/2}\|\mathbf{f}\|(\mathcal{F}(0))^{1/2}t + 2\mathcal{F}(0)t^2. \tag{II.103}$$

Remark. Suppose that $\mathcal{E}(0) > -\tilde{k}$, where $\tilde{k} > 0$. Then $\mathcal{F}(0) > 0$ provided that

$$\tilde{k} < k_1 \sup_{[0,\infty)} \|\mathbf{K}(t)\|_{\mathcal{L}_S(H_+,H_-)} + k_2 \sup_{[0,\infty)} \|\mathbf{K}_t(t)\|_{\mathcal{L}_S H_+,H_-)}. \tag{II.104}$$

In view of (II.98), (II.104) will be satisfied if

$$\sup_{[0,\infty)} \|\mathbf{K}_t(t)\|_{\mathcal{L}_S(H_+,H_-)} > \frac{\tilde{k}}{\gamma k_1 T + k_2}. \tag{II.105}$$

Thus, if $\mathbf{u} \in \mathcal{N}$ is a solution of (II.23)–(II.25) with $\mathcal{E}(0) > -\tilde{k}$, for some $\tilde{k} > 0$, and $\langle \mathbf{f}, \mathbf{g}\rangle > (2\mathcal{F}(0))^{1/2}\|\mathbf{f}\|$, then $\|\mathbf{u}(t)\|$ satisfies the growth estimate (II.102) on $[0, T)$ whenever $\mathbf{K}(t)$ satisfies (II.55), (II.105); if $\langle \mathbf{f}, \mathbf{g}\rangle = (2\mathcal{F}(0))^{1/2}\|\mathbf{f}\|$ then $\|\mathbf{u}(t)\|$ satisfies the polynomial growth estimate (II.103), on $[0, T)$, whenever $\mathbf{K}(t)$ satisfies (II.55), (II.105).

Remark. A wide variety of growth estimates based on the differential inequality (II.56) may be found in [79]; for example, if $\mathcal{F}(0) > 0$ and $\langle \mathbf{f}, \mathbf{g}\rangle > 2(2+\varepsilon)^{1/2}(\mathcal{F}(0))^{1/2}\|\mathbf{f}\|$, for $\varepsilon > 0$ an appropriately chosen positive constant, then it is shown that exponential growth estimates for $\|\mathbf{u}(t)\|$ hold on $[0, T)$. Specifically, if $\varepsilon > 0$ is such that for given β, $t_0 \geqq 0$,

$$\varepsilon\beta t_0^2 - 2t_0\langle \mathbf{f}, \mathbf{g}\rangle + (2+\varepsilon)\|\mathbf{f}\|^2 < 0, \tag{II.106}$$

then

(II.107)
$$\|\mathbf{u}(t)\|^2 + \beta(t+t_0)^2 \geqq (\|\mathbf{f}\| + \beta t_0^2)\left(\frac{t}{t_0}+1\right)^{2+\varepsilon} \times \exp\left\{\frac{2\langle \mathbf{f}, \mathbf{g}\rangle + 2\beta t_0}{\|\mathbf{f}\|^2 + \beta t_0^2} - \frac{(2+\varepsilon)}{t_0}t\right\}$$

on $[0, T)$. Thus, for given $\varepsilon > 0$ such that $\langle \mathbf{f}, \mathbf{g}\rangle > 2(2+\varepsilon)^{1/2}(\mathcal{F}(0))^{1/2}\|\mathbf{f}\|$, the growth estimate (II.107) would be valid on $[0, T)$ for β, $t_0 \geqq 0$ satisfying (II.106); a further analysis may be found in [79].

Example. We present an application of some of the growth estimates obtained above to a problem in one dimensional viscoelasticity. We consider the viscoelastic model governed by the evolution equation (II.5) with $\rho = 1$, $a = 0$, $b = 1$ and homogeneous relaxation function $g_0(t) = e^{-\lambda t}$, $\lambda > 0$. Thus our initial-history boundary value problem is of the form

$$\frac{\partial^2 u}{\partial t^2} - \frac{\partial^2 u}{\partial x^2} + \lambda \int_{-\infty}^{t} e^{-\lambda(t-\tau)} \frac{\partial^2 u(x,\tau)}{\partial x^2}\, d\tau = 0, \tag{II.108}$$

for $(x, t) \in [0, 1] \times [0, T)$ and

$$\begin{aligned} &u(0,t) = u(1,t) = 0, \qquad t \in (-\infty, T),\\ &\dot{u}(x,0) = f(x), \quad \left(\frac{\partial u}{\partial t}\right)(x,0) = g(x), \qquad x \in [0,1], \end{aligned} \tag{II.109}$$

$$u(x,\tau) = U(x,\tau), \qquad (x,\tau) \in [0,1] \times (-\infty, 0). \tag{II.110}$$

Clearly $H = L_2(0, 1)$, $H_+ = H_0^1(0, 1)$ and $H_- = H^{-1}(0, 1)$ and the basic operators are just

$$\begin{aligned} &\mathbf{G}(t) = e^{-\lambda t} \frac{\partial^2}{\partial x^2},\\ &\mathbf{N} = \frac{\partial^2}{\partial x^2}, \qquad \mathbf{K}(t-\tau) = \lambda\, e^{-\lambda(t-\tau)} \frac{\partial^2}{\partial x^2}. \end{aligned} \tag{II.111}$$

We must assume that the past history U satisfies

$$\int_{-\infty}^{0} \left(\int_0^1 \left(\frac{\partial U(x,\tau)}{\partial x} \right)^2 dx \right)^{1/2} d\tau < \infty. \tag{II.112}$$

The conditions on the operator $\mathbf{K}(t)$ in this situation reduce, as already shown, to (II.82) with $\rho_0 = 1$; also for $u(\cdot, t) \in H_0^1(0, 1)$

$$u(x,t) = \int_0^x \frac{\partial u}{\partial y}(y,t)\, dy, \qquad x \in [0,1],$$

so

$$u^2(x,t) = \left(\int_0^x \frac{\partial u}{\partial y}(y,t)\, dy \right)^2 \leqq x \int_0^x \left(\frac{\partial u}{\partial y} \right)^2 dy$$

and therefore, for $0 \leqq x \leqq 1$,

$$u^2(x,t) \leqq \int_0^1 \left(\frac{\partial u}{\partial x} \right)^2 dx \Rightarrow \|u\| \leqq \|u\|_+,$$

i.e., $\gamma = 1$ for the embedding of $H_0^1(0, 1)$ into $L_2(0, 1)$. Thus, we must satisfy $g_0'(0) \leqq -\kappa$ with $\kappa \geqq T \sup_{[0,\infty)} |g_0''(t)|$ and, with $g_0(t) = e^{-\lambda t}$; these conditions are easily seen to be equivalent to requiring that $\lambda \leqq 1/T$. Suppose now that $\lambda \leqq 1/T$; then

$$\hat{F}(t; \beta, t_0) = \int_0^1 u^2(x,t)\, dx + \beta(t + t_0)^2 \tag{II.113}$$

satisfies (II.56) where

$$\mathscr{F}(t)\equiv\frac{1}{2}\int_0^1\left[\left(\frac{\partial u(x,t)}{\partial t}\right)^2+\left(\frac{\partial u(x,t)}{\partial x}\right)^2\right]dx+\hat{k}_1\lambda+\hat{k}_2\lambda^2 \tag{II.114}$$

with

$$\begin{aligned}\hat{k}_1&=\frac{1}{2}\left[T\hat{N}^2+\left(\hat{N}^2+2\left[\int_0^1\left(\frac{\partial f}{\partial x}\right)^2dx\right]^{1/2}\right)\times\int_{-\infty}^0\left[\int_0^1\left(\frac{\partial U}{\partial x}\right)^2dx\right]^{1/2}d\tau\right],\\ \hat{k}_2&=T\hat{N}^2\int_{-\infty}^0\left[\int_0^1\left(\frac{\partial U}{\partial x}\right)^2dx\right]^{1/2}d\tau,\end{aligned} \tag{II.115}$$

and $\hat{N}$ serves to delineate the class of bounded functions

$$\hat{\mathcal{N}}=\left\{w\in C^2([0,T);H_0^1(0,1))\,\middle|\,\sup_{[0,T)}\left[\int_0^1\left(\frac{\partial w(x,t)}{\partial x}\right)^2dx\right]^{1/2}\leqq\hat{N}^2\right\} \tag{II.116}$$

in which the estimates implied by (II.56) are valid for the present one-dimensional situation. For the problem at hand

$$\mathscr{E}(0)=\frac{1}{2}\int_0^1\left[\left(\frac{\partial f}{\partial x}\right)^2+g^2(x)\right]dx\geqq 0, \tag{II.117}$$

so that $\mathscr{F}(0)$ is always positive; in this situation just one of the estimates from Case III above is applicable and we have the following growth estimate:

PROPOSITION (Bloom [22]). *Let $u\in\hat{\mathcal{N}}$ be a solution of the initial-history boundary value problem* (II.108)–(II.110) *where $0<\lambda\leqq 1/T$. If*

(II.118)

$$\left\{\int_0^1 f(x)g(x)\,dx=\left[\int_0^1 f^2(x)\,dx\right]^{1/2}(2\mathscr{F}(0))^{1/2}\right\}\Leftrightarrow f(x)\equiv 0\quad\text{a.e. on }[0,1],$$

where $\mathscr{F}(0)$ is given by (II.114), (II.115), *then for all t, $0\leqq t<T$*

$$\int_0^1 u^2(x,t)\,dx\geqq\left(\int_0^1 g^2(x)\,dx\right)t^2. \tag{II.119}$$

The equivalence in (II.118) *follows trivially from the estimates*

$$\begin{aligned}\left[\int_0^1 f^2(x)\,dx\right]^{1/2}\left[\int_0^1 g^2(x)\,dx\right]^{1/2}&\geqq\int_0^1 f(x)g(x)\,dx\\ &=\left[\int_0^1 f^2(x)\,dx\right]^{1/2}(2\mathscr{F}(0))^{1/2}\\ &\geqq\left[\int_0^1 f^2(x)\,dx\right]^{1/2}\left[\int_0^1(g^2(x)+f'^2(x))\,dx\right]^{1/2}\end{aligned} \tag{II.120}$$

and the fact that $f(x)$ must vanish at $x=0$ and $x=1$.

For certain cases, such as the one we present below for $\mathscr{E}(0)$ sufficiently negative, growth estimates for $\|\mathbf{u}(t)\|^2$ may be established by a particular differential inequality argument which requires even weaker hypotheses than those which led to the differential inequality (II.86); we are, in particular, interested in weakening the assumption (II.55), for the case of sufficiently negative initial-energy. To this end, we will augment our hypotheses concerning the past history $\mathbf{U}$ by assuming that $\exists \tau_\infty > 0$ such that $\mathbf{U}(\tau) = 0$, $-\infty < \tau < -\tau_\infty$ and we also assume that $M^2 = \sup_{[-\tau_\infty, 0)} \|\mathbf{U}(\tau)\|_+ < \infty$. If we set

(II.121)
$$\mathscr{K}(t) = \|\mathbf{K}(t)\|_{\mathscr{L}_S(H_+, H_-)},$$
$$\hat{\mathscr{K}}(t) = \int \|\mathbf{K}_t(t)\|_{\mathscr{L}_S(H_+, H_-)}\, dt,$$

then we can prove the following:

THEOREM (Bloom [11]). *Let* $\mathbf{u} \in \mathscr{N}$ *be a solution of* (II.23)–(II.25) *with prescribed past history satisfying the above smoothness assumptions. Assume that* $\mathbf{K}(\cdot)$ *satisfies*

(II.122) $$\mathscr{K}(\cdot) \in L_1[0, \infty), \quad \hat{\mathscr{K}}(\cdot) \in L_1[0, \infty) \quad \text{with } \hat{\mathscr{K}}(0) = 0$$

and

(II.123) $$-\langle \mathbf{v}, \mathbf{K}(0)\mathbf{v}\rangle \geqq 0 \quad \forall \mathbf{v} \in H_+.$$

Then if $\mathscr{E}(0) < 0$ *with*

(II.124) $$|\mathscr{E}(0)| \geqq \gamma (M^2 + N^2)^2 (\tfrac{3}{2}\|\mathscr{K}\|_{L_1[0,\infty)} + \|\hat{\mathscr{K}}\|_{L_1[0,\infty)}),$$

$F(t) = \|\mathbf{u}(t)\|^2$ *satisfies the differential inequality*

(II.125) $$FF'' - \left(\frac{\beta + 1}{2\beta + 1}\right) F'^2 \geqq 0, \qquad 0 \leqq t < T, \quad 0 < \beta < \infty.$$

Proof. By the definition of $F(t)$, $F'(t) = 2\langle \mathbf{u}, \mathbf{u}_t\rangle$ and $F''(t) = 2\|\mathbf{u}_t\|^2 + 2\langle \mathbf{u}, \mathbf{u}_{tt}\rangle$. A direct computation then yields the identity

(II.126) $$FF'' - (\beta + 1)F'^2 = 4(\beta + 1)Q_\beta^2 + 2F[\langle \mathbf{u}, \mathbf{u}_t\rangle - (2\beta + 1)\|\mathbf{u}_t\|^2],$$

valid for any β, $0 < \beta < \infty$, where

(II.127) $$Q_\beta^2(t) = \|\mathbf{u}\|^2 \|\mathbf{u}_t\|^2 - \langle \mathbf{u}, \mathbf{u}_t\rangle^2 \geqq 0,$$

by the Schwarz inequality. Therefore, for any $\beta > 0$, $F(t) = \|\mathbf{u}(t)\|^2$ satisfies the differential inequality

(II.128) $$FF'' - (\beta + 1)F'^2 \geqq 2F\mathscr{R}_\beta, \qquad t > 0,$$

where

$$\mathcal{R}_\beta(t) = \langle \mathbf{u}, \mathbf{N}\mathbf{u}\rangle - (2\beta+1)\|\mathbf{u}_t\|^2 - \left\langle \mathbf{u}, \int_{-\infty}^{t} \mathbf{K}(t-\tau)\mathbf{u}(\tau)\,d\tau \right\rangle$$

$$(\text{II.129}) \qquad = -(2\beta+1)[\|\mathbf{u}_t\|^2 - \langle \mathbf{u}, \mathbf{N}\mathbf{u}\rangle] - 2\beta\langle \mathbf{u}, \mathbf{N}\mathbf{u}\rangle - \left\langle \mathbf{u}, \int_{-\infty}^{t} \mathbf{K}(t-\tau)\mathbf{u}(\tau)\,d\tau \right\rangle$$

by virtue of the governing integrodifferential evolution equation (II.23). Combining (II.129) with the definition of the total energy $\mathscr{E}(t)$ then yields

$$(\text{II.130}) \quad \mathcal{R}_\beta(t) = -2(2\beta+1)\mathscr{E}(t) - 2\beta\langle \mathbf{u}, \mathbf{N}\mathbf{u}\rangle - \left\langle \mathbf{u}, \int_{-\infty}^{t} \mathbf{K}(t-\tau)\mathbf{u}(\tau)\,d\tau \right\rangle.$$

Now, by (II.63),

$$\mathscr{E}(t) - \mathscr{E}(0) = -\int_0^t \left\langle \mathbf{u}_\tau, \int_{-\infty}^{\tau} \mathbf{K}(t-\lambda)\mathbf{u}(\lambda)\,d\lambda \right\rangle d\tau,$$

and therefore (II.130) is equivalent to

$$(\text{II.131}) \quad \begin{aligned} \mathcal{R}_\beta(t) = {} & -2(2\beta+1)\mathscr{E}(0) - 2\beta\langle \mathbf{u}, \mathbf{N}\mathbf{u}\rangle \\ & + 2(2\beta+1)\int_0^t \left\langle \mathbf{u}_\tau, \int_{-\infty}^{\tau} \mathbf{K}(t-\lambda)\mathbf{u}(\lambda)\,d\lambda \right\rangle d\tau \\ & - \left\langle \mathbf{u}, \int_{-\infty}^{t} \mathbf{K}(t-\tau)\mathbf{u}(\tau)\,d\tau \right\rangle. \end{aligned}$$

If we now take the inner-product in H of (II.23) with $\mathbf{u}$ and use the definition of $F(t)$ we easily obtain the identity

$$(\text{II.132}) \qquad \frac{1}{2}F'' = \|\mathbf{u}_t\|^2 + \langle \mathbf{u}, \mathbf{N}\mathbf{u}\rangle - \left\langle \mathbf{u}, \int_{-\infty}^{t} \mathbf{K}(t-\tau)\mathbf{u}(\tau)\,d\tau \right\rangle.$$

We now substitute for $-2\beta\langle \mathbf{u}, \mathbf{N}\mathbf{u}\rangle$ from (II.132) into (II.131), collect terms, and drop the nonnegative term proportional to $\|\mathbf{u}_t\|^2$ so as to obtain the lower bound

$$(\text{II.133}) \quad \begin{aligned} \mathcal{R}_\beta(t) \geqq {} & -\beta F'' - 2(2\beta+1)\mathscr{E}(0) - (2\beta+1)\left\langle \mathbf{u}, \int_{-\infty}^{t} \mathbf{K}(t-\tau)\mathbf{u}(\tau)\,d\tau \right\rangle \\ & + 2(2\beta+1)\int_0^t \left\langle \mathbf{u}_\tau, \int_{-\infty}^{t} \mathbf{K}(t-\lambda)\mathbf{u}(\lambda)\,d\lambda \right\rangle d\tau. \end{aligned}$$

Finally, we combine the estimate given by (II.133) with the differential inequality (II.128) and we obtain

$$
\begin{aligned}
FF'' - \left(\frac{\beta+1}{2\beta+1}\right)F'^2 \geqq{} & -4F\mathscr{E}(0) \\
& -2F\left\langle \mathbf{u}, \int_{-\infty}^{t} \mathbf{K}(t-\tau)\mathbf{u}(\tau)\,d\tau \right\rangle \\
& +4F\int_0^t \left\langle \mathbf{u}_\tau, \int_{-\infty}^{t} \mathbf{K}(t-\tau)\mathbf{u}(\lambda)\,d\lambda \right\rangle d\tau.
\end{aligned} \tag{II.134}
$$

As we have yet to make use of the hypothesis that $\mathbf{u} \in \mathcal{N}$, the differential inequality (II.134) is valid for $t>0$. Using the hypothesis that $\mathscr{E}(0)<0$ we can rewrite (II.134) in the form

$$
\begin{aligned}
FF'' - \left(\frac{\beta+1}{2\beta+1}\right)F'^2 \geqq 2F\Big\{ & 2|\mathscr{E}(0)| - \left\langle \mathbf{u}, \int_{-\infty}^{t} \mathbf{K}(t-\tau)\mathbf{u}(\tau)\,d\tau \right\rangle \\
& +2\int_0^t \left\langle \mathbf{u}_\tau, \int_{-\infty}^{t} \mathbf{K}(\tau-\lambda)\mathbf{u}(\lambda)\,d\lambda \right\rangle d\tau \Big\}.
\end{aligned} \tag{II.135}
$$

We now make use of an identity employed in the proof of the previous theorem, namely,

$$
\begin{aligned}
\left\langle \mathbf{u}_\tau, \int_{-\infty}^{\tau} \mathbf{K}(\tau-\lambda)\mathbf{u}(\lambda)\,d\lambda \right\rangle = {} & \frac{d}{d\tau}\left\langle \mathbf{u}(\tau), \int_{-\infty}^{\tau} \mathbf{K}(\tau-\lambda)\mathbf{u}(\lambda)\,d\lambda \right\rangle \\
& -\left\langle \mathbf{u}(\tau), \int_{-\infty}^{\tau} \mathbf{K}_\tau(\tau-\lambda)\mathbf{u}(\lambda)\,d\lambda \right\rangle \\
& -\left\langle \mathbf{u}(\tau), \mathbf{K}(0)\mathbf{u}(\tau) \right\rangle,
\end{aligned}
$$

and (II.135) may be recast in the form

$$
\begin{aligned}
FF'' - \left(\frac{\beta+1}{2\beta+1}\right)F'^2 \geqq 2F\Big\{ & 2|\mathscr{E}(0)| - 2\int_0^t \langle \mathbf{u}(\tau), \mathbf{K}(0)\mathbf{u}(\tau)\rangle\,d\tau \\
& -2\int_0^t \left\langle \mathbf{u}(\tau), \int_{\infty}^{\tau} \mathbf{K}_\tau(\tau-\lambda)\mathbf{u}(\lambda)\,d\lambda \right\rangle d\tau \\
& -2\left\langle \mathbf{f}, \int_{-\infty}^{0} \mathbf{K}(-\tau)\mathbf{u}(\tau)\,d\tau \right\rangle + \left\langle \mathbf{u}, \int_{-\infty}^{t} \mathbf{K}(t-\tau)\mathbf{u}(\tau)\,d\tau \right\rangle \Big\} \\
\geqq 2F\Big\{ & 2|\mathscr{E}(0)| - 2\left\langle \mathbf{f}, \int_{-\infty}^{0} \mathbf{K}(-\tau)\mathbf{u}(\tau)\,d\tau \right\rangle \\
& -2\int_0^t \left\langle \mathbf{u}(\tau), \int_{-\infty}^{\tau} \mathbf{K}_\tau(\tau-\lambda)\mathbf{u}(\lambda)\,d\lambda \right\rangle d\tau \\
& +\left\langle \mathbf{u}, \int_{-\infty}^{t} \mathbf{K}(t-\tau)\mathbf{u}(\tau)\,d\tau \right\rangle \Big\}
\end{aligned} \tag{II.136}
$$

in view of our hypothesis (II.123). It should be clear that the theorem will be proven if we can verify that, under the hypotheses on $\mathscr{E}(0)$, the right-hand side of (II.136$_1$) is nonnegative on $[0, T)$ for $\mathbf{u}\in\mathcal{N}$; to this end we exhibit the following series of estimates:

$$\left|\left\langle \mathbf{f}, \int_{-\infty}^{0} \mathbf{K}(-\tau)\mathbf{u}(\tau)\,d\tau\right\rangle\right|$$

$$\text{(II.137)}\qquad \begin{aligned} &\leqq \gamma\|\mathbf{f}\|_{+} \sup_{[-\tau_\infty,0)} \|\mathbf{U}\|_{+} \int_{-\infty}^{0} \|\mathbf{K}(-\tau)\|_{\mathscr{L}_S(H_+,H_-)}\,d\tau \\ &= \gamma\|\mathbf{f}\|_{+} \sup_{[-\tau_\infty,0)} \|\mathbf{U}\|_{+}\|\mathscr{K}\|_{L_1[0,\infty)} \end{aligned}$$

so that

$$\text{(II.138)}\qquad -\left\langle \mathbf{f}, \int_{-\infty}^{0} \mathbf{K}(-\tau)\mathbf{u}(\tau)\,d\tau\right\rangle \geqq -\gamma M^4\|\mathscr{K}\|_{L_1[0,\infty)}.$$

Also,

$$\left|\left\langle \mathbf{u}, \int_{-\infty}^{t} \mathbf{K}(t-\tau)\mathbf{u}(\tau)\,d\tau\right\rangle\right|$$

$$\text{(II.139)}\qquad \begin{aligned} &\leqq \gamma\Big(\sup_{[-\tau_\infty,T)} \|\mathbf{u}\|_{+}\Big)^2 \int_{-\infty}^{t} \|\mathbf{K}(t-\tau)\|_{\mathscr{L}_S(H_+,H_-)}\,d\tau \\ &\leqq \gamma\Big(\sup_{[-\tau_\infty,0)} \|\mathbf{U}\|_{+} + N^2\Big)^2 \int_{0}^{\infty} \|\mathbf{K}(\rho)\|_{\mathscr{L}_S(H_+,H_-)}\,d\rho \\ &\leqq \gamma(M^2+N^2)^2\|\mathscr{K}\|_{L_1[0,\infty)} \end{aligned}$$

so that

$$\text{(II.140)}\qquad \left\langle \mathbf{u}, \int_{-\infty}^{t} \mathbf{K}(t-\tau)\mathbf{u}(\tau)\,d\tau\right\rangle \geqq -\gamma(M^2+N^2)\|\mathscr{K}\|_{L_1[0,\infty)}.$$

Finally, we have

$$\left|\int_0^t \left\langle \mathbf{u}(\tau), \int_{-\infty}^{\tau} \mathbf{K}_\tau(\tau-\lambda)\mathbf{u}(\lambda)\,d\lambda\right\rangle d\tau\right|$$

$$\text{(II.141)}\qquad \begin{aligned} &\leqq \gamma\Big(\sup_{[-\tau_\infty,T)} \|\mathbf{u}\|_{+}\Big)^2 \int_0^T \int_{-\infty}^{\tau} \|\mathbf{K}_\tau(\tau-\lambda)\|_{\mathscr{L}_S(H_+,H_-)}\,d\lambda\,d\tau \\ &\leqq \gamma(M^2+N^2)^2 \int_0^\infty \int_0^{\tau+\tau_\infty} \|\mathbf{K}_\rho(\rho)\|_{\mathscr{L}_S(H_+,H_-)}\,d\rho\,d\tau \\ &\leqq \gamma(M^2+N^2)^2 \int_0^\infty \big(\hat{\mathscr{K}}(\rho)\big|_0^{\tau+\tau_\infty}\big)\,d\tau \\ &= \gamma(M^2+N^2)^2 \int_{\tau_\infty}^\infty \hat{\mathscr{K}}(\lambda)\,d\lambda \leqq \gamma(M^2+N^2)\|\hat{\mathscr{K}}\|_{L_1[0,\infty)} \end{aligned}$$

in view of our hypothesis that $\hat{\mathcal{K}}(0)=0$. Thus

$$\text{(II.142)}\qquad -\int_0^t \left\langle \mathbf{u}(\tau), \int_\infty^\tau \mathbf{K}_\tau(\tau-\lambda)\mathbf{u}(\lambda)\,d\lambda \right\rangle d\tau \geqq -\gamma(M^2+N^2)^2\|\mathcal{K}\|_{L_1[0,\infty)}.$$

It is easy now to see that the differential inequality (II.125) follows by combining the estimates (II.138), (II.140), (II.142) with the second inequality in (II.136) and our hypothesis (II.124) relative to the magnitude $|\mathscr{E}(0)|$ of the initial energy. Q.E.D.

An immediate consequence of the theorem above is the growth estimate given by the following:

COROLLARY. *Under the same hypotheses which prevail in the theorem above, any solution* $\mathbf{u}\in\mathcal{N}$ *of* (II.23)–(II.25) *must satisfy*

$$\text{(II.143)}\qquad \|\mathbf{u}(t)\|^2 \geqq \|\mathbf{f}\|^2\left[1+2(1-\alpha)\left(\frac{\langle \mathbf{f},\mathbf{g}\rangle}{\|\mathbf{f}\|^2}\right)t\right]^{1/(1-\alpha)},\qquad 0\leqq t<T,$$

for all α, $\frac{1}{2}<\alpha<1$.

Proof. For any β, $0<\beta<\infty$, set $\alpha=(\beta+1)/(2\beta+1)$; clearly α ranges over $(\frac{1}{2},1)$ as β ranges over $(0,\infty)$. By the previous theorem we have

$$\text{(II.144)}\qquad FF''-\alpha F'^2\geqq 0,\qquad 0\leqq t<T,\quad 0<\alpha<\tfrac{1}{2},$$

where $F(t)=\|\mathbf{u}(t)\|^2$. However, for any $t>0$

$$\text{(II.145)}\qquad [F^{(1-\alpha)}]''(t)=(1-\alpha)F^{-\alpha-1}(t)[F(t)F''(t)-\alpha F'^2(t)],$$

and therefore (II.144) implies that

$$\text{(II.146)}\qquad [F^{(1-\alpha)}]''(t)\geqq 0,\qquad 0\leqq t<T,\quad \tfrac{1}{2}<\alpha<1.$$

Integration now yields

$$\text{(II.147)}\qquad [F^{(1-\alpha)}]'(t)\geqq(1-\alpha)F^{-\alpha}(0)F'(0),\quad \tfrac{1}{2}<\alpha<1,$$

and by a second integration we obtain

$$\text{(II.148)}\qquad \begin{aligned} F^{(1-\alpha)}(t) &\geqq F^{(1-\alpha)}(0)+(1-\alpha)F^{-\alpha}(0)F'(0)t \\ &= F^{(1-\alpha)}(0)\left[1+(1-\alpha)\left(\frac{F'(0)}{F(0)}\right)t\right] \end{aligned}$$

or, as $1-\alpha>0$,

$$\text{(II.149)}\qquad F(t)\geqq F(0)\left[1+(1-\alpha)\left(\frac{F'(0)}{F(0)}\right)t\right]^{1/(1-\alpha)},\qquad 0\leqq t<T.$$

The estimate (II.143) now follows directly from (II.149) via the definition of $F(t)$. Q.E.D.

Remark. The estimate (II.143) actually includes, and is, therefore, stronger than the exponential growth estimate (II.90) (obtained for $\|\mathbf{u}(t)\|$ from the differential inequality (II.56) for the case of $\mathscr{E}(0)<0$ and sufficiently large in magnitude, i.e., for $\mathscr{E}(0)$ satisfying either (II.84) or (II.85)). In fact if we take the limit in (II.143) as $\alpha \to 1^-$ then the elementary fact that $\lim_{\lambda\to 0^+} [1+\lambda x]^{1/\lambda} = e^x$ establishes (II.90) for $0 \leqq t < T$.

Remark. In the theorem above, and the resulting corollary, we have weakened the hypothesis on $\mathbf{K}(0)$ from (II.55) to (II.23) and added the two additional hypotheses on the functions $\mathscr{K}(\cdot)$, $\hat{\mathscr{K}}(\cdot)$, respectively, which are given by (II.122). In terms of conditions on the relaxation tensor $g_{ijkl}(\mathbf{x}, t)$ in our three-dimensional viscoelastic model we are, therefore, imposing the conditions

$$\text{(II.150)} \qquad \int_\Omega \dot{g}_{ijkl}(\mathbf{x}, 0) \frac{\partial v_i}{\partial x_j} \frac{\partial v_k}{\partial x_l}\, d\mathbf{x} \leqq 0, \quad \forall \mathbf{v} \in (H_0^1(\Omega))^3,$$

and

$$\text{(II.151)} \qquad \begin{aligned} &\int_0^\infty \sup_\Omega |\dot{g}_{ijkl}(\mathbf{x}, t)|\, dt < \infty, \\ &\int_0^\infty \sup_\Omega \left(\int_0^t \left| \ddot{g}_{ijkl}(\mathbf{x}, \tau) \right| d\tau \right) dt < \infty, \end{aligned}$$

with $\int \ddot{g}_{ijkl}(\mathbf{x}, \tau)\, d\tau|_{\tau=0} = 0$. For the one-dimensional homogeneous viscoelastic body with relaxation function $g_0(t)$, (II.150)–(II.151) reduce to

$$\text{(II.152)} \qquad \begin{aligned} &g_0'(0) \leqq 0, \qquad \int g_0''(\tau)\, d\tau|_{\tau=0} = 0, \\ &\int_0^\infty |g_0'(t)|\, dt < \infty, \qquad \int_0^\infty \int_0^t |g_0''(\tau)|\, d\tau\, dt < \infty. \end{aligned}$$

We will close out this section on growth estimates, for solutions $\mathbf{u} \in \mathscr{N}$, by showing how a modification of the convexity argument based on the differential inequality (II.56) can lead to exponential growth estimates for $\|\mathbf{u}(t)\|$, even in the case where the initial energy $\mathscr{E}(0)$ satisfies (II.99) so that $\mathscr{F}(0) > 0$. For the case where $\mathscr{F}(0) > 0$ with the initial data satisfying $\langle \mathbf{f}, \mathbf{g} \rangle \geqq (2\mathscr{F}(0))^{1/2} \|\mathbf{f}\|$ we have obtained the growth estimates (II.102) ($\langle \mathbf{f}, \mathbf{g} \rangle > (2\mathscr{F}(0))^{1/2} \|\mathbf{f}\|$) and (II.103) ($\langle \mathbf{f}, \mathbf{g} \rangle = (2\mathscr{F}(0))^{1/2} \|\mathbf{f}\|$); we are assuming, of course, the hypotheses (II.55) relative to $\mathbf{K}(0)$. Suppose now that we retain the hypothesis (II.99) relative to $\mathscr{E}(0)$ but drop the requirement that $\langle \mathbf{f}, \mathbf{g} \rangle \geqq (2\mathscr{F}(0)^{1/2} \|\mathbf{f}\|$; then as in Bloom [18] we may proceed as follows: First of all, we define

$$\text{(II.153)} \qquad \lambda(\beta; t_0) = \frac{2(2\mathscr{F}(0) + \beta)}{\beta t_0^2} > 0 \quad \forall \beta, \quad t_0 \geqq 0.$$

Directly from the definition (II.54) of $F(t; \beta, t_0)$ it is clear that

$$\text{(II.154)} \qquad \lambda \beta t_0^2 \leqq \lambda F(t; \beta, t_0) \quad \forall t \geqq 0,$$

for any positive λ and all $\beta, t_0 \geqq 0$. Setting $\lambda = \lambda(\beta; t_0)$ we have, therefore,

$$2(2\mathcal{F}(0)+\beta) \leqq \lambda(\beta; t_0)F(t; \beta, t_0) \quad \forall t \geqq 0, \tag{II.155}$$

and all nonnegative β, t_0. From (II.56) we then obtain the differential inequality

$$FF''-F'^2 \geqq -\lambda(\beta; t_0)F^2, \qquad 0 \leqq t < T. \tag{II.156}$$

Now, as $F(t; \beta, t_0) > 0$ (unless $\mathbf{u}(t) = \mathbf{0}$, $0 \leqq t < T$, and $\beta = 0$), (II.156) implies that for $t \in [0, T)$

$$\frac{d^2}{dt^2}[\ln F(t; \beta, t_0)] + \lambda(\beta; t_0) \geqq 0 \tag{II.157}$$

or

$$\frac{d^2}{dt^2}[\ln \{\exp(\tfrac{1}{2}\lambda(\beta; t_0)t^2)F(t; \beta, t_0)\}] \geqq 0 \tag{II.158}$$

for $t \in [0, T)$, $\beta, t_0 \geqq 0$. We now set

$$G(t; \beta, t_0) = \ln[\exp(\tfrac{1}{2}\lambda(\beta; t_0)t^2)F(t; \beta, t_0)]. \tag{II.159}$$

Then by (II.157) and our smoothness hypotheses, it follows that

$$G(t; \beta, t_0) \geqq G(0; \beta, t_0) + \dot{G}(0; \beta, t_0)t, \qquad 0 \leqq t < T. \tag{II.160}$$

Substituting for $G(t; \beta, t_0)$ from (II.159) and simplifying we easily find that the estimate

$$F(t; \beta, t_0) \geqq F(0; \beta, t_0)\exp(-\lambda(\beta; t_0)t^2)\exp\left\{\frac{\dot{F}(0; \beta, t_0)}{F(0; \beta, t_0)}t\right\} \tag{II.161}$$

is valid on $[0, T)$ for all $\beta, t_0 \geqq 0$, with

$$F(0; \beta, t_0) = \|\mathbf{f}\|^2 + \beta t_0^2, \qquad \dot{F}(0; \beta, t_0) = 2\langle \mathbf{f}, \mathbf{g}\rangle + 2\beta t_0. \tag{II.162}$$

Up to this point, the nonnegative constants β, t_0 have been abitrary. Now, let $\varepsilon > 0$ be an arbitrary positive constant, set $\beta = \varepsilon/t_0^2$ and let $t_0 \to +\infty$ in (II.153), (II.161), (II.162); we easily obtain from (II.161) the estimate

$$\|\mathbf{u}(t)\|^2 + \varepsilon \geqq (\|\mathbf{f}\|^2 + \varepsilon)\exp(-\bar{\lambda}t^2)\exp\left(\frac{2\langle \mathbf{f}, \mathbf{g}\rangle t}{\|\mathbf{f}\|^2+\varepsilon}\right) \tag{II.163}$$

valid on $[0, T)$, where $\bar{\lambda} = 4\mathcal{F}(0)/\varepsilon > 0$. For $\mathbf{f} \neq \mathbf{0}$ we may set $\varepsilon = \|\mathbf{f}\|^2$ in (II.163) and obtain, for $0 \leqq t < T$, the growth estimate

$$\frac{\|\mathbf{u}(t)\|^2 + \|\mathbf{f}\|^2}{2} \geqq \|\mathbf{f}\|^2 \exp\left\{\frac{-4\mathcal{F}(0)t^2 + \langle \mathbf{f}, \mathbf{g}\rangle t}{\|\mathbf{f}\|^2}\right\}, \tag{II.164}$$

valid for all initial data $\mathbf{f}$, $\mathbf{g} \in H_+$, $\mathbf{f} \neq \mathbf{0}$. Various special growth estimates are immediate consequences of (II.164); for example, if $\mathbf{g}=\mathbf{0}$ there results

$$\|\mathbf{u}(t)\|^2 \geqq \|\mathbf{f}\|^2\left[2\exp\left\{\frac{-4\mathcal{F}(0)t^2}{\|\mathbf{f}\|^2}\right\}-1\right], \quad 0\leqq t<T_0, \tag{II.165}$$

where $T_0=\min(T, [\frac{1}{4}\mathcal{F}^{-1}(0)\ln 2]^{1/2}\|\mathbf{f}\|)$. Also, without a lower bound on $\langle \mathbf{f}, \mathbf{g}\rangle$ of the type represented by the earlier condition $\langle \mathbf{f}, \mathbf{g}\rangle \geqq (2\mathcal{F}(0))^{1/2}\|\mathbf{f}\|$, it is clear that (II.164) yields a monotonically increasing exponential lower bound for $\|\mathbf{u}(t)\|^2$ on some interval $[0, T_1)$, with T_1 sufficiently small, provided the initial data satisfies $\langle \mathbf{f}, \mathbf{g}\rangle>0$. In fact, the right-hand side of (II.164) is a monotonically increasing exponential function of t for all $t\in[0, \frac{1}{8}\mathcal{F}^{-1}(0)\langle \mathbf{f}, \mathbf{g}\rangle]$. Our results may be summed up in the following:

THEOREM (Bloom [18]). *Let* $\mathbf{u}\in\mathcal{N}$ *be any solution of* (II.23)–(II.25) *for which* $\mathcal{E}(0)$ *satisfies* (II.99) (*or equivalent conditions which guarantee* $\mathcal{F}(0)>0$) *and* $\mathbf{K}(t)$ *satisfies the hypotheses* (II.55). *If* $\langle \mathbf{f}, \mathbf{g}\rangle>0$ *then* $\exists\psi(t)$, *a real-valued monotonically increasing function on* $[0, T_1)$, *with* $\psi(0)=0$ *and* $T_1=\min(T, \mathcal{F}^{-1}(0)\langle \mathbf{f}, \mathbf{g}\rangle)$, *such that*

$$\|\mathbf{u}(t)\|^2 \geqq 2\|\mathbf{f}\|^2(\exp\psi(t)-\tfrac{1}{2}), \qquad 0\leqq t<T_1. \tag{II.166}$$

Remarks. Upper bounds on the growth of solutions $\mathbf{u}\in\mathcal{N}$ are easily obtainable from (II.158). In fact, if we integrate (II.158) according to the secant property of convex functions (i.e., the estimate immediately following (I.11)) we obtain

$$G(t;\beta,t_0)\leqq G(0;\beta,t_0)+\frac{t}{T}[G(T;\beta,t_0)-G(0;\beta,t_0)] \tag{II.167}$$

on $[0, T)$, for all β, $t_0\geqq 0$, where $G(t;\beta,t_0)$ is given by (II.159). Substituting for G in (II.167) we easily obtain the estimate

$$F(t;\beta,t_0)\leqq e^{-\lambda(\beta;t_0)t^2}F(0;\beta,t_0)^{1-t/T}(e^{\lambda(\beta;t_0)T^2}F(T;\beta,t_0))^{t/T} \tag{II.168}$$

on $[0, T)$, again valid for all β, $t_0\geqq 0$. We again set $\beta=\varepsilon/t_0^2$, for $\varepsilon>0$ arbitrary, and take the limit as $t_0\to+\infty$, this time obtaining

$$\|\mathbf{u}(t)\|^2\leqq e^{-4\mathcal{F}(0)t^2/\varepsilon}(\|\mathbf{f}\|^2+\varepsilon)^{1-t/T}(e^{4\mathcal{F}(0)T^2/\varepsilon}[\|\mathbf{u}(T)\|^2+\varepsilon])^{t/T} \tag{II.169}$$

for all t, $0\leqq t<T$. As $\mathbf{u}\in\mathcal{N}$, by hypothesis,

$$\|\mathbf{u}(T)\|^2\leqq\left(\sup_{[0,T)}\|\mathbf{u}(t)\|\right)^2\leqq\gamma^2\left(\sup_{[0,T)}\|\mathbf{u}(t)\|_+\right)^2\leqq\gamma^2N^2.$$

Therefore, if we again choose $\varepsilon=\|\mathbf{f}\|^2$, $\mathbf{f}\neq\mathbf{0}$, and pick $M=M(T)$ so large that

$$\gamma^2N^4+\|\mathbf{f}\|^2\leqq M(T)\exp\left(\frac{-4\mathcal{F}(0)T^2}{\|\mathbf{f}\|^2}\right),$$

then we obtain from (II.169) the estimate

$$\|\mathbf{u}(t)\|^2 \leqq C(T)\|\mathbf{f}\|^{2\delta(t)} \exp\left(\frac{-4\mathcal{F}(0)t^2}{\|\mathbf{f}\|^2}\right) \tag{II.170}$$

valid on $[0, T)$, where $\delta(t) = 1 - t/T$ and

$$C(T) = \sup_{[0,T)} [2^{\delta(t)} M(T)^{1-\delta(t)}].$$

The growth estimate represented by (II.170) shows that under the hypotheses of the above theorem, $\|\mathbf{u}(t)\|^2$ is bounded from above by a monotonically decreasing exponential function of t on $[0, T)$.

Example. The results delineated above in the estimates (II.166), (II.170) are immediately applicable to the one-dimensional, homogeneous viscoelastic initial-history boundary value problem (II.108), (II.109), (II.110). The operators $\mathbf{G}$, $\mathbf{N}$, $\mathbf{K}(t)$ are again given by (II.111) and we again assume that the past history $U(x, t)$ satisfies (II.112) and that the constitutive parameter λ which appears in (II.108), (II.111) satisfies $\lambda \leqq 1/T$ so that the hypotheses (II.55) relative to $\mathbf{K}(t)$ are satisfied. In view of (II.117) we always satisfy the hypothesis (II.99) on $\mathscr{E}(0)$, in this case, and, in particular $\mathcal{F}(0)$ is always positive.. For solutions $\mathbf{u} \in \hat{\mathcal{N}}$ of (II.108)–(II.110), with $\hat{\mathcal{N}}$ the class of bounded functions defined by (II.116), we have, therefore, the following results:

(i) If $\int_0^1 f(x)g(x)\,dx > 0$, and $T_1 = \min(T, (\mathcal{F}(0)^{-1} \int_0^1 fg\,dx)$ where

$$\mathcal{F}(0) = \frac{1}{2}\int_0^1 (f^2(x) + g^2(x))\,dx + \hat{k}_1\lambda + \hat{k}_2\lambda^2,$$

($\hat{k}_1$, $\hat{k}_2$ defined by (II.115)) then

$$\int_0^1 u^2(x,t)\,dx \geqq 2\int_0^1 f^2(x)\,dx\,[\exp\hat{\psi}(t) - \tfrac{1}{2}], \qquad 0 \leqq t < T_1, \tag{II.171}$$

$$\hat{\psi}(t) = \frac{1}{\int_0^1 f^2(x)\,dx}\left(\left[\int_0^1 f(x)g(x)\,dx\right]t - 4\mathcal{F}(0)t^2\right). \tag{II.172}$$

(ii) If $f(x) \neq 0$, a.e. on $[0, 1]$ then there exists a constant $\hat{B} = \hat{B}(T)$ such that for all t, $0 \leqq t < T$,

$$\int_0^1 u^2(x,t)\,dx \leqq \hat{B}\left(\int_0^1 f^2(x)\,dx\right)^{1-t/T} \exp\left[\frac{-4\mathcal{F}(0)t^2}{\int_0^1 f^2(x)\,dx}\right]. \tag{II.173}$$

4. Continuous data dependence for ill-posed initial-history boundary value problems in linear viscoelasticity.

We want, in this section, to consider the problem of continuous dependence upon initial-data, initial-geometry, etc., for solutions of the ill-posed initial-history value problem (II.23)–(II.25). Under appropriate conditions on the viscoelastic relaxation tensor $g_{ijkl}(\mathbf{x}, t)$ our results will be immediately applicable to the problem of continuous dependence upon data for solutions of the viscoelastic initial-boundary value problem (II.2′), (II.3), (II.4) and we will, in fact, conclude the section by giving some specific continuous dependence estimates for solutions of an initial-history boundary value problem for a homogeneous, one-dimensional viscoelastic body. Our approach in this section will be based on a logarithmic convexity argument not unlike that presented in Chapter I for the initial-value problem associated with the abstract equation (I.13); the relevant spaces and operators are precisely the same as those used in the formulation of the initial-history value problem (II.23)–(II.25).

We will begin by looking at the nonhomogeneous version of the evolution equation, i.e.,

$$\text{(II.23}'\text{)}\qquad \mathbf{u}_{tt}-\mathbf{N}\mathbf{u}+\int_0^t \mathbf{K}(t-\tau)\mathbf{u}(\tau)\,d\tau=\mathcal{F}(t),\qquad 0\leqq t<T,$$

with $\mathcal{F}(\cdot)\in L^2([0,T);H_-)$ such that $\mathcal{F}(0)\neq\mathbf{0}$, and associated initial data $\mathbf{f},\mathbf{g}:J\to H_+$ where J, the domain of $\mathbf{u}(\cdot,t)$, $t\in[0,T)$, is an arbitrary topological space which is endowed with a positive measure μ; the prescribed past history $\mathbf{U}$, which appears in the expression $\int_{-\infty}^0 \mathbf{K}(t-\tau)\mathbf{U}(\tau)\,d\tau$ has been absorbed into $\mathcal{F}(t)$. For any $\mathbf{v}\in C([0,T);H_+)$ we define

$$\text{(II.174)}\qquad \|\mathbf{v}\|_t^2=\int_0^t \|\mathbf{v}(\tau)\|^2\,d\tau,\qquad 0\leqq t<T,$$

and

$$\text{(II.175)}\qquad \mathcal{N}_T=\{\mathbf{v}\in C^2([0,T);H_+)\,|\,\mathbf{v}_{tt}(0)\neq\mathbf{0},\ \|\mathbf{v}\|_T^2\leqq N^2\}$$

for some real $N\neq 0$. Now let $\mathbf{v}\in\mathcal{N}_T$ with $\mathbf{v}(0)=\mathbf{0}$, $\mathbf{v}_t(0)=\mathbf{0}$ and define, for $0\leqq t<T$,

$$\text{(II.176)}\qquad \begin{aligned} Q_{\mathbf{v}}(t)&\equiv\frac{t\int_0^t \|\mathbf{v}(\tau)\|_+^2\,d\tau}{\int_0^t\int_0^\eta \|\mathbf{v}(\tau)\|_+^2\,d\tau\,d\eta},\qquad \mathbf{v}\neq 0\\ &=\frac{tq_{\mathbf{v}}(t)}{\int_0^t q_{\mathbf{v}}(\tau)\,d\tau},\end{aligned}$$

where $q_{\mathbf{v}}(t)=\int_0^t \|\mathbf{v}(\tau)\|_+^2\,d\tau$. By virtue of the monotonicity of $q_{\mathbf{v}}$ on $[0,T)$, and the mean-value theorem for integrals, $Q_{\mathbf{v}}\geqq 1$ on $[0,T)$ for $\mathbf{v}\neq\mathbf{0}$. We state the following Lemma, a proof of which can be found in Bloom [21]:

LEMMA. *If* $\mathbf{v}\in\mathcal{N}_T$, *with* $\mathbf{v}(0)=\mathbf{v}_t(0)=\mathbf{0}$, *then* $\sup_{[0,T)} Q_{\mathbf{v}}(t)<+\infty$.

Remark. As $\mathbf{v}\in\mathcal{N}_T$ the only problem is to show that

$$\lim_{t\to+0}\left[\frac{tq_{\mathbf{v}}(t)}{\int_0^t q_{\mathbf{v}}(\tau)\,d\tau}\right]<+\infty\quad \text{when } \mathbf{v}(0)=\mathbf{v}_t(0)=\mathbf{0}.$$

Thus, for any $K>0$ we may define a subset $\mathscr{P}_K \subset \mathscr{N}_T$ as follows:

(II.177) $$\mathscr{P}_K = \{\mathbf{v} \in \mathscr{N}_T \mid \mathbf{v}(0) = \mathbf{v}_t(0) = 0, \sup_{[0,T)} Q_{\mathbf{v}}(t) \leqq 2K\}.$$

Examples are given in [21] to show that $\mathscr{P}_K \neq \varnothing$ (empty set) for given $K>0$. The set $\mathscr{P}_K$ will define the basic class of bounded functions in which our logarithmic convexity arguments will be valid, more precisely, we have the following:

THEOREM (Bloom [21]). *Let* $\mathbf{u} \in \mathscr{P}_K$, $K>0$, *be any solution of* (II.23') [*thus*, $\mathbf{u}(0) \equiv \mathbf{f} = \mathbf{0}$ *and* $\mathbf{u}_t(0) \equiv \mathbf{g} = \mathbf{0}$]. If $\mathbf{K}(t)$ *satisfies*

$$-\langle \mathbf{v}, \mathbf{K}(0)\mathbf{v}\rangle \geqq \kappa \|\mathbf{v}\|_+^2 \quad \forall \mathbf{v} \in H_+,$$

with

(II.178) $$\kappa \geqq K\left[\sup_{[0,T)} \|\mathbf{K}(t)\|_{\mathscr{L}_S(H_+,H_-)} + 2T \sup_{[0,T)} \|\mathbf{K}_t(t)\|_{\mathscr{L}_S(H_+,H_-)}\right],$$

then $\exists P, Q \geqq 0$ *such that for all* t, $0 \leqq t < T$,

(II.179) $$\|\mathbf{u}\|_t^2 \leqq P Q^{2\delta} \|\mathscr{F}\|^{2(1-\delta)}, \qquad 0 \leqq t < T, \quad \delta(t) = \frac{t}{T}.$$

Proof. We consider the real-valued function

(II.180) $$F(t) = \|\mathbf{u}\|_t^2 + T^4 \|\mathscr{F}\|_T^2, \qquad 0 \leqq t < T.$$

Direct computation using the definition of $\|(\cdot)\|_t$ yields

(II.181) $$F'(t) = \|\mathbf{u}(t)\|^2 = 2\int_0^t \langle \mathbf{u}(\eta), \mathbf{u}_\eta(\eta)\rangle \, d\eta$$

as $\mathbf{u}(0) = \mathbf{u}_t(0) = \mathbf{0}$. Also

(II.182) $$\begin{aligned} F''(t) &= 2\langle \mathbf{u}(t), \mathbf{u}_t(t)\rangle \\ &= 2\int_0^t \{\langle \mathbf{u}_\eta(\eta), \mathbf{u}_\eta(\eta)\rangle + \langle \mathbf{u}(\eta), \mathbf{u}_{\eta\eta}(\eta)\rangle\} \, d\eta \end{aligned}$$

which, in view of (II.23'), we can rewrite as

(II.183) $$\begin{aligned} F''(t) = 2\int_0^t \|\mathbf{u}_\eta\|^2 \, d\eta + 2\int_0^t \langle \mathbf{u}(\eta), \mathbf{N}\mathbf{u}(\eta)\rangle \, d\eta \\ + 2\int_0^t \langle \mathbf{u}(\eta), \mathscr{F}(\eta)\rangle \, d\eta - 2\int_0^t \left\langle \mathbf{u}(\eta), \int_0^\eta \mathbf{K}(\eta-\tau)\mathbf{u}(\tau)\, d\tau \right\rangle d\eta. \end{aligned}$$

If we apply the Schwarz inequality twice in succession to (II.181) we easily obtain

(II.184) $$[F'(t)]^2 \leqq 4\|\mathbf{u}\|_t^2 \int_0^t \|\mathbf{u}_\eta\|^2 \, d\eta$$

and thus we have the differential inequality for $F(t)$

$$
\begin{aligned}
FF''-F'^2 &\geqq 2F\int_0^t \|\mathbf{u}_\eta\|^2\,d\eta+2F\int_0^t \langle \mathbf{u}(\eta),\mathbf{N}\mathbf{u}(\eta)\rangle\,d\eta+2F\int_0^t \langle \mathbf{u}(\eta),\mathscr{F}(\eta)\rangle\,d\eta\\
&\quad -2F\int_0^t \left\langle \mathbf{u}(\eta),\int_0^\eta \mathbf{K}(\eta-\tau)\mathbf{u}(\tau)\,d\tau\right\rangle d\eta-4\|\mathbf{u}\|_t^2\int_0^t \|\mathbf{u}_\eta\|^2\,d\eta\\
&= -2F\left\{\int_0^t \|\mathbf{u}_\eta\|^2\,d\eta-\int_0^t \langle \mathbf{u}(\eta),\mathbf{N}\mathbf{u}(\eta)\rangle\,d\eta\right\}+2F\int_0^t \langle \mathbf{u}(\eta),\mathscr{F}(\eta)\rangle\,d\eta\\
&\quad -2F\int_0^t \left\langle \mathbf{u}(\eta),\int_0^\eta \mathbf{K}(\eta-\tau)\mathbf{u}(\tau)\,d\tau\right\rangle d\eta+4T^4\|\mathscr{F}\|_T^2\int_0^t \|\mathbf{u}_\eta\|^2\,d\eta,
\end{aligned}
\tag{II.185}
$$

where we have made use of (II.180) to substitute for $\|\mathbf{u}\|_t^2$ in the first estimate. Now, if we take the inner-product of (II.23′) with $\mathbf{u}_\tau$ and use the symmetry of $\mathbf{N}$ we obtain

$$
\frac{d}{d\tau}\langle \mathbf{u}_\tau,\mathbf{u}_\tau\rangle-\frac{d}{d\tau}\langle \mathbf{u},\mathbf{N}\mathbf{u}\rangle=2\langle \mathbf{u}_\tau,\mathscr{F}(\tau)\rangle-2\left\langle \mathbf{u}_\tau,\int_0^\tau \mathbf{K}(\tau-\lambda)\mathbf{u}(\lambda)\,d\lambda\right\rangle.
$$

Integrating this last identity with respect to τ over $[0,\eta]$ and then with respect to η over $[0,t]$, and using the homogeneous initial conditions $\mathbf{u}(0)=\mathbf{u}_t(0)=\mathbf{0}$, we arrive at the identity

$$
\begin{aligned}
&\int_0^t \|\mathbf{u}_\eta\|^2\,d\eta-\int_0^t \langle \mathbf{u}(\eta),\mathbf{N}\mathbf{u}(\eta)\rangle\,d\eta\\
&\quad =2\int_0^t (t-\eta)\langle \mathbf{u}_\eta,\mathscr{F}(\eta)\rangle\,d\eta-2\int_0^t\int_0^\eta\left\langle \mathbf{u}_\tau,\int_0^\tau \mathbf{K}(\tau-\lambda)\mathbf{u}(\lambda)\,d\lambda\right\rangle d\tau\,d\eta.
\end{aligned}
\tag{II.186}
$$

Substituting from (II.186) into the estimate (II.185) we obtain

$$
\begin{aligned}
FF''-F'^2 \geqq& -4F\int_0^t (t-\eta)\langle \mathbf{u}_\eta,\mathscr{F}(\eta)\rangle\,d\eta\\
&+2F\int_0^t \langle \mathbf{u}(\eta),\mathscr{F}(\eta)\rangle\,d\eta\\
&+4T^4\|\mathscr{F}\|_T^2\int_0^t \|\mathbf{u}_\eta\|^2\,d\eta\\
&-2F\int_0^t \left\langle \mathbf{u}(\eta),\int_0^\eta \mathbf{K}(\eta-\tau)\mathbf{u}(\tau)\,d\tau\right\rangle d\eta\\
&+4F\int_0^t\int_0^\eta\left\langle \mathbf{u}_\tau,\int_0^\tau \mathbf{K}(\tau-\lambda)\mathbf{u}(\lambda)\,d\lambda\right\rangle d\tau\,d\eta.
\end{aligned}
\tag{II.187}
$$

We now need the following:

LEMMA. *For $F(t)$ as defined by* (II.180)

$$\text{(II.188)}\quad \begin{aligned} &2F\int_0^t \langle \mathbf{u}(\lambda), \mathcal{F}(\lambda)\rangle\, d\lambda \geqq -T^{-2}F^2(t),\\ &-4F\int_0^t (t-\eta)\langle \mathbf{u}_\eta, \mathcal{F}(\eta)\rangle\, d\eta \geqq -T^{-2}F^2(t) - 4T^4\|\mathcal{F}\|_T^2 \int_0^t \|\mathbf{u}_\eta\|^2\, d\eta. \end{aligned}$$

Remark. Proofs of the estimates (II.188) depend only on applications of the Schwarz and arithmetic-geometric mean inequalities and follow the analogous results in [77].

Use of the estimates (II.188) now allows us to reduce the differential inequality (II.187) to

$$\text{(II.189)}\quad \begin{aligned} FF'' - F'^2 \geqq -T^{-2}F^2 &\\ &-2F\int_0^t \left\langle \mathbf{u}(\eta), \int_0^\eta \mathbf{K}(\eta-\tau)\mathbf{u}(\tau)\, d\tau\right\rangle d\eta\\ &+4F\int_0^t\int_0^\eta \left\langle \mathbf{u}_\tau, \int_0^\tau \mathbf{K}(\tau-\lambda)\mathbf{u}(\lambda)\, d\lambda\right\rangle d\tau\, d\eta. \end{aligned}$$

We can, once again, make use of the identity

$$\begin{aligned} \left\langle \mathbf{u}_\tau, \int_0^\tau \mathbf{K}(\tau-\lambda)\mathbf{u}(\lambda)\, d\lambda\right\rangle = \frac{d}{d\tau}&\left\langle \mathbf{u}(\tau), \int_0^\tau \mathbf{K}(\tau-\lambda)\mathbf{u}(\lambda)\, d\lambda\right\rangle\\ &-\left\langle \mathbf{u}(\tau), \int_0^\tau \mathbf{K}_\tau(\tau-\lambda)\mathbf{u}(\lambda)\, d\lambda\right\rangle\\ &-\langle \mathbf{u}(\tau), \mathbf{K}(0)\mathbf{u}(\tau)\rangle, \end{aligned}$$

this time to rewrite (II.189) in the form

$$\text{(II.190)}\quad \begin{aligned} FF'' - F'^2 \geqq -T^{-2}F^2 - 4F\int_0^t\int_0^\eta \langle \mathbf{u}(\tau), \mathbf{K}(0)\mathbf{u}(\tau)\rangle\, d\tau\, d\eta &\\ -4F\int_0^t\int_0^\eta \left\langle \mathbf{u}(\tau), \int_0^\tau \mathbf{K}_\tau(\tau-\lambda)\mathbf{u}(\lambda)\, d\lambda\right\rangle d\tau\, d\eta&\\ +2F\int_0^t \left\langle \mathbf{u}(\eta), \int_0^\eta \mathbf{K}(\eta-\tau)\mathbf{u}(\tau)\, d\tau\right\rangle d\eta.& \end{aligned}$$

As in our previous logarithmic convexity arguments, we now seek to bound several terms in the differential inequality (II.190). First of all

$$\text{(II.191)}\quad \begin{aligned} &\int_0^t \left\langle \mathbf{u}(\eta), \int_0^\eta \mathbf{K}(\eta-\tau)\mathbf{u}(\tau)\, d\tau\right\rangle d\eta\\ &\quad\leqq \sup_{[0,T)} \|\mathbf{K}(t)\|_{\mathcal{L}_S(H_+,H_-)} \int_0^t \|\mathbf{u}(\eta)\|_+ \left(\int_0^\eta \|\mathbf{u}(\tau)\|_+\, d\tau\right) d\eta\\ &\quad\leqq \left(\sup_{[0,T)} \|\mathbf{K}(t)\|_{\mathcal{L}_S(H_+,H_-)}\right) t \int_0^t \|\mathbf{u}(\eta)\|_+^2\, d\eta \end{aligned}$$

for all $t \geqq 0$; we have, therefore, the estimate

$$2F\int_0^t \left\langle \mathbf{u}(\eta), \int_0^\eta \mathbf{K}(\eta-\tau)\mathbf{u}(\tau)\,d\tau\right\rangle d\eta \geqq -2\pi_1 tF\int_0^t \|\mathbf{u}(\eta)\|_+^2\,d\eta, \tag{II.192}$$

where $\pi_1 = \sup_{[0,T)} \|\mathbf{K}(t)\|_{\mathcal{L}_S(H_+,H_-)}$. In an analogous fashion we easily establish that

$$-4F\int_0^t\int_0^\eta \left\langle \mathbf{u}(\tau), \int_0^\tau \mathbf{K}_\tau(\tau-\lambda)\mathbf{u}(\lambda)\,d\lambda\right\rangle d\tau\,d\eta \geqq -4\pi_2 tF\int_0^t \|\mathbf{u}(\eta)\|_+^2\,d\eta, \tag{II.193}$$

where $\pi_2 = T\sup_{[0,T)} \|\mathbf{K}_t(t)\|_{\mathcal{L}_S(H_+,H_-)}$. Combining the estimates (II.192), (II.193) with the differential inequality (II.190) we have

$$\begin{aligned} FF''-F'^2 &\geqq -T^{-2}F^2 \\ &\quad -2(\pi_1+2\pi_2)tF\int_0^t \|\mathbf{u}(\tau)\|_+^2\,d\tau \\ &\quad -4F\int_0^t\int_0^\eta \langle \mathbf{u}(\tau), \mathbf{K}(0)\mathbf{u}(\tau)\rangle\,d\tau\,d\eta \\ &\geqq -T^{-2}F^2(\pi_1+2\pi_2)tF\int_0^t \|\mathbf{u}(\tau)\|_+^2\,d\tau \\ &\quad +4\kappa F\int_0^t\int_0^\eta \|\mathbf{u}(\tau)\|_+^2\,d\tau\,d\eta, \end{aligned} \tag{II.194}$$

by virtue of the first part of our hypothesis (II.178); using the second part of that hypothesis it is easily seen that (II.194) implies that

$$\begin{aligned} FF''-F'^2 \geqq -T^{-2}F^2 &- 2(\pi_1+2\pi_2)tF\int_0^t \|\mathbf{u}(\tau)\|_+^2\,d\tau \\ &+4K(\pi_1+2\pi_1)F\int_0^t\int_0^\eta \|\mathbf{u}(\tau)\|_+^2\,d\tau\,dn \end{aligned} \tag{II.195}$$

So,

$$\begin{aligned} FF''-F'^2 &\geqq 2(\pi_1+2\pi_2)F\left[2K\int_0^t\int_0^\eta \|\mathbf{u}(\tau)\|_+^2\,d\tau\,d\eta - t\int_0^t \|\mathbf{u}(\tau)\|_+^2\,d\tau\right] + T^{-2}F^2 \\ &\geqq T^{-2}F^2, \qquad 0 \leqq t < T, \end{aligned} \tag{II.196}$$

in view of our assumption that $\mathbf{u} \in \mathcal{P}_K$. The last inequality in (II.195) may, however, be rewritten [compare Chapter I, eq. (I.30)] in the form

$$\frac{d^2}{dt^2}\left(\ln\left[\exp\left(\frac{t^2}{T^2}\right)F(t)\right]\right) \geqq 0, \qquad 0 \leqq t < T. \tag{II.197}$$

The estimate (II.179) now follows by integrating (II.197) according to the secant property of convex functions and noting that, by the definition of $F(t)$, $F(t) \geqq \|\mathbf{u}\|_t^2$, $0 \leqq t < T$. Q.E.D.

By employing the stability estimate (II.179), valid for solutions $\mathbf{u} \in \mathcal{P}_K$, $K > 0$, of (II.23′), when $\mathbf{K}(t)$ satisfies (II.178), we may obtain a number of continuous data dependence results for the initial-history value problem (II.23)–(II.25). Some of these results are delineated below while others may be found in [21].

(a) *Continuous dependence on initial data.* Let $\mathcal{M}_T = \{\mathbf{v} \in C^2([0, T); H_+) \mid \|\mathbf{v}\|_T^2 \leqq M^2\}$ for some $M \neq 0$. Suppose that $\mathbf{U}(\tau) = \mathbf{0}$, $-\infty < \tau < 0$ and consider the system

(II.198)
$$\mathbf{u}_{tt} - \mathbf{N}\mathbf{u} + \int_0^t \mathbf{K}(t-\tau)\mathbf{u}(\tau)\, d\tau = \mathbf{0}, \qquad 0 \leqq t < T,$$
$$\mathbf{u}(0) = \mathbf{f}, \qquad \mathbf{u}_t(0) = \mathbf{g},$$

where it is assumed that $\mathbf{Nf} \neq \mathbf{0}$ and $\mathbf{u}_{tt}(0) \neq \mathbf{0}$. For any t, $0 \leqq t \leqq T$, define $\hat{\mathbf{u}} \in C^2([0, T); H_+)$ by

(II.199)
$$\hat{\mathbf{u}}(\cdot, t) = \mathbf{u}(\cdot, t) - t\mathbf{u}_t(\cdot, 0) - \mathbf{u}(\cdot, 0)$$

so that $\hat{\mathbf{u}}(0) = \hat{\mathbf{u}}_t(0) = \mathbf{0}$, $\hat{\mathbf{u}}_{tt}(0) = \mathbf{Nf} \neq 0$. We note that

(II.200)
$$\|\hat{\mathbf{u}}\|_T^2 \leqq 2\|\mathbf{u}\|_T^2 + 2\|t\mathbf{u}_t(0) + \mathbf{u}(0)\|_T^2$$
$$\leqq 2\|\mathbf{u}\|_T^2 + 2k(T) \max\{\|\mathbf{f}\|^2, \|\mathbf{g}\|^2\},$$

where $k(T) = T^3/3 + T^2 + T$, for any $\mathbf{u} \in \mathcal{M}_T$. If we now choose N such that

(II.201)
$$N^2 \geqq 2M^2 + 2k(T) \max\{\|\mathbf{f}\|^2, \|\mathbf{g}\|^2\},$$

and define $\mathcal{N}_T$ as in (II.175), then every $\hat{\mathbf{u}}$ of the form (II.199) lies in $\mathcal{N}_T$ when $\mathbf{u} \in \mathcal{M}_T$. Also, if $\mathbf{u}(\cdot, t)$ is any solution of the system (II.198), then it is easily verified that $\hat{\mathbf{u}}(\cdot, t)$ satisfies (II.23′), with

(II.202)
$$\mathcal{F}(t) = \left(t\mathbf{N} - \int_0^t \tau\mathbf{K}(t-\tau)\, d\tau\right)\mathbf{g}$$
$$+ \left(\mathbf{N} - \int_0^t \mathbf{K}(t-\tau)\, d\tau\right)\mathbf{f}$$
$$\equiv \mathbf{A}(t)\mathbf{g} + \mathbf{B}(t)\mathbf{f},$$

where $\mathbf{A}(\cdot), \mathbf{B}(\cdot) \in L^2([0, T); \mathcal{L}_S(H_+, H_-))$. Now suppose that $\hat{\mathbf{u}} \in \mathcal{P}_{K_0}$ for some $K_0 > 0$ and $\mathbf{K}(0)$ satisfies (II.178) with $K \geqq K_0$; then the estimate (II.179) may be applied to $\hat{\mathbf{u}}(\cdot, t)$, with $\mathcal{F}$ given by (II.202), so as to conclude that for $t \in [0, T)$

(II.203)
$$\|\hat{\mathbf{u}}\|_t^2 \leqq PQ^{2\delta}\|\mathbf{A}(t)\mathbf{g} + \mathbf{B}(t)\mathbf{f}\|_T^{2(1-\delta)}, \qquad \delta = t/T.$$

However, by (II.199)

$$\begin{aligned}\|\hat{\mathbf{u}}\|_t &\leqq \|\mathbf{u}\|_t + \|t\mathbf{u}_t(0) + \mathbf{u}(0)\|_T \\ &\leqq \|\hat{\mathbf{u}}\|_t + \sqrt{k(T)\max\{\|\mathbf{f}\|^2, \|\mathbf{g}\|^2\}}\end{aligned} \tag{II.204}$$

and thus

$$\|\mathbf{u}\|_t \leqq \sqrt{P}\,Q^\delta \|\mathbf{A}(t)\mathbf{g} + \mathbf{B}(t)\mathbf{f}\|_T^{1-\delta} + \sqrt{k(T)\max\{\|\mathbf{f}\|^2, \|\mathbf{g}\|^2\}} \tag{II.205}$$

for $0 \leqq t < T$. A simple computation yields the estimate

$$\begin{aligned}\|\mathbf{A}(t)\mathbf{g} + \mathbf{B}(t)\mathbf{f}\|_T^2 \leqq 2T\,\{&\sup_{[0,T)} \|\mathbf{A}(t)\|^2_{\mathscr{L}_S(H_+,H_-)}\|\mathbf{g}\|_+^2 \\ &+ \sup_{[0,T)} \|\mathbf{B}(t)\|^2_{\mathscr{L}_S(H_+,H_-)}\|\mathbf{f}\|_+^2\}.\end{aligned} \tag{II.206}$$

If we set

$$a = \sup_{[0,T)} \|\mathbf{A}\|^2_{\mathscr{L}_S(H_+,H_-)}, \qquad b = \sup_{[0,T)} \|\mathbf{B}(t)\|^2_{\mathscr{L}_S(H_+,H_-)}$$

and combine (II.206) with (II.205) we are led to the estimate

$$\begin{aligned}\|\mathbf{u}\|_t &\leqq \mu(t)[a\|\mathbf{g}\|_+^2 + b\|\mathbf{f}\|_+^2]^{1/2-\delta/2} \\ &\quad + [\gamma k(T)]^{1/2}(\max\{\|\mathbf{f}\|_+^2, \|\mathbf{g}\|_+^2\})^{1/2}\end{aligned} \tag{II.207}$$

on $[0, T)$, where $\mu(t) = \sqrt{P}\,Q^\delta(2T)^{1/2-\delta/2}$. Continuous dependence in the $\|(\cdot)\|_t$ norm, for solutions $\mathbf{u} \in \mathcal{M}_T$ of (II.198), on the initial data $\mathbf{f}$, $\mathbf{g}$, follows immediately from the estimate (II.207); this estimate is valid under the assumption that $\hat{\mathbf{u}}(\cdot, t)$, as defined by (II.199), lies in $\mathscr{P}_{K_0} \subset \mathcal{N}_T$ for some $K_0 > 0$, (where $\mathcal{N}_T$ is defined by the real number N satisfying (II.201)) and that $\mathbf{K}(0)$ satisfies (II.178) with $K \geqq K_0$. Uniqueness of solutions $\mathbf{u} \in \mathcal{M}_T$ to the initial-value problem (II.198) follows directly from (II.207), and the definition of $\|(\cdot)\|_t$, and continuous dependence on initial data for the initial-history value problem (II.23)–(II.25) follows from the basic estimate (II.179) by obvious modifications of the above argument.

(b) *Continuous dependence on initial geometry.* Suppose that $\chi: J \to R^+$ is a continuous nonnegative function on J such that $\sup_{\mathbf{x}\in J} |\chi(\mathbf{x})| < \varepsilon$ for some $\varepsilon > 0$. We consider solutions $\mathbf{u}^\chi(\cdot, t)$ of (II.23′) for which the associated initial datum $\mathbf{f}, \mathbf{g}$ are prescribed on the surface $t = -\chi(\mathbf{x})$, $\mathbf{x} \in J$, as in (I.39). Let $\mathbf{u}(\cdot, t)$ be any solution of (II.23′) taking on the initial datum $\mathbf{f}, \mathbf{g}$ on the initial hyperplane $t = 0$ and assume that $\mathbf{u}(\cdot, t) \in \mathcal{M}_T$ for some $M \neq 0$. If, as in Chapter I, we set $\mathbf{u}^\varepsilon = \mathbf{u}^\chi - \mathbf{u}$ then, clearly, $\mathbf{u}^\varepsilon$ is a solution of (II.23′) with $\mathscr{F} = \mathbf{0}$ and $\mathbf{u}_t^\varepsilon(\cdot, 0)$, $\mathbf{u}^\varepsilon(\cdot, 0)$ given by (I.40) and (I.41), respectively. Thus, $\mathbf{u}^\varepsilon(\cdot, t)$ is a solution of the system (II.198) with

$$\text{(II.208)}\qquad \begin{aligned} \mathbf{f} \to \mathbf{f}^\varepsilon &\equiv \chi(\cdot)\mathbf{g}(\cdot) + \int_0^{-\chi(\cdot)} \eta \mathbf{u}^\chi_{\eta\eta}\, d\eta, \\ \mathbf{g} \to \mathbf{g}^\varepsilon &\equiv \int_{-\chi(\cdot)}^0 \mathbf{u}^\chi_{\eta\eta}\, d\eta. \end{aligned}$$

If we assume that $\mathbf{u}^\varepsilon(\cdot, t) \in \mathcal{M}_T$ for ε sufficiently small, and that $\mathbf{N}\mathbf{f}^\varepsilon \neq \mathbf{0}$, then our previous continuous dependence estimate (II.207) may be applied to $\mathbf{u}^\varepsilon$ provided

$$\text{(II.209)}\qquad \hat{\mathbf{u}}^\varepsilon \equiv \mathbf{u}^\varepsilon - t\mathbf{g}^\varepsilon - \mathbf{f}^\varepsilon \in \mathcal{P}_{K_0} \subset \mathcal{N}$$

for some $K_0 > 0$ and $\mathbf{K}(0)$ satisfies (II.178) with $K \geqq K_0$; note that (II.209) need only be satisfied for sufficiently small $\varepsilon > 0$. Applying (II.207) to $\mathbf{u}^\varepsilon(\cdot, t)$ we then obtain

$$\text{(II.210)}\qquad \begin{aligned} \|\mathbf{u}^\varepsilon\|_t \leqq \mu(t)[a\|\mathbf{u}^\varepsilon_t(0)\|^2_+ + b\|\mathbf{u}^\varepsilon(0)\|^2_+]^{1/2-\delta/2} \\ + [\gamma k(T)]^{1/2}(\max\{\|\mathbf{u}^\varepsilon(0)\|^2_+, \|\mathbf{u}_t(0)\|^2_+\})^{1/2} \end{aligned}$$

for $0 \leqq t < T$. The following lemma (see Theorem IV, [21]) then establishes the desired continuous dependence result:

LEMMA. *If $\|\chi\mathbf{g}\|_+ \geqq (1+\varepsilon)\|\mathbf{u}^\varepsilon_t(0)\|_+$ at each $x \in J$, for ε sufficiently small, then for such ε*

$$\text{(II.211)}\qquad \begin{aligned} \|\mathbf{u}^\varepsilon(0)\|_+ &\leqq \varepsilon(\|\mathbf{g}\|_+ + \|\mathbf{u}^\varepsilon_t(0)\|_+), \\ \|\mathbf{u}^\varepsilon_t(0)\|_+ &\leqq \frac{\varepsilon}{1-\varepsilon}\|\mathbf{g}\|_+. \end{aligned}$$

From (II.211) it readily follows that $\|\boldsymbol{u}^\varepsilon(0)\|_+ \to 0$, $\|\mathbf{u}^\varepsilon_t(0)\|_+ \to 0$, as $\varepsilon \to 0$ and, thus, by virtue of the estimate (II.210), $\|\mathbf{u}^\varepsilon\|_t \to 0$, $0 \leqq t < T$, as $\varepsilon \to 0$.

(c) *Joint continuous dependence on perturbations of the past history and the initial data.* We now return and consider the initial-history value problem (II.23)–(II.25), i.e., the abstract model for the viscoelastic initial-history boundary value problem (II.2′), (II.3), (II.4); with obvious modifications the results of this subsection will be valid if a term corresponding to an external forcing function appears on the right-hand side of (II.2′). For the continuous dependence estimate to be derived here we retain all our basic assumptions about the operators $\mathbf{N}$, $\mathbf{K}(t)$ and the past history $\mathbf{U}$ (i.e., (i)–(iii) following (II.25)) and append the additional smoothness requirements that

$$\text{(II.212)}\qquad \frac{\partial^k \mathbf{K}(t)}{\partial t^k}, \quad k = 2, 3, 4, \quad \text{exists a.e. on } [0, T)$$

(in the sense of strong operator derivatives) and belongs to $\mathcal{L}_S(H_+, H_-)$ and

$$\text{(II.213)}\qquad \int_{-\infty}^0 \mathbf{K}(-\tau)\mathbf{U}(\tau)\, d\tau - \mathbf{N}\mathbf{f} \neq 0.$$

The last assumption guarantees, of course, that any solution $\mathbf{u}$ of (II.23)–(II.25) satisfies $\mathbf{u}_{tt}(0) \neq 0$. We will now derive an estimate which yields joint continuous dependence on perturbations of the past history $\mathbf{U}(\tau)$, $-\infty<\tau<0$, and the initial data $\mathbf{f}$, $\mathbf{g}$ for solutions $\mathbf{u}$ of (II.23)–(II.25) which lie in suitable classes of bounded functions. To this end, suppose that $\mathbf{u} \in \mathcal{M}_T$ is a solution of (II.23)–(II.25) and define

$$\mathbf{v} \in C^2([0, T); H_+) \cap C^1((-\infty, 0); H_+)$$

by

$$\text{(II.214)} \qquad \mathbf{v}(t) = \begin{cases} \mathbf{u}(t) - t\mathbf{u}_t(0) - \mathbf{u}(0), & 0 \leqq t < T, \\ \mathbf{U}(t), & -\infty < t < 0. \end{cases}$$

Obviously, $\mathbf{v}(0) = \mathbf{v}_t(0) = \mathbf{0}$, and it is easily verified that $\mathbf{v}$ satisfies the evolution equation

$$\text{(II.215)} \qquad \mathbf{v}_{tt} - \mathbf{N}\mathbf{v} + \int_0^t \mathbf{K}(t-\tau)\mathbf{v}(\tau)\,d\tau = \mathbf{A}(t)\mathbf{g} + \mathbf{B}(t)\mathbf{f} - \int_{-\infty}^0 \mathbf{K}(t-\tau)\mathbf{U}(\tau)\,d\tau$$

for each t, $0 \leqq t < T$, where $\mathbf{A}(t)$, $\mathbf{B}(t)$ are given by

$$\mathbf{A}(t) = t\mathbf{N} - \int_0^t \tau\mathbf{K}(t-\tau)\,d\tau,$$

$$\mathbf{B}(t) = \mathbf{N} - \int_0^t \mathbf{K}(t-\tau)\,d\tau.$$

From (II.215), (II.213), we have $\mathbf{v}_{tt}(0) \neq \mathbf{0}$ and if we choose $N \neq 0$ so as to satisfy (II.201) then $\mathbf{v} \in \mathcal{N}_T$. Suppose now that $\mathbf{v} \in \mathcal{P}_{K_1} \subset \mathcal{N}_T$ for some $K_1 > 0$ and that $\mathbf{K}(0)$ satisfies (II.178) with $K \geqq K_1$. Then the stability estimate (II.179) may be applied to $\mathbf{v}(\cdot, t)$ with

$$\text{(II.216)} \qquad \mathcal{F} \to \mathcal{F}_{\mathbf{U}}(\mathbf{f}, \mathbf{g}) \equiv \mathbf{A}(t)\mathbf{g} + \mathbf{B}(t)\mathbf{f} - \int_{-\infty}^0 \mathbf{K}(t-\tau)\mathbf{U}(\tau)\,d\tau;$$

i.e.,

$$\text{(II.217)} \qquad \|\mathbf{v}\|_t \leqq \sqrt{P}\,Q^\delta \|\mathcal{F}_{\mathbf{U}}(\mathbf{f}, \mathbf{g})\|_T^{1-\delta}, \qquad 0 \leqq t < T.$$

However, from (II.214), (II.204) with $\hat{\mathbf{u}} \to \mathbf{v}$

$$\|\mathbf{u}\|_t \leqq \|\mathbf{v}\|_t + (k(T) \max\{\|\mathbf{f}\|^2, \|\mathbf{g}\|^2\})^{1/2}$$

and thus

$$\text{(II.218)} \qquad \|\mathbf{u}\|_t \leqq \sqrt{P}\,Q^\delta \|\mathcal{F}_{\mathbf{U}}(\mathbf{f}, \mathbf{g})\|_T^{1-\delta} + \sqrt{k(T) \max(\|\mathbf{f}\|^2, \|\mathbf{g}\|^2)}.$$

We now observe the following set of elementary estimates

$$\text{(II.219)} \qquad \|\mathcal{F}_{\mathbf{U}}(\mathbf{f}, \mathbf{g})\|_T^2 \leqq 2\|\mathbf{A}(t)\mathbf{g} + \mathbf{B}(t)\mathbf{f}\|_T^2 + 2\left\|\int_{-\infty}^0 \mathbf{K}(t-\tau)\mathbf{U}(\tau)\,d\tau\right\|_T^2,$$

$$\left\| \int_{-\infty}^{0} \mathbf{K}(t-\tau)\mathbf{U}(\tau)\, d\tau \right\|_T^2 \leqq \int_0^T \left(\int_{-\infty}^{0} \|\mathbf{K}(t-\tau)\|^2_{\mathscr{L}s(H_+,H_-)}\, d\tau \right) dt \int_{-\infty}^{0} \|\mathbf{U}(\tau)\|_+^2\, d\tau \tag{II.220}$$

so that

$$\left\| \int_{-\infty}^{0} \mathbf{K}(t-\tau)\mathbf{U}(\tau)\, d\tau \right\|_T^2 \leqq \left[T \sup_{[0,T)} \int_{-\infty}^{0} \|\mathbf{K}(t-\tau)\|^2_{\mathscr{L}s(H_+,H_-)}\, d\tau \right] \int_{-\infty}^{0} \|\mathbf{U}(\tau)\|_+^2\, d\tau. \tag{II.221}$$

Also, by (II.206) and the definitions of a, b immediately following, we have

$$\|\mathbf{A}(t)\mathbf{g}+\mathbf{B}(t)\mathbf{f}\|_T^2 \leqq 4T \max(a, b) \max(\|\mathbf{f}\|_+^2, \|\mathbf{g}\|_+^2). \tag{II.222}$$

By combining the estimates (II.219)–(II.222) with (II.218), we obtain

$$\begin{aligned} \|\mathbf{u}\|_t \leqq \sqrt{P}\, Q^{\delta} \Bigg[\left(2T \sup_{[0,T)} \int_{-\infty}^{0} \|\mathbf{K}(t-\tau)\|^2_{\mathscr{L}s(H_+,H_-)}\, d\tau \right) \int_{-\infty}^{0} \|\mathbf{U}(\tau)\|_+^2\, d\tau \\ + 8T \max(a, b) \max(\|\mathbf{f}\|_+^2, \|\mathbf{g}\|_+^2) \Bigg]^{1/2-\delta/2} \\ + (\gamma\sqrt{k(T)})[\max(\|\mathbf{f}\|_+^2, \|\mathbf{g}\|_+^2)]^{1/2} \end{aligned} \tag{II.223}$$

from which it follows that $\|\mathbf{u}\|_t \to 0$, $0 \leqq t < T$ as

$$\max\left[\int_{-\infty}^{0} \|\mathbf{U}\|_+^2\, d\tau, \max(\|\mathbf{f}\|_+^2, \|\mathbf{g}\|_+^2) \right] \to 0. \tag{II.224}$$

Remarks. For the application we have in mind, i.e., to the viscoelastic initial-history boundary value problem (II.2′), (II.3), (II.4), the basic hypothesis of this section, i.e., (II.178) will be satisfied provided the viscoelastic relaxation tensor $\mathscr{g}_{ijkl}(\mathbf{x}, t)$ satisfies (II.80) with

$$\begin{aligned} \kappa \geqq K\left(\sup_{\Omega} \frac{1}{\rho} \right) [\max_{i,j,k,l} \sup_{[0,T)} \sup_{\Omega} |\dot{\mathscr{g}}_{ijkl}(\mathbf{x}, t)| \\ + 2T \sup_{[0,T)} \sup_{\Omega} |\ddot{\mathscr{g}}_{ijkl}(\mathbf{x}, t)|], \end{aligned} \tag{II.225}$$

where $K > 0$ is determined by the governing class of bounded perturbations $\mathscr{P}_K$ arising in each of the various continuous dependence estimates we have presented.

Example. For the one-dimensional homogeneous viscoelastic material governed by the integrodifferential evolution equation (II.5), with homogeneous relaxation function $\mathscr{g}_0(t)$ and constant density ρ_0, the basic hypothesis

(II.178) is satisfied provided $\mathscr{g}_0'(0) \leqq -\kappa$ with

(II.226)
$$\kappa \geqq \frac{K}{\rho_0}\left[\sup_{[0,T)} |\mathscr{g}_0'(t)| + 2T \sup_{[0,T)} |\mathscr{g}_0''(t)|\right],$$

K being determined by the type of continuous dependence relation we have in mind. For example, suppose $\Omega = [0, 1]$ and we associate with (II.5) (with $\mathscr{g} = \mathscr{g}_0(t)$, $\rho = \rho_0$) the data (II.109), (II.110); if

(i) $\displaystyle\int_{-\infty}^{0} \mathscr{g}_0'(-\tau)\frac{\partial^2 U(x,\tau)}{\partial x^2}\,d\tau \neq \mathscr{g}_0(0)\frac{\partial^2 f}{\partial x^2}, \qquad x \in [0, 1],$

(ii) $v(x, t) = u(x, t) - tg(x)$, $x \in [0, 1]$ satisfies $v \in \mathscr{P}_{\bar{K}}$ for some $\bar{K} > 0$,

and

(iii) $\mathscr{g}_0'(0) \leqq -\kappa$, with κ satisfying (II.226) for some $K \geqq \bar{K}$,

then it follows that

(II.227)
$$\int_0^t \int_0^1 u^2(x, \tau)\,dx\,d\tau \to 0, \qquad 0 \leqq t < T,$$

as

(II.228)
$$\max\left[\int_{-\infty}^{0}\int_0^1 \left(\frac{\partial U(x,\tau)}{\partial x}\right)^2 dx\,d\tau, \max\left(\int_0^1 \left(\frac{\partial f}{\partial x}\right)^2 dx, \int_0^1 \left(\frac{\partial g}{\partial x}\right)^2 dx\right)\right] \to 0.$$

Essentially, what (II.227), (II.228), and the conditions preceding them indicate is that the uniform boundedness of $u - tg - f$ (in the sense of (ii) above) and the fact that $\mathscr{g}_0(t)$ is decreasing sufficiently fast at $t = 0$ (in the sense of (iii) above) imply the joint continuous dependence of $u(x, t)$ on perturbations of the past history and the initial datum. We indicate here that other continuous data dependence theorems have been derived in the recent literature, for solutions of ill-posed initial-history boundary value problems in three-dimensional isothermal viscoelasticity, which do not make use of differential inequality arguments; in this vein we mention, in particular, the work of Beevers [9], [10] which makes use of the so-called weighted energy arguments of Murray and Protter [118] and the work of Brún [30] which employs the Lagrange identity technique.

5. Growth estimates for solutions of initial-history boundary value problems in one-dimensional nonlinear viscoelasticity.

In recent years considerable effort has been expended on the problem of proving existence, uniqueness, stability, and asymptotic stability of solutions to initial-history boundary value problems in nonlinear viscoelasticity. Most of the work to date has dealt with one-dimensional situations and, in this connection we note, in particular, the papers of Dafermos [42], Browne [28], Dafermos and Nohel [43], Slemrod [139]–[141], and MacCamy [104]; the particular models of nonlinear viscoelastic response employed vary from paper to paper in the references listed above. In this section we will consider one particular model of nonlinear viscoelastic response, that of MacCamy [104], and for this model we will obtain certain growth estimates by employing a variant of concavity argument that has been used by Knops [72] to prove global nonexistence of smooth solutions to initial-boundary value problems in one-dimensional nonlinear elasticity; we also comment on the recent work of Slemrod [139], Dafermos [42], and Dafermos and Nohel [43]. In the last chapter logarithmic convexity arguments are used to obtain growth estimates for yet another one-dimensional viscoelastic model (Bloom [16]) when perturbations are restricted to lie in a suitable class of bounded functions.

In [104] MacCamy considered the following model for one-dimensional nonlinear viscoelasticity:

$$u_{tt} = \mathscr{g}(0)\sigma(u_x)_x - \int_0^t \mathscr{g}_\tau(t-\tau)\sigma(u_x)_x \, d\tau + \mathscr{F} \tag{II.229}$$

on $[0, 1]\times[0, \infty)$ subject to initial and boundary data of the form (II.109) (with $(-\infty, T)\rightarrow[0, \infty)$); the term involving the specification of the past history, i.e., $\int_{-\infty}^0 \mathscr{g}_\tau(t-\tau)\sigma(U_x)_x \, d\tau$, has been absorbed into the nonhomogeneous term $\mathscr{F}(x, t)$. The nonlinear evolution equation (II.229) generalizes the homogeneous linear one-dimensional evolution equation ((II.5) with $\mathscr{g} = \mathscr{g}_0(t)$, $\rho \equiv 1$) in an obvious manner. By using Riemann invariants and an energy estimate argument, MacCamy was able to show that the initial-boundary value problem corresponding to (II.229) has a unique classical solution for all t when $\mathscr{F}$ is suitably restricted and the data f, g are sufficiently small; furthermore, the solution is asymptotically stable. The essential hypotheses in [104] are that $\mathscr{g}(t) = \mathscr{g}_\infty + \mathscr{G}(t)$, where $\mathscr{g}_\infty > 0$, $\mathscr{G} \in L_1[0, \infty)$, $(-1)^k \mathscr{g}^{(k)}(t) \geqq 0$, $k = 0, 1, 2$, $\sigma(0) = 0$, $\sigma'(\zeta) \geqq \varepsilon > 0$, as well as various smoothness assumptions relative to σ, f, g, and $\mathscr{F}$ and boundedness and growth conditions on $\mathscr{F}$; without any loss of generality it may be assumed that $\mathscr{g}(0) = 1$.

The model considered by MacCamy [104] is intermediate between two extremes; there are the cases where (II.229) is replaced, respectively, by the evolution equations

$$u_{tt} = \sigma(u_x)_x, \qquad (x, t) \in (0, 1)\times[0, \infty), \tag{II.230}$$

and

$$u_{tt} = \frac{\partial}{\partial x}(\sigma(u_x) + \lambda(u_x)u_{xt}), \qquad (x, t) \in (0, 1)\times[0, \infty). \tag{II.231}$$

Equation (II.230) corresponds to one-dimensional nonlinear elastic response and it is well known that if σ is genuinely nonlinear the corresponding initial boundary value problem does not have a smooth global solution for any nonzero data f, g; in fact, Knops [72] has shown that global existence fails for the initial-boundary value problem corresponding to (II.230) when there exists a strain-energy function Σ (i.e., $\sigma(\zeta)=\Sigma'(\zeta)$, for all ζ) such that for some $\alpha>2$, $\alpha\Sigma(\zeta)>\zeta\Sigma'(\zeta)$ for all ζ. On the other hand, the initial-boundary value problems associated with (II.231) always have global smooth solutions which are asymptotically stable no matter how large the initial data are; this has been demonstrated by MacCamy [106] and by Greenberg, MacCamy, and Mizel [58]. It has been conjecutred by MacCamy in [104] that global existence of smooth solutions fails for the initial-boundary value problems associated with (II.229) if the data are too large. To date, this author is not aware of the existence of a proof of the conjecture of non well-posedness for the model governed by (II.229), in the presence of sufficiently large data, but we will comment below on some recent global nonexistence results of Slemrod [139] for a closely related nonlinear one-dimensional model of viscoelastic response; we will also briefly review the work of Dafermos [42] on a nonlinear one-dimensional viscoelastic initial-boundary value problem which is closely related to the problems considered in [58] and [104].[1]

We now want to demonstrate that it is possible to derive some growth estimates for solutions of initial-boundary value problems associated with the nonlinear integrodifferential equation (II.229), under relatively mild assumptions on $\mathscr{g}(\cdot)$ and the nonlinearity $\sigma(\cdot)$, by means of a concavity argument of the kind employed in Chapter I for the nonlinear heat conduction problem (I.52). Our basic assumptions are as follows: First of all we assume that $\sigma(0)=0$, $\mathscr{g}(0)=1$, $F\equiv 0$, and that

$$\sigma(\zeta)=\Sigma'(\zeta), \quad \text{with } \alpha\Sigma(\zeta)\geqq\zeta\Sigma'(\zeta), \quad \text{for all } \zeta \text{ and some } \alpha>2. \tag{II.232}$$

For $T>0$ we introduce the class of functions

$$\mathscr{C}=\{u:[0,T)\to H_0^1[0,1] \mid \sup_{[0,T)}\|u\|_{H_0^1}\leqq C\}, \quad \text{for some } C>0. \tag{II.233}$$

we make the smoothness assumptions that $\mathscr{g}(\cdot)\in C^2([0,T))$ with

$$\begin{gathered}\sup_{[0,T)}\int_0^t |\mathscr{g}'(t-\tau)|\,d\tau<\infty,\\ \int_0^T\left(\int_0^t |\mathscr{g}''(t-\tau)|^2\,d\tau\right)^{1/2} dt<\infty.\end{gathered} \tag{II.234}$$

---

[1] [Note added in proof] The MacCamy conjecture has now been positively affirmed by H. Hattori in his PhD thesis, *Breakdown of Smooth Solutions in Dissipative Nonlinear Hyperbolic Equations,* RPI, 1981.

Finally, we assume that there exists $\bar{\sigma}$, $0<\bar{\sigma}<\infty$, such that

$$|\sigma'(\zeta)|<\bar{\sigma} \quad \forall \zeta. \tag{II.235}$$

No sign definiteness assumptions are imposed on the derivatives $g^{(k)}(t)$, $k=0, 1, 2$ as in [104]. Before proceeding with the statement and proof of our growth estimate, we need the result in the following:

LEMMA (Bloom [12]). *Let the total energy $\mathscr{E}$ associated with the one-dimensional nonlinear viscoelastic model governed by* (II.229) *be given by*

$$\mathscr{E}(t)\equiv\frac{1}{2}\int_0^1 [\dot{u}(x,t)]^2\,dx+\int_0^1 \Sigma(u_x(x,t))\,dx. \tag{II.236}$$

Then

$$\begin{aligned}\mathscr{E}(t)=\mathscr{E}(0)&-\int_0^1 u(x,t)\int_0^t \tilde{g}(t-\tau)\sigma(u_x(x,\tau))_x\,d\tau\,dx\\&+\int_0^t\left(\int_0^1 u(x,\tau)\int_0^\tau \tilde{g}_\tau(\tau-\lambda)\sigma(u_x(x,\lambda))_x\,d\lambda\,dx\right)d\tau\\&-\dot{g}(0)\int_0^t\int_0^1 u(x,\tau)\sigma(u_x(x,\tau))_x\,dx\,d\tau,\end{aligned} \tag{II.237}$$

where $\tilde{g}(t-\tau)=g_\tau(t-\tau)$.

Proof. By direct computation based on (II.236),

$$\begin{aligned}\dot{\mathscr{E}}(t)&=\int_0^1 u\ddot{u}\,dx+\int_0^1 \Sigma'(u_x)\dot{u}_x\,dx\\&=\int_0^1 \dot{u}\sigma(u_x)_x\,dx+\int_0^1\Sigma'(u_x)\dot{u}_x\,dx\\&\quad-\int_0^1 \dot{u}\int_0^t g_\tau(t-\tau)\sigma(u_x)_x\,d\tau\,dt,\end{aligned} \tag{II.238}$$

or, in view of the fact that $\Sigma'(\zeta)=\sigma(\zeta)$, for all ζ, and the definition of $\tilde{g}$

$$\begin{aligned}\dot{\mathscr{E}}(t)&=\int_0^1 \dot{u}\frac{\partial}{\partial x}(\Sigma'(u_x))\,dx+\int_0^1\Sigma'(u_x)\dot{u}_x\,dx\\&\quad-\int_0^1\dot{u}\int_0^t\tilde{g}(t-\tau)\frac{\partial}{\partial x}(\Sigma'(u_x))\,d\tau\,dx\\&=\int_0^1\frac{\partial}{\partial x}(\dot{u}\Sigma'(u_x))\,dx\\&\quad-\int_0^1\dot{u}\int_0^t\tilde{g}(t-\tau)\frac{\partial}{\partial x}(\Sigma'(u_x))\,d\tau\,dx.\end{aligned} \tag{II.239}$$

Thus

$$\dot{\mathscr{E}}(t) = -\int_0^1 \dot{u}(x,t)\int_0^t \tilde{g}(t-\tau)\frac{\partial}{\partial x}(\Sigma'(u_x(x,\tau)))\,d\tau\,dx, \tag{II.240}$$

by virtue of the boundary conditions in (II.109); i.e., $u(0,t) = u(1,t) = 0$, $t \geqq 0$. Integrating (II.240) we have

$$\mathscr{E}(t) - \mathscr{E}(0) = -\int_0^t \left(\int_0^1 \dot{u}(x,\tau)\int_0^\tau \tilde{g}(\tau-\lambda)\frac{\partial}{\partial x}\Sigma'(u_x(x,\lambda))\,d\lambda\,dx\right)d\tau. \tag{II.241}$$

However,

$$\begin{aligned}
&\frac{d}{d\tau}\int_0^1 u(x,\tau)\int_0^\tau \tilde{g}(\tau-\lambda)\frac{\partial}{\partial x}\Sigma'(u_x(x,\lambda))\,d\lambda\,dx \\
&\quad= \frac{d}{d\tau}\int_0^1 u(x,\tau)\int_0^\tau \tilde{g}(\tau-\lambda)\sigma(u_x(x,\lambda))_x\,d\lambda\,dx \\
&\quad= \int_0^1 \frac{d}{d\tau}\left(u(x,\tau)\int_0^\tau \tilde{g}(\tau-\lambda)\sigma(u_x(x,\lambda))_x\,d\lambda\right)dx \\
&\quad= \int_0^1 \dot{u}(x,\tau)\int_0^\tau \tilde{g}(\tau-\lambda)\sigma(u_x(x,\lambda))_x\,d\lambda\,dx \\
&\qquad+ \int_0^1 u(x,\tau)\int_0^\tau \tilde{g}(\tau-\lambda)\sigma(u_x(x,\lambda))_x\,d\lambda\,dx \\
&\qquad+ \tilde{g}(0)\int_0^1 u(x,\tau)\sigma(u_x(x,\tau))_x\,dx,
\end{aligned} \tag{II.242}$$

Also,

$$\tilde{g}(0) = \tilde{g}(t-\tau)|_{t=\tau} = \frac{\partial}{\partial\tau}g(t-\tau)|_{t=\tau} = -g'(0), \tag{II.243}$$

so

$$\begin{aligned}
&\frac{d}{d\tau}\int_0^1 u(x,\tau)\int_0^\tau \tilde{g}(\tau-\lambda)\frac{\partial}{\partial x}\Sigma'(u_x(x,\lambda))\,d\lambda\,dx \\
&\quad= \int_0^1 \dot{u}(x,\tau)\int_0^\tau \tilde{g}(\tau-\lambda)\sigma(u_x(x,\lambda))_x\,d\lambda\,dx \\
&\qquad+ \int_0^1 u(x,\tau)\int_0^\tau \tilde{g}_\tau(\tau-\lambda)\sigma(u_x(x,\lambda))_x\,d\lambda\,dx \\
&\qquad- \dot{g}(0)\int_0^1 u(x,\tau)\sigma(u(x,\tau))_x\,dx.
\end{aligned} \tag{II.244}$$

The lemma now follows if we substitute for $\int_0^1 \dot{u}(x,\tau)\int_0^\tau \tilde{g}(\tau-\lambda)$ $\sigma(u_x(x,\lambda))_x\,d\lambda\,dx$ from (II.244) into (II.241) and perform the indicated integration over $[0,t)$. Q.E.D.

We now set

(II.245)
$$\varkappa_T = |\dot{g}(0)|T + \left(1 - \frac{1}{\alpha}\right) \sup_{[0,T)} \int_0^t |\dot{g}(t-\tau)|\, d\tau + \sqrt{T} \int_0^T \left(\int_0^t \ddot{g}^2(t-\tau)\, d\tau \right)^{1/2} dt,$$

(II.246)
$$\delta = \max\,(\mathscr{E}(0), \bar{\sigma}\varkappa_T C^2).$$

Then we have the following.

THEOREM. *Consider the nonlinear one-dimensional viscoelastic initial-boundary value problem* (II.229), (II.109) *with* $\sigma(0) = 0$, $g(0) = 1$, *and* $\mathscr{F} = 0$. *If* $u \in C^2([0, 1] \times [0, T)) \cap \mathscr{C}$ *is a solution of* (II.229), (II.109) *with initial data satisfying*

(II.247)
$$\int_0^1 f(x)g(x)\, dx > \nu \left(\int_0^1 f^2(x)\, dx \right)^{1/2}, \qquad \nu = \frac{2\alpha\delta}{\alpha - 1}.$$

Then for $0 \leqq t < T$ *we have the quadratic growth estimate*

(II.248)
$$\|u\|_{L_2}^2 \geqq \|f\|_{L_2}^2 + 2\nu \|f\|_{L_2} t + \nu^2 t^2.$$

Proof. Let $U(t) = \int_0^1 u^2(x, t)\, dx$. Then by direct computation

(II.249)
$$\dot{U}(t) = 2 \int_0^1 u\dot{u}\, dx, \qquad \ddot{U}(t) = 2 \int_0^1 u\ddot{u}\, dx + 2 \int_0^1 \dot{u}^2\, dx,$$

or, in view of (II.229),

(II.250)
$$\begin{aligned}
\ddot{U}(t) &= 2 \int_0^1 u \frac{\partial}{\partial x} \Sigma'(u_x)\, dx + 2 \int_0^1 \dot{u}^2\, dx \\
&\quad - 2 \int_0^1 u \int_0^t \tilde{g}(t-\tau) \frac{\partial}{\partial x} \Sigma'(u_x)\, d\tau\, dx \\
&= 2 \int_0^1 \frac{\partial}{\partial x} (u\Sigma'(u_x))\, dx - 2 \int_0^1 u_x \Sigma'(u_x)\, dx \\
&\quad - 2 \int_0^1 \frac{\partial}{\partial x} \left(\int_0^t \tilde{g}(t-\tau) u(x, t) \Sigma'(u_x(x, \tau))\, d\tau \right) dx \\
&\quad + 2 \int_0^1 \int_0^t \tilde{g}(t-\tau) u_x(x, t) \Sigma'(u_x(x, \tau))\, d\tau\, dx + 2 \int_0^1 \dot{u}^2\, dx \\
&= -2 \int_0^1 u_x \Sigma'(u_x)\, dx + 2 \int_0^1 \dot{u}^2\, dx \\
&\quad + 2 \int_0^1 \int_0^t \tilde{g}(t-\tau) u_x(x, t) \Sigma'(u_x(x, \tau))\, d\tau\, dx,
\end{aligned}$$

in view of the homogeneous boundary data in (II.109). By adding and subtracting the expression $2\alpha \int_0^1 \Sigma(u_x(x,t))\,dx$, $\alpha > 2$, in the last equality in (II.250) we obtain

$$
\begin{aligned}
\ddot{U}(t) &= 2\int_0^1 [\alpha\Sigma(u_x(x,t)) - u_x(x,t)\Sigma'(u_x(x,t))]\,dx \\
&\quad + 2\int_0^1 \dot{u}^2(x,t)\,dx - 2\alpha\int_0^1 \Sigma(u_x(x,t))\,dx \\
&\quad + 2\int_0^1\int_0^t \tilde{g}(t-\tau)u_x(x,t)\Sigma'(u_x(x,\tau))\,d\tau\,dx \\
&\geqq 2\int_0^1 \dot{u}^2(x,t)\,dx - 2\alpha\int_0^1 \Sigma(u_x(x,t))\,dx \\
&\quad + 2\int_0^1\int_0^t \tilde{g}(t-\tau)u_x(x,t)\Sigma'(u_x(x,\tau))\,d\tau\,dx,
\end{aligned}
\tag{II.251}
$$

where we have used the hypothesis (II.232). By combining the lower bound in (II.251) with the definition of the energy $\mathscr{E}(t)$ we obtain

$$
\begin{aligned}
\ddot{U}(t) &\geqq 2\int_0^1 \dot{u}^2(x,t)\,dx - 2\alpha\left(\mathscr{E}(t) - \frac{1}{2}\int_0^1 \dot{u}^2(x,t)\,dx\right) \\
&\quad + 2\int_0^1\int_0^t \tilde{g}(t-\tau)u_x(x,\tau)\Sigma'(u_x(x,\tau))\,d\tau\,dx \\
&= (2+\alpha)\int_0^1 \dot{u}^2(x,t)\,dx - 2\alpha\mathscr{E}(t) \\
&\quad + 2\int_0^1\int_0^t \tilde{g}(t-\tau)u_x(x,t)\Sigma'(u_x(x,\tau))\,d\tau\,dx.
\end{aligned}
\tag{II.252}
$$

At this point we can make efficient use of the previous lemma to substitute in (II.252) for $\mathscr{E}(t)$; in this manner, we easily obtain

$$
\begin{aligned}
\ddot{U}(t) &\geqq (2+\alpha)\int_0^1 \dot{u}^2(x,t)\,dx - 2\alpha\mathscr{E}(0) \\
&\quad + 2\alpha\int_0^1 u(x,t)\int_0^t \tilde{g}(t-\tau)\sigma(u_x(x,\tau))_x\,d\tau\,dx \\
&\quad + 2\alpha\dot{g}(0)\int_0^t\int_0^1 u(x,\tau)\sigma(u_x(x,\tau))_x\,dx\,d\tau \\
&\quad - 2\alpha\int_0^t\left(\int_0^1 u(x,\tau)\int_0^\tau \tilde{g}(\tau-\lambda)\sigma(u_x(x,\lambda))_x\,d\lambda\,dx\right)d\tau \\
&\quad + 2\int_0^1\int_0^t \tilde{g}(t-\tau)u_x(x,t)\Sigma'(u_x(x,\tau))\,d\tau\,dx.
\end{aligned}
\tag{II.253}
$$

Now, by (II.249) $\dot{U}^2(t)=4(\int_0^1 u\dot{u}\,dx)^2$ and therefore (II.253) easily leads us to the differential inequality

$$\begin{aligned}
&\ddot{U}(t)U(t)-\left(\frac{\alpha+2}{4}\right)\dot{U}^2(t)\\
&\geqq(2+\alpha)\left[\int_0^1 u^2\,dx\int_0^1 \dot{u}^2\,dx-\left(\int_0^1 u\dot{u}\,dx\right)^2\right]\\
&\quad-2\alpha U(t)\left[\int_0^t\left(\int_0^1 u(x,\tau)\int_0^\tau \tilde{g}(\tau-\lambda)\sigma(u_x(x,\lambda))_x\,d\lambda\,dx\right)d\tau\right.\\
&\qquad-\int_0^1\int_0^t \tilde{g}(t-\tau)u(x,t)\sigma(u_x(x,\tau))_x\,d\tau\,dx\\
&\qquad+\dot{g}(0)\int_0^t\int_0^1 u(x,\tau)\sigma(u_x(x,\tau))_x\,dx\,d\tau\\
&\qquad\left.-\frac{1}{\alpha}\int_0^1\int_0^t \tilde{g}(t-\tau)u_x(x,t)\sigma(u_x(x,\tau))\,d\tau\,dx\right]-2\alpha U(t)\mathscr{E}(0)\\
&\geqq 2\alpha U(t)\left[-(\mathscr{E}(0))-\int_0^1\left(\int_0^t u(x,\tau)\int_0^\tau \tilde{g}(\tau-\lambda)\sigma(u_x(x,\lambda))_x\,d\lambda\,d\tau\right)dx\right.\\
&\qquad+\left(\frac{1}{\alpha}\right)\int_0^1\int_0^t \tilde{g}(t-\tau)u_x(x,t)\sigma(u_x(x,\tau))\,d\tau\,dx\\
&\qquad+\int_0^1\int_0^t \tilde{g}(t-\tau)u(x,t)\sigma(u_x(x,\tau))_x\,d\tau\,dx\\
&\qquad\left.+\dot{g}(0)\int_0^t\int_0^1 u(x,\tau)\sigma(u_x(x,\tau))_x\,dx\,d\tau\right],
\end{aligned}\tag{II.254}$$

by virtue of the Schwarz inequality and our hypothesis relative to the initial energy $\mathscr{E}(0)$. We now note that, by virtue of the homogeneity of the boundary conditions (II.109),

$$\begin{aligned}
&\int_0^1\int_0^t \tilde{g}(t-\tau)u(x,t)\sigma(u_x(x,\tau))_x\,d\tau\,dx\\
&\quad=\int_0^1\frac{\partial}{\partial x}\left(\int_0^t \tilde{g}(t-\tau)u(x,t)\sigma(u_x(x,\tau))\,d\tau\right)dx\\
&\qquad-\int_0^1\int_0^t \tilde{g}(t-\tau)u_x(x,t)\sigma(u_x(x,\tau))\,d\tau\,dx\\
&\quad=-\int_0^1\int_0^t \tilde{g}(t-\tau)u_x(x,t)\sigma(u_x(x,\tau))\,d\tau\,dx,
\end{aligned}\tag{II.255}$$

and thus the differential inequality (II.254) may be rewritten in the form

$$\ddot{U}(t)U(t)-\left(\frac{\alpha+2}{4}\right)\dot{U}^2(t)$$

$$\geqq -2\alpha U(t)\left[\mathscr{E}(0)-\dot{g}(0)\int_0^t\int_0^1 u(x,\tau)\sigma(u_x(x,\tau))_x\,dx\,d\tau\right.$$

$$\text{(II.256)}\qquad +\int_0^1\left(\int_0^t u(x,\tau)\int_0^\tau \tilde{g}_\tau(\tau-\lambda)\sigma(u_x(x,\lambda))_x\,d\lambda\,d\tau\right)dx$$

$$\left.+\left(1-\frac{1}{\alpha}\right)\int_0^1\int_0^t \tilde{g}(t-\tau)u_x(x,t)\sigma(u_x(x,\tau))\,d\tau\,dx\right].$$

Our aim at this point in the proof is to bound, from above, the three integrals on the right-hand side of (II.256). To this end we have the following series of estimates (beginning with the third integral expression):

$$\left|\int_0^1\int_0^t \tilde{g}(t-\tau)u_x(x,t)\sigma(u_x(x,\tau))\,d\tau\,dx\right|$$

$$\leqq \int_0^t |\tilde{g}(t-\tau)|\int_0^1 |u_x(x,t)||\sigma(u_x(x,\tau))|\,dx\,d\tau$$

$$\text{(II.257)}\qquad \leqq \int_0^t |\tilde{g}(t-\tau)|\left(\int_0^1 u_x^2(x,t)\,dx\right)^{1/2}\left(\int_0^1 \sigma^2(u_x(x,\tau))\,dx\right)^{1/2}d\tau$$

$$\leqq \|u(\cdot,t)\|_{H_0^1}\int_0^t |\tilde{g}(t-\tau)|\left(\int_0^1 \sigma^2(u_x(x,\tau))\,dx\right)^{1/2}d\tau.$$

However, in view of (II.235) and our assumption that $\sigma(0)=0$,

$$\int_0^1 \sigma^2(u_x(x,\tau))\,dx \leqq \bar{\sigma}^2\int_0^1 u_x^2(x,\tau)\,dx,$$

and, therefore,

$$\left|\int_0^1\int_0^t \tilde{g}(t-\tau)u_x(x,t)\sigma(u_x(x,\tau))\,d\tau\,dx\right|$$

$$\leqq \bar{\sigma}\|u(\cdot,t)\|_{H_0^1}\int_0^t |\tilde{g}(t-\tau)|\|u(\cdot,t)\|_{H_0^1}\,d\tau$$

$$\text{(II.258)}\qquad \leqq \bar{\sigma}\left(\sup_{[0,T)}\|u(\cdot,t)\|_{H_0^1}\right)^2\int_0^t |\tilde{g}(t-\tau)|\,d\tau$$

$$\leqq \bar{\sigma}\left(\sup_{[0,T)}\|u(\cdot,t)\|_{H_0^1}\right)^2\sup_{[0,T)}\int_0^t |\dot{g}(t-\tau)|\,d\tau.$$

Next, if we integrate by parts in the second integral on the right-hand side of (II.256) and estimate as in (II.257), (II.258) we obtain

(II.259)
$$\begin{aligned}
&\left|\int_0^1 \left[\int_0^t u(x,\tau)\int_0^\tau \tilde{g}_\tau(\tau-\lambda)\sigma(u_x(x,\lambda))_x\,d\lambda\,d\tau\right]dx\right| \\
&\quad=\left|\int_0^1 \left[\int_0^t u_x(x,\tau)\int_0^\tau \tilde{g}_\tau(\tau-\lambda)\sigma(u_x(x,\lambda))\,d\lambda\,d\tau\right]dx\right| \\
&\quad=\left|\int_0^t \left[\int_0^1 u_x(x,\tau)\left(\int_0^\tau \tilde{g}_\tau(\tau-\lambda)\sigma(u_x(x,\lambda))\,d\lambda\right)dx\right]d\tau\right| \\
&\quad\leqq \int_0^t \left|\int_0^1 u_x(x;\tau)\left(\int_0^\tau \tilde{g}_\tau(\tau-\lambda)\sigma(u_x(x,\lambda)\,d\lambda\right)dx\right|d\tau \\
&\quad\leqq \int_0^t \left(\int_0^1 u_x^2(x,\tau)\,dx\right)^{1/2}\left(\int_0^1\left(\int_0^\tau \tilde{g}_\tau(\tau-\lambda)\sigma(u_x(x,\lambda))\,d\lambda\right)^2 dx\right)^{1/2} d\tau \\
&\quad\leqq \sup_{[0,T)}\|u(\cdot,t)\|_{H_0^1}\int_0^t\left(\int_0^1\left(\int_0^\tau \tilde{g}_\tau(\tau-\lambda)\sigma(u_x(x,\lambda))\,d\lambda\right)^2 dx\right)^{1/2} d\tau.
\end{aligned}$$

However,

$$\begin{aligned}
&\int_0^1\left(\int_0^\tau \tilde{g}_\tau(\tau-\lambda)\sigma(u_x(x,\lambda))\,d\lambda\right)^2 dx \\
&\quad\leqq \int_0^\tau \tilde{g}_\tau^2(\tau-\lambda)\,d\lambda\int_0^1\int_0^\tau \sigma^2(u_x(x,\lambda))\,d\lambda\,dx \\
&\quad\leqq \left(\int_0^\tau \tilde{g}_\tau^2(\tau-\lambda)\,d\lambda\right)\bar{\sigma}^2\int_0^1\int_0^\tau u_x^2(x,\lambda)\,d\lambda\,dx \\
&\quad=\bar{\sigma}^2\int_0^\tau \|u(\cdot,\lambda)\|_{H_0^1}^2\,d\lambda\int_0^\tau \tilde{g}_\tau(\tau-\lambda)\,d\lambda,
\end{aligned}$$

and therefore

(II.260)
$$\begin{aligned}
&\left|\int_0^1\left[\int_0^t u(x,\tau)\int_0^\tau \tilde{g}_\tau(\tau-\lambda)\sigma(u_x(x,\lambda))_x\,d\lambda\,d\tau\right]dx\right| \\
&\quad\leqq \bar{\sigma}\sup_{[0,T)}\|u(\cdot,t)\|_{H_0^1}\int_0^t\left(\int_0^\tau \tilde{g}_\tau^2(\tau-\lambda)\,d\lambda\right)^{1/2}\left(\int_0^\tau \|u(\cdot,\lambda)\|_{H_0^1}^2\,d\lambda\right)^{1/2} d\tau \\
&\quad\leqq \bar{\sigma}\sqrt{T}\left(\sup_{[0,T)}\|u(\cdot,t)\|_{H_0^1}\right)^2\int_0^T\left(\int_0^t \tilde{g}_\tau^2(t-\tau)\,d\tau\right)^{1/2} dt \\
&\quad=\bar{\sigma}\sqrt{T}\left(\sup_{[0,T)}\|u(\cdot,t)\|_{H_0^1}\right)^2\int_0^T\left(\int_0^t \ddot{g}^2(t-\tau)\,d\tau\right)^{1/2} dt.
\end{aligned}$$

Finally,

$$\left|\int_0^t \int_0^1 u(x,\tau)\sigma(u_x(x,\tau))_x\,dx\,d\tau\right|$$

$$= \left|\int_0^t \int_0^1 u_x(x,\tau)\sigma(u_x(x,\tau))\,dx\,d\tau\right|$$

(II.261)
$$\leqq \int_0^t \left(\int_0^1 u_x^2(x,\tau)\,dx\right)^{1/2} \left(\int_0^1 \sigma^2(u_x(x,\tau))\,dx\right)^{1/2} d\tau$$

$$\leqq \bar{\sigma} \sup_{[0,T)} \|u(\cdot,t)\|_{H_0^1} \int_0^t \left(\int_0^1 u_x^2(x,\tau)\,dx\right)^{1/2} d\tau$$

$$\leqq \bar{\sigma} T\Big(\sup_{[0,T)} \|u(\cdot,t)\|_{H_0^1}\Big)^2.$$

Combining the estimates (II.258), (II.260), (II.261) with (II.256) then yields the differential inequality

(II.262)
$$\ddot{U}(t)U(t) - \left(\frac{\alpha+2}{4}\right)\dot{U}^2(t) \geqq -2\alpha U(t) \times \left(\mathscr{E}(0) + \bar{\sigma}\varkappa_T\Big[\sup_{[0,T)} \|u(\cdot,t)\|_{H_0^1}\Big]^2\right)$$

in view of the definition of $\varkappa_T$. However, by assumption $u \in \mathscr{C}$ and, therefore, it follows that

(II.263)
$$\ddot{U}(t)U(t) - \left(\frac{\alpha+2}{4}\right)\dot{U}^2(t) \geqq -2\alpha U(t)(\mathscr{E}(0) + \bar{\sigma}\varkappa_T C^2);$$

if we introduce δ, as in (II.246), then our differential inequality becomes

(II.264)
$$\ddot{U}(t)U(t) - \left(\frac{\alpha+2}{4}\right)\dot{U}^2(t) \geqq -2\alpha\delta U(t), \qquad 0 \leqq t < T.$$

We note here, in passing, that as a consequence of our hypotheses (II.232), (II.235), $\mathscr{E}(t) \geqq 0$ $\forall t \geqq 0$ (and thus $\mathscr{E}(0) \geqq 0$); this fact will be substantiated in the Remarks following the derivation of the growth estimate (II.248). For the sake of convenience we now set

(II.265)
$$\gamma = \frac{\alpha-2}{4} \qquad \text{(so that } \gamma > 0\text{)},$$

and define

(II.266)
$$\nu^2 = \frac{2\alpha\delta}{(2\gamma+1)} = \frac{2\alpha\delta}{\alpha-1}.$$

Then (II.264) has the equivalent form

$$\ddot{U}(t)U(t)-(\gamma+1)\dot{U}^2(t) \geqq -2\nu^2(2\gamma+1)U(t), \tag{II.267}$$

for $0 \leqq t < T$, a differential inequality which has appeared several times in the recent literature on ill-posed initial-boundary value problems associated with nonlinear partial differential equations of hyperbolic and parabolic type; indeed, by (II.247) and the definition of $U(t)$, $\dot{U}(0)>0$ and, thus, $\dot{U}(t)>0$ for $t \in [0, \eta)$. Following the analysis in Levine [94] we multiply both sides of (II.267) by

$$-\gamma(U^{-\gamma}(t))'(U^{-(\gamma+2)}(t))'' \qquad (\text{for } t \in [0, \eta)) \tag{II.268}$$

and then integrate both sides of the resulting inequality over $[0, t)$ so as to obtain (compare with [94, II-15] for $t \in [0, \eta)$:

$$[(U^{-\gamma}(t)']^2 - 4\gamma^2\nu^2 U^{-(2\gamma+1)}(t) \geqq [(U^{-\gamma}(0))']^2 - 4\gamma^2\nu^2 U^{-(2\gamma+1)}(0) > 0, \tag{II.269}$$

where the right-hand side is positive by virtue of the definition of $U(t)$ and our hypothesis (II.247) relative to the initial data. Proceeding as in [94] we factor both sides of (II.269) and rewrite the inequality as

$$\begin{aligned}&(U^{-\gamma}(t)' - 2\gamma\nu U^{-(\gamma+1/2)}(t))(U^{-\gamma}(t)' + 2\gamma\nu U^{-(\gamma+1/2)}(t))\\ &\qquad \geqq (U^{-\gamma'}(0) - 2\gamma U^{-(\gamma+1/2)}(0))(U^{-\gamma'}(0) + 2\gamma\nu U^{-(\gamma+1/2)}(0)).\end{aligned} \tag{II.270}$$

Again, by (II.247) the factor

$$U^{-\gamma'}(0) + 2\gamma\nu U^{-(\gamma+1/2)}(0) < 0, \tag{II.271}$$

and thus, since neither factor in (II.270) can change sign on $[0, \eta)$ (by virtue of our smoothness assumptions relative to $u(x, t)$), we have

$$U^{-\gamma}(t)' < -2\gamma\nu U^{-(\gamma+1/2)}(t), \qquad 0 \leqq t < \eta. \tag{II.272}$$

Now suppose that $\dot{U}(\eta)=0$. Then, by (II.269),

$$[[U^{-\gamma}(t)']^2 - 4\gamma^2\nu^2 U^{-(2\gamma+1)}(t)]_{t=\eta} = -4\gamma^2\nu^2 U^{-(2\gamma+1)}(\eta) < 0,$$

contradicting the fact that (II.269) holds at $t=\eta$. Thus, $\dot{U}(t)>0$ for $t \in [0, T)$ and, therefore, (II.272) is valid for $t \in [0, T)$. Clearly (II.272) on $[0, T)$ is equivalent to

$$U'(t) > 2\nu U^{1/2}(t), \qquad 0 \leqq t < T. \tag{II.273}$$

Integration of (II.273) then yields the estimate

$$U(t) \geqq (\nu t + U^{1/2}(0))^2, \qquad 0 \leqq t < T, \tag{II.274}$$

which by virtue of the definition of $U(t)$ is easily seen to be equivalent to (II.248). Q.E.D.

Remarks. We have indicated, in the course of the above derivation, that the hypotheses (II.232), (II.235) together imply that

$$\text{(II.275)} \qquad \mathscr{E}(0)=\frac{1}{2}\int_0^1 g^2(x)\,dx+\int_0^1 \Sigma(f'(x))\,dx>0$$

for all choices of initial data $f(x)$, $g(x)$. In fact $\mathscr{E}(t)\geqq 0\ \forall t\geqq 0$, and this result is a direct consequence of the fact that

$$\text{(II.276)} \qquad \text{Hypotheses (II.232), (II.235)}\Rightarrow \Sigma(\zeta)>0 \quad \forall\zeta\in R^1.$$

In order to verify (II.276) we begin by noting that as a consequence of (II.232), $\Sigma(0)>0$. Also, for $\zeta\neq 0$,

$$\text{(II.277)} \qquad \frac{d}{d\zeta}\left(\frac{\Sigma(\zeta)}{\zeta^\alpha}\right)=\frac{\zeta\Sigma'(\zeta)-\alpha\Sigma(\zeta)}{\zeta^{\alpha+1}}.$$

Therefore, for $\zeta>0$, it follows from (II.232) that

$$\text{(II.278)} \qquad \frac{d}{d\zeta}\left(\frac{\Sigma(\zeta)}{\zeta^\alpha}\right)>0 \qquad (\zeta>0),$$

so that $\Sigma(\zeta)/\zeta^\alpha$ is nonincreasing on $(0,\infty)$. Also

$$\text{(II.279)} \qquad \lim_{\zeta\to 0^+}\left(\frac{\Sigma(\zeta)}{\zeta^\alpha}\right)\geqq 0.$$

By (II.235) and the definition of Σ in terms of σ, we have

$$\text{(II.280)} \qquad \begin{aligned}|\sigma(\zeta)|&\leqq|\sigma(0)|+\bar\sigma|\zeta|,\\ |\Sigma(\zeta)|&\leqq|\Sigma(0)|+|\sigma(0)||\zeta|+\bar\sigma|\zeta|^2.\end{aligned}$$

Since $\alpha>2$, therefore, we have

$$\text{(II.281)} \qquad \left|\frac{\Sigma(\zeta)}{\zeta^\alpha}\right|\leqq\frac{|\Sigma(0)|+|\sigma(0)||\zeta|+\bar\sigma|\zeta|^2}{\zeta^\alpha}\to 0, \qquad \zeta\to+\infty.$$

Thus, $\Sigma(\zeta)/\zeta^\alpha$ is nonincreasing on $(0,\infty)$ and satisfies $\underline{\lim}_{\zeta\to 0^+}|\Sigma(\zeta)/\zeta^\alpha|\geqq 0$ and $\lim_{\zeta\to+\infty}|\Sigma(\zeta)/\zeta^\alpha|=0$; hence $\Sigma(\zeta)/\zeta^\alpha\geqq 0$ for $\zeta\geqq 0$, which implies that $\Sigma(\zeta)\geqq 0$ for $\zeta\geqq 0$. We now define

$$\text{(II.282)} \qquad \Lambda(\zeta)=\Sigma(-\zeta) \quad \forall\zeta\in R^1.$$

Then $\Lambda'(\zeta)=-\Sigma'(-\zeta)$ and, by (II.232),

$$\text{(II.283)} \qquad \zeta\Lambda'(\zeta)=-\zeta\Sigma'(-\zeta)\geqq\alpha\Sigma(-\zeta)=\alpha\Lambda(\zeta).$$

Repeating the argument given above yields the conclusion that $\Lambda(\zeta)\geqq 0\ \forall\zeta\geqq 0$ and, thus, $\Sigma(\zeta)\geqq 0\ \ \forall\zeta\leqq 0$. We conclude that $\Sigma(\zeta)\geqq 0\ \ \forall\zeta\in R^1$, and thus $\mathscr{E}(t)\geqq 0\ \forall t\geqq 0$. Q.E.D.

Remarks. M. Slemrod has recently published [139] global nonexistence results for a problem closely related to (II.229), (II.109). Slemrod considers steady shearing flows in a nonlinear viscoelastic fluid with shearing stress given by

$$\sigma\left(\int_0^\infty e^{-\alpha\tau} v_x(x,t-\tau)\,d\tau\right), \tag{II.284}$$

where σ is a nonlinear, odd, real analytic function, α a positive constant, and $v(x,t)=u(x,t)$ the fluid velocity; using the constitutive hypothesis (II.284) the corresponding evolution equation takes the form

$$u_{tt}=-\sigma\left(\int_0^\infty e^{-\alpha\tau}\frac{\partial}{\partial\tau}u_x(x,t-\tau)\,d\tau\right)_x, \tag{II.285}$$

if we set the density $\rho=1$. If we assume in (II.229) the same kind of decaying exponential behavior for the relaxation function g as appears in the constitutive assumption (II.229), take $\mathcal{F}\equiv\int_{-\infty}^0 g_\tau(t-\tau)\sigma(U_x)_x\,d\tau$ in (II.284), where $U(x,\tau)$, $(x,\tau)\in[0,1]\times(-\infty,0]$ is the prescribed past history, integrate by parts with respect to τ, and then make an obvious change of variables in the integral, (II.229) reduces to

$$u_{tt}=-\int_0^\infty e^{-\alpha\tau}\frac{\partial}{\partial\tau}\sigma(u_x(x,t-\tau))_x\,d\tau \tag{II.286}$$

which is to be compared with the evolution equation in [139], i.e., (II.285). The particular choice of history dependence chosen in [139], i.e., $g(t)=e^{-\alpha\tau}$, and, more essentially, the particular choice of how the nonlinearity $\sigma(\zeta)$ enters the constitutive theory, allows for the initial-history boundary value problem in [139] to be transformed into an initial-boundary value problem for a nonlinear hyperbolic conservation law with linear damping; hyperbolicity for the transformed problem in [139] is shown to be equivalent to the condition that $\sigma'(\zeta)>0$. In fact, the transformation in [139] is effected as follows: Clearly (II.285) is equivlent to

$$\begin{aligned} v_t &= \sigma(w)_x, \\ w(x,t) &\equiv \int_0^\infty e^{-\alpha s} v_x(x,t-s)\,ds. \end{aligned} \tag{II.287}$$

If we set

$$r(x,t)\equiv\int_0^\infty e^{-\alpha s} v_t(x,t-s)\,ds, \tag{II.288}$$

then

$$r(x,t)=v(x,t)-\alpha\int_0^\infty e^{-\alpha s} v(x,t-s)\,ds \tag{II.289}$$

via integration by parts. Then

$$r_t(x, t) = v_t(x, t) - \alpha \int_0^\infty e^{-\alpha s} v_t(x, t-s)\, ds \tag{II.290}$$
$$= v_t(x, t) - \alpha r(x, t).$$

Thus, the evolution equation (II.285) is equivalent to the system

$$\begin{aligned} w_t(x, t) &= r_x(x, t), \\ r_t(x, t) &= \sigma(w(x, t)_x - \alpha r(x, t). \end{aligned} \tag{II.291}$$

which is a strictly hyperbolic system if $\sigma'(\zeta) > 0$; a similar reduction does not seem possible for the evolution equation (II.286). Once (II.285) has been transformed into the (hyperbolic) system (II.291), it is shown in [139] to be amenable to analysis based on the use of Riemann invariants[2] and an argument due to Lax [86], to prove that singularities develop in the solutions of nonlinear hyperbolic systems, in finite time, for sufficiently large initial data; the manner in which the nonlinearity σ enters into (II.286), as opposed to (II.285), seems to preclude the possibility of carrying over the global nonexistence results of [139], for initial-history boundary value problems associated with (II.285), to analogous problems for (II.286). For the viscoelastic model based on (II.229), or even its specialization (II.286) to the case of an exponentially decaying viscoelastic relaxation function, the original conjecture of MacCamy [104] concerning global nonexistence of smooth solutions for sufficiently large initial data still represents, to the best of this author's knowledge, an open problem.

Remarks. Other models of one-dimensional nonlinear viscoelastic response, besides those represented by the models governed by the evolution equations (II.285), (II.286), have appeared in the recent literature. Greenberg, MacCamy, and Mizel [58] considered the semilinear situation where the stress is of the form

$$\sigma(u_x, \dot{u}_x) \equiv \phi(u_x) + \dot{u}_x \tag{II.292}$$

with ϕ a strictly increasing function (the evolution equation which results is then a special case of (II.231) for which $\lambda \equiv 1$). Dafermos [42], on the other hand, considered the fully nonlinear situation in which the stress is of the form $\sigma(u_x, \dot{u}_x)$; the dependence of $\sigma(u_x, \dot{u}_x)$ on $\dot{u}_x$ is restricted in [42] by the requirement that the viscosity be bounded away from zero while the dependence on u_x is essentially unrestricted except for certain boundedness assumptions. More specifically, Dafermos [42] assumes that $\sigma(p, q)$ is continuously differentiable in p, q, that $\sigma_p(p, q)$ and $\sigma_q(p, q)$ are locally Hölder continuous in

[2] See Chapter IV for a brief discussion of the procedure involved. Slemrod was awarded the 1980 Munroe Martin Prize in Applied Mathematics, primarily for the contribution represented by reference [139].

R^2 with exponent μ, $0<\mu<1$, that $\sigma(0,0)=0$, and that $\exists K, L>0$ such that

$$\sigma_q(p,q) \leqq K, \qquad |\sigma_p(p,q)| \leqq L(\sigma_q(p,q))^{1/2}, \tag{II.293}$$

for all p, q. The evolution equation in [42] assumes the form

$$\rho u_{tt} = \sigma(u_x, \dot{u}_x)_x + \mathscr{F}(x,t), \tag{II.294}$$

and there are associated initial and boundary data of the form

$$\sigma(u_x(0,t), \dot{u}_x(0,t)) = \sigma_0(t), \quad 0 \leqq t < T, \qquad \sigma(u_x(1,t), \dot{u}_x(1,t)) = \sigma_1(t), \quad 0 \leqq t < T, \tag{II.295}$$

$$u(x,0) = f(x), \; u_t(x,0) = g(x). \tag{II.296}$$

Actually, for the existence problem in [42], the analysis is first restricted to the special case in which the boundary conditions (II.295) reduce to

$$u_x(0,t)=0, \qquad u_x(1,t)=0, \qquad 0 \leqq t < T. \tag{II.297}$$

Dafermos shows [42] that the viscoelastic part of the constitutive hypothesis, under the assumptions (II.293), actually dominates the elastic part and guarantees the existence of a unique smooth solution on $[0,1]\times[0,T)$ for all $T>0$; the proof combines certain energy estimates with known a priori bounds for parabolic equations and involves invoking the Leray-Schauder fixed point theorem. An additional assumption on σ, namely that $\exists J$ (constant) such that

$$\int_0^p \sigma(\zeta, 0)\, ds \equiv W(p) \geqq J, \qquad p \in (-\infty, \infty), \tag{II.298}$$

suffices to guarantee the asymptotic stability of solutions to (II.294)–(II.296); $\sigma(p,0)$ can be interpreted as the elastic part of the stress and $W(p)$ as the elastic energy of the body.

Additional remarks [solutions of the nonlinear viscoelastic integro-differential equation (II.229)].

Before departing this chapter, some additional remarks are in order concerning the nature of the solutions to initial-history boundary value problems associated with the model of nonlinear viscoelastic response (II.229); some of our remarks concern work which came to this author's attention after an initial draft of this work had been prepared.

(α) As we have already indicated, it does not appear possible to transform (II.229) into an equivalent damped nonlinear hyperbolic system of the form (II.291) as was the case for the Slemrod model, i.e., (II.285). It is possible, however, to transform (II.229) into an equivalent damped, *nonhomogeneous*, nonlinear hyperbolic system and, indeed, the analysis in MacCamy [104] is based, in large measure, on such a transformation. In fact, MacCamy, employing results established in [105], shows that (II.229) is equivalent to a nonlinear, nonhomogeneous, hyperbolic equation of the form

$$u_{tt}+k(0)u_t-\sigma(u_x)_x=G(x,t), \tag{II.299}$$

where $k(0)>0$ *and*

$$\begin{aligned} G(x,t)\equiv\mathscr{F}(x,t)+L_k[\mathscr{F}]+k(t)g(x) \\ -\dot{k}(0)u(x,t)+\dot{k}(t)f(x)+L_k[u(x,\cdot)]. \end{aligned} \tag{II.300a}$$

By $L_k[v]$ in (II.300a) we understand the convolution of $k(t)$ with $v(t)$, i.e.,

$$L_k[v](t)=\int_0^t k(t-\tau)v(\tau)\,d\tau\equiv(k*v)(t). \tag{II.300b}$$

Furthermore, it is shown in [104] that $k(\cdot)$ satisfies, besides the condition $k(0)>0$, $t^jk^{(n)}(t)\in L_1[0,\infty)$ for $j\leqq 4$, $n=0, 1, 2$ and for $T>0$,

$$\int_0^T\int_0^1 v(x,t)\frac{d}{dt}L_k[v(x,t)]\,dx\,dt\geqq 0. \tag{II.301}$$

If we set $\alpha=k(0)$, $r=u_t$, $w=u_x$, then (II.299) is easily seen to be equivalent to the damped, nonhomogeneous system

$$\begin{aligned} w_t(x,t)&=r_x(x,t), \\ r_t(x,t)&=\sigma(w(x,t))_x-\alpha r(x,t)+G(x,t), \end{aligned} \tag{II.302}$$

which is of the form (II.291) if and only if $G(x,t)\equiv 0$. One arrives at the representation (II.299) by employing a Laplace transform argument; the proof of existence, uniqueness, and asymptotic stability of solutions to initial-history boundary value problems associated with (II.229), for sufficiently small initial data, is then based on the derivation of energy estimates for the solutions of the equivalent nonhomogeneous, damped, nonlinear hyperbolic equation (II.299) and a modification of an argument due to Nishida [121] which is based on the Riemann Invariants concept that we will discuss in Chapter IV in connection with Slemrod's work [139]–[141]; Nishida dealt with the homogeneous damped system (II.291), and thus with the equivalent homogeneous, damped, nonlinear hyperbolic equation, and proved that global smooth solutions to the corresponding initial-boundary value problems always exist provided the initial data is sufficiently small. The arguments which are presented by Slemrod in [139], for solutions of the damped homogeneous system (II.291), do not, unfortunately, carry over to the nonhomogeneous system (II.302) and, thus, can not be used to deduce any information about the possible breakdown of smooth solutions to initial-history boundary-value problems associated with the model governed by (II.229), when the initial data are not small.

We will discuss, in Chapter IV, the application of Riemann invariant arguments to the damped homogeneous system (II.291). At this point, however, it seems worthwhile to give some indication of how Laplace transform arguments may be used to bring the evolution equation (II.229) into the form of a damped, nonhomogeneous, nonlinear hyperbolic equation (and, by extension, into an equivalent nonhomogeneous, damped, nonlinear hyperbolic system of the form (II.302)). Instead of using here the general argument[3] as it was first presented in MacCamy [105], we will review, instead, the presentation put forth in the recent survey article by Slemrod [141] in which (II.229), by virtue of an obvious choice for the relaxation function $g(t)$, and with the assumption that $\mathcal{F}(x, t) \equiv 0$, assumes the form

$$u_{tt} = \kappa_0 \sigma(u_x)_x - \sigma(\kappa_0 - \kappa_1) \int_0^t e^{-\lambda(t-\tau)} \sigma(u_x)_x \, d\tau, \tag{II.229'}$$

where $\lambda > 0$, $0 < \kappa_1 < \kappa_0$ are constants. From (II.229') it follows directly that

$$\frac{\partial}{\partial t} u_t(x, t) - \int_0^t [(\kappa_0 - \kappa_1) e^{-\lambda(t-\tau)} + \kappa_1] \sigma(u_x)_x \, d\tau = 0$$

and thus

$$u_t(x, t) = \int_0^t [(\kappa_0 - \kappa_1) e^{-\lambda(t-\tau)} + \kappa_1] \sigma(u_x)_x \, d\tau + g(x). \tag{II.303}$$

Following [141] we now let $\mathcal{L}$ denote the Laplace transform and ω the transform variable. From (II.303) we obtain directly

$$\mathcal{L}[u_t - g] = \frac{\kappa_0 \omega + \kappa_1 \lambda}{(\lambda + \omega)\omega} \mathcal{L}[\sigma(u_x(x, t))_x]. \tag{II.304}$$

An application of the convolution theorem then yields

$$\begin{aligned} \int_0^t e^{-\lambda(t-\tau)} \sigma(u_x(x, \tau))_x \, d\tau &= \mathcal{L}^{-1}\left\{\frac{\mathcal{L}[\sigma(u_x(s, t))_x]}{\lambda + \omega}\right\} \\ &= \mathcal{L}^{-1}\left\{\frac{\omega \mathcal{L}(u_t - g]}{\kappa_0 \omega + \kappa_1 \lambda}\right\}, \end{aligned} \tag{II.305}$$

where (II.304) has been used to arrive at (II.205$_2$). From (II.305$_2$) it now

[3] The more general argument will be presented below in connection with the discussion of the work of Dafermos and Nohel [43].

follows that

$$
\begin{aligned}
\int_0^t e^{-\lambda(t-\tau)}\sigma(u_x(x,\tau))_x\,d\tau
&= \frac{1}{\kappa_0}\mathscr{L}^{-1}\left\{\left(1-\frac{\kappa_1\lambda}{\kappa_0\omega+\kappa_1\lambda}\right)\mathscr{L}[u_t-g]\right\}\\
&= \frac{u_t-g(x)}{\kappa_0}-\frac{\kappa_1\lambda}{\kappa_0}\mathscr{L}^{-1}\{\mathscr{L}[e^{-\kappa_1\lambda t/\kappa_0}]\mathscr{L}[u_t-g]\}\\
&= \frac{1}{\kappa_0}(u_t(x,t)-g(x))-\frac{\kappa_1\lambda}{\kappa_0^2}\int_0^t e^{-\lambda\kappa_1(t-\tau)/\kappa_0}(u_\tau(x,\tau)-g(x))\,d\tau.
\end{aligned}
\tag{II.306}
$$

Integration by parts, applied to (II.306$_3$), now yields

$$
\begin{aligned}
\int_0^t e^{-\lambda(t-\tau)}\sigma(u_x(x,\tau))_x\,d\tau &= \frac{1}{\kappa_0}(u_t(x,t)-g(x))\\
&\quad+\frac{1}{\kappa_0}\left(\frac{\kappa_1\lambda}{\kappa_0}\right)^2\int_0^t e^{-\kappa_1\lambda(t-\tau)/\kappa_0}u(x,\tau)\,d\tau\\
&\quad-\frac{\kappa_1}{\kappa_0}\lambda u(x,t)+\frac{\kappa_1\lambda}{\kappa_0^2}e^{-\kappa_1\lambda t/\kappa_0}f(x)\\
&\quad+\frac{1}{\kappa_0}(1-e^{-\kappa_1\lambda t/\kappa_0})g(x).
\end{aligned}
\tag{II.307}
$$

Following Slemrod, we now define

$$
\hat{\sigma}(\xi)=\kappa_0\sigma(\zeta),\qquad \beta=\lambda(\kappa_0-\kappa_1),
\tag{II.308}
$$

and

$$
\begin{aligned}
\Phi(x,t)=-\lambda(\kappa_0-\kappa_1)\Bigg[&\frac{1}{\kappa_0}\left(\frac{\kappa_1\lambda}{\kappa_0}\right)^2\int_0^t e^{-\kappa_1\lambda(t-\tau)/\kappa_0}u(x,\tau)\,d\tau\\
&-\frac{\kappa_1}{\kappa_0}\lambda u(x,t)+\frac{\kappa_1\lambda}{\kappa_0^2}e^{-\kappa_1\lambda t/\kappa_0}f(x)-\frac{1}{\kappa_0}e^{-\lambda\kappa_1 t/\kappa_0}g(x)\Bigg].
\end{aligned}
\tag{II.309}
$$

From (2.29′), (II.307), (II.308)–(II.309) it then follows directly that

$$
u_{tt}+\beta u_t=\hat{\sigma}(u_x)_x+\Phi(x,t),
\tag{II.310}
$$

which is the desired result.

(β) In a recent major piece of work [43], Dafermos and Nohel have reconsidered the problem of global existence, uniqueness, and asymptotic stability of solutions to the pure initial-history value problem associated with (II.229), with $g(0)=1$, as well as to the associated Dirichlet and Neumann initial-boundary history value problems; their work extends and improves upon the earlier work of MacCamy in as much as MacCamy considered only the problem with associated Dirichlet boundary data. Also the approach in [43] depends, essentially, only upon energy estimates and a contraction mapping argument (in conjunction with the Banach fixed point theorem), while the work in [104] employs energy estimates in combination with Nishida's [121] Riemann invariant argument; as Dafermos and Nohel point out, therefore, MacCamy's approach is essentially limited to one-dimensional situations whereas the energy methods employed in [43], which seem to be based on some recent work of Matsumura [111], on the multi-space-dimensional nonlinear wave equation with frictional damping, may be extended to problems in more than one space-dimension. It should also be pointed out that in the same paper [43], and by essentially the same method, Dafermos and Nohel treat the problem of global existence, uniqueness, and asymptotic stability of solutions to first-order, nonlinear, integrodifferential initial-history, and initial-history boundary value problems of the type considered by MacCamy in [105]. Both in the work of Dafermos and Nohel [43] and in MacCamy's work [104], [105], the data f, g, $\mathcal{F}$ are assumed to be sufficiently small (in an appropriate sense).

Although our interest in this book is in the direction of ill-posed initial-history value and initial-history boundary value problems, we will sketch, in detail, the basic ideas behind the approach taken in [43]. The first basic idea in [43] is to note that the pure initial-history value problem associated with (II.229) may be brought into the equivalent form

$$\begin{aligned} &u_{tt}+\frac{\partial}{\partial t}\left(\int_0^t k(t-\tau)u_t(x,\tau)\,d\tau\right) \\ \text{(II.311)}\qquad &\qquad =\sigma(u_x(x,t))_x+\Gamma(x,t), \qquad 0<t<\infty, \quad -\infty\leqq x<\infty, \\ &u(x,0)=f(x), \qquad u_t(x,0)=g(x), \quad -\infty\leqq x<\infty, \end{aligned}$$

where $k(t)$ is the resolvent kernel associated with $g'(t)$ and

$$\text{(II.312)}\qquad \Gamma(x,t)=\mathcal{F}(x,t)+(k*\mathcal{F})(x,t)+k(t)g(x).$$

We note that (II.311), (II.312) are essentially equivalent to (II.299), (II.300a) in MacCamy [104]. In fact (II.311) is easily established by noting that a Laplace transform argument, similar to that employed for the special kernel considered in (α), above, may be used to show that the unique solution of the Volterra equation

$$\text{(II.313)}\qquad y(t)+(g'*y)(t)=\phi(t), \qquad 0\leqq t<\infty,$$

$\phi(\cdot)\in L^1_{\text{loc}}(0,\infty)$, is given by

$$\text{(II.314)}\qquad y(t)=\phi(t)+(k*\phi)(t), \qquad 0\leqq t<\infty.$$

This observation is then applied to (II.229), with $g(0)=1$, and $y(t)=\sigma(u_x(x,t))_x$, $\phi(t)=u_{tt}(x,t)-\mathcal{F}(x,t)$ so as to yield

$$\text{(II.315)} \qquad \begin{aligned} &u_{tt}(x,t)+(k*u_{tt})(x,t) \\ &\quad =\sigma(\acute{u}_x(x,t))_x+\mathcal{F}(x,t)+(k*\mathcal{F})(x,t). \end{aligned}$$

The evolution equation in (II.311) now results by integrating, by parts, the convolution term on the left-hand side of (II.315). Having established the equivalence of the pure initial-history value problems for (II.229) and (II.311) Dafermos and Nohel then impose the following conditions on the relaxation function $g(t)$:

$$\text{(II.316)} \qquad \begin{aligned} &\text{(i)} && g(t)\in C^3[0,\infty),\ g(t),\ g'(t),\ g''(t),\ g'''(t) \text{ all bounded on } [0,\infty) \\ &\text{(ii)} && g(t)=g_\infty+\mathcal{G}(t),\ g_\infty>0,\ g(0)=1 \\ &\text{(iii)} && (-1)^{(k)}\mathcal{G}^{(k)}(t)\geqq 0,\ 0\leqq t<\infty,\ k=0,1,2 \\ &\text{(iv)} && \mathcal{G}'(t)\not\equiv 0 \\ &\text{(v)} && t^j\mathcal{G}^{(k)}(t)\in L^1[0,\infty),\ j=0,1,2,3,\ k=0,1,2,3. \end{aligned}$$

Under the assumptions in (II.316), it is then possible to prove that $k(t)$, the resolvent kernel associated with $g'(t)$ must satisfy

$$\text{(II.317)} \qquad \begin{aligned} &\text{(i)} && k(t)\in C^2[0,\infty);\ k(t),\ k'(t),\ k''(t) \text{ all bounded on } [0,\infty). \\ &\text{(ii)} && k^{(l)}(t)\in L^1(0,\infty),\ l=0,1,2. \\ &\text{(iii)} && \int_0^T v(t)\frac{d}{dt}(k*v)(t)\,dt\geqq 0 \text{ for any } T>0,\ \forall v(\cdot)\in L^2(0,T). \end{aligned}$$

Concerning the nonlinearity σ, Dafermos and Nohel [43] assume that

$$\text{(II.318)} \qquad \sigma\in C^3(-\infty,\infty), \quad \sigma(0)=0, \quad \sigma'(0)>0$$

and, vis à vis the forcing term $\mathcal{F}(x,t)$, that

$$\text{(II.319)} \qquad \begin{aligned} &\mathcal{F},\ \mathcal{F}_t\in L^1([0,\infty);L^2(-\infty,\infty)), \\ &\mathcal{F}_x,\ \mathcal{F}_{tt},\ \mathcal{F}_{xt}\in L^2([0,\infty);L^2(-\infty,\infty)). \end{aligned}$$

Finally, the initial data f, g are assumed to satisfy

$$\text{(II.320)} \qquad \begin{aligned} &f'(\cdot),\ f''(\cdot),\ f'''(\cdot)\in L^2(-\infty,\infty), \\ &g(\cdot),\ g'(\cdot),\ g''(\cdot)\in L^2(-\infty,\infty). \end{aligned}$$

Under the above stated assumptions it follows that $\Gamma(\cdot,t)$ in (II.311) satisfies

$$\text{(II.321)} \qquad \begin{aligned} &\Gamma,\ \Gamma_t\in L^1([0,\infty);L^2(-\infty,\infty)), \\ &\Gamma_x,\ \Gamma_{tt},\ \Gamma_{xt}\in L^2([0,\infty);L^2(-\infty,\infty)). \end{aligned}$$

We note that yet another alternative form of the integrodifferential evolution equation (II.229) (besides (II.311)) is required for the derivation of the energy estimates employed in [43] and this will be indicated below. At this point, however, we note that the authors in [43] employ the form of the evolution equation as given by (II.311) to prove a local existence and uniqueness theorem for the pure initial-history value problem associated with (II.229). To accomplish this they must introduce the stricter requirements on the nonlinearity σ,

$$\text{(II.318}'\text{)} \qquad \sigma \in C^3(-\infty, \infty); \qquad \sigma(0) = 0, \quad \begin{cases} \sigma'(w) \geqq p_0 > 0, \\ \forall w \in (-\infty, \infty). \end{cases}$$

The theorem they prove says, in essence, that there exists a unique, maximally defined solution of (II.311), $u(x, t) \in C^2([0, T_0) \times (-\infty, \infty))$ such that for $T \in [0, T_0)$, $T_0 \leqq \infty$,

$$\text{(II.322)} \quad u_t, u_x, u_{tt}, u_{tx}, u_{xx}, u_{ttt}, u_{ttx}, u_{txx}, u_{xxx} \in L^\infty([0, T); L^2(-\infty, \infty))$$

and if $T_0 < \infty$ then

$$\text{(II.323)} \quad \begin{aligned} \int_{-\infty}^{\infty} [u_t^2(x, t) &+ u_x^2(x, t) + u_{tt}^2(x, t) + u_{tx}^2(x, t) \\ &+ u_{xx}^2(x, t) + u_{ttt}^2(x, t) + u_{ttx}^2(x, t) \\ &+ u_{txx}^2(x, t) + u_{xxx}^2(x, t)]\, dx \to \infty \qquad \text{as } t \to T_0. \end{aligned}$$

In brief outline, the approach taken in the proof of the above stated local existence and uniqueness theorem is as follows: Let $M, T > 0$ and let $\chi(M, T)$ denote the set of all functions $u(x, t) \in C^2([0, T); (-\infty, \infty))$ such that $u(x, 0) = f(x)$, $u_t(x, 0) = g(x)$ and (II.322) is satisfied, as well as,

$$\text{(II.324)} \quad \sup_{[0,T)} \int_{-\infty}^{\infty} [u_t^2 + u_x^2 + u_{tt}^2 + u_{tx}^2 + u_{xx}^2 + u_{ttt}^2 + u_{ttx}^2 + u_{txx}^2 + u_{xxx}^2]\, dx \leqq M^2.$$

Clearly $\chi(M, T) \neq \emptyset$ if M is sufficiently large. Also, if $u \in \chi(M, T)$ then

$$\text{(II.325)} \qquad \sup_{[0,T)\times(-\infty,\infty)} \{|u_t(x, t)|, |u_x(x, t)|, |u_{tt}(x, t)|, |u_{tx}(x, t)|, |u_{xx}(x, t)|\} \leqq M.$$

Dafermos and Nohel [43] now construct a map S of $\chi(M, T)$ into $C^2([0, T) \times (-\infty, \infty))$ by defining, for $v(x, t) \in \chi(M, T)$, $u(x, t) = S(v(x, t))$ to be the solution of the initial value problem

$$\text{(II.326)} \quad \begin{aligned} u_{tt}(x, t) + k(0)u_t(x, t) &= \sigma'(v_x(x, t))u_{xx}(x, t) \\ &\quad + \Gamma(x, t) - (k' * v_t)(x, t), \qquad 0 \leqq t < \infty, \quad -\infty < x < \infty, \\ u(x, 0) = f(x), &\qquad u_t(x, 0) = g(x), \qquad -\infty < x < \infty. \end{aligned}$$

Remark. The motivation for defining $u(x, t)$ as the solution of (II.326) is clear:

(i) The evolution equation in (II.311) may be rewritten in the form

$$
\text{(II.327)}\qquad \begin{aligned} u_{tt}(x, t)+k(0)u_t(x, t) &= \sigma'(u_x(x, t))u_{xx}(x, t) \\ &\quad +\Gamma(x, t)-\int_0^t k'(t-\tau)u_\tau(x, \tau)\, d\tau \end{aligned}
$$

(compare with II.299), if we carry out the indicated differentiation in (II.311), and use the fact that $\sigma(u_x)_x = \sigma'(u_x)u_{xx}$, and

(ii) If we can show that $S:\chi(M, T)\to C^2([0, T)\times(-\infty, \infty))$ has a fixed point $\hat{u}(x, t)$ then by virtue of (II.326) $\hat{u}(x, t)$ will be a solution of (II.327) and hence a solution of (II.311).

In order to show that $S:\chi(M, T)\to C^2([0, T)\times(-\infty, \infty))$ has such a fixed point $\hat{u}(x, t)$ Dafermos and Nohel [43] first demonstrate that for M sufficiently large and T sufficiently small, $S:\chi(M, T)\to\chi(M, T)$. This is accomplished by using energy estimates, established for the solutions of (II.326), to show that any such solution $u(x, t)\in\chi(M, T)$, under the stated hypotheses on M, T; more precisely, by employing the evolution equation in (II.326) (and equations obtained from it by successive differentiations with respect to t and x) multiplying by u_t (respectively, u_{tt}, u_{tx}, etc.) and employing integration by parts in each case, the authors obtain an estimate, for solutions $u(x, t)$ of (II.326), of the form

$$
\text{(II.328)}\qquad \begin{aligned} \int_{-\infty}^{\infty} [u_t^2(x, t)&+u_x^2(x, t)+u_{tt}^2(x, t)+u_{tx}^2(x, t) \\ &+u_{xx}^2(x, t)+u_{ttt}^2(x, t)+u_{ttx}^2(x, t) \\ &\quad +u_{txx}^2(x, t)+u_{xxx}^2(x, t)]\, dx \\ &\leqq A(f, g, \Gamma)+B(M)M^2T, \qquad 0\leqq t< T, \end{aligned}
$$

where A is a constant which can be estimated in terms of the $L^2(-\infty, \infty)$ norms of f', f'', f''', g, g', and g'', the $L^2([0, \infty); L^2(-\infty, \infty))$ norms of Γ, Γ_t, Γ_x, Γ_{tt}, Γ_{tx}, the constant p_0 of (II.318′), and the bounds on σ', σ'', and k'. The constant B may be estimated in terms of M and the bounds on σ', σ'', σ''', k' and k''. Thus, if we choose

$$
\text{(II.329)}\qquad T\leqq[2B(M)]^{-1}, \qquad M\geqq[2A(f, g, \Gamma)]^{1/2},
$$

it follows that the right-hand side of (II.328) is bounded from above by M^2 and thus, by the defining property (II.234), $u(x, t)\in\chi(M, T)$.

Having established that $S:\chi(M, T)\to\chi(M, T)$ with, M, T chosen so as to satisfy (II.329), Dafermos and Nohel [43] then introduce on $\chi(M, T)$ the metric ρ defined by

(II.330)
$$\rho(u, \bar{u}) \equiv \max_{[0,T)} \left[\int_{-\infty}^{\infty} [(u_t(x, t) - \bar{u}_t(x, t))^2 + (u_x(x, t) - \bar{u}_x(x, t))^2] \, dx \right]^{1/2},$$

under which $\chi(M, T)$ becomes a complete (bounded) metric space. If $v(x, t)$, $\bar{v}(x, t) \in \chi(M, T)$ and we set $u(x, t) = S(v(x, t))$, $\bar{u}(x, t) = S(\bar{v}(x, t))$, $V(x, t) = v(x, t) - \bar{v}(x, t)$, $U(x, t) = u(x, t) - \bar{u}(x, t)$ it follows that $U(x, t)$ is a solution of the initial-value problem

(II.331)
$$\begin{aligned} U_{tt}(x, t) + k(0)U_t(x, t) &= \sigma'(v_x(x, t))U_{xx}(x, t) \\ &\quad + \Lambda(x, t)\bar{u}_{xx}(x, t)V_x(x, t) - (k' * v_t)(x, t), \\ &\quad 0 \leqq t \leqq T, \quad -\infty < x < \infty, \\ U(x, 0) = 0, \quad U_t(x, 0) &= 0, \quad -\infty < x < \infty, \end{aligned}$$

where $\Lambda(x, t)$, defined by

(II.332)
$$\Lambda(x, t) = \begin{cases} \dfrac{\sigma'(v_x(x, t)) - \sigma'(\bar{v}_x(x, t))}{v_x(x, t) - \bar{v}_x(x, t)}, & v_x \neq \bar{v}_x, \\ \sigma''(v_x(x, t)), & v_x = \bar{v}_x \end{cases}$$

is a bounded, continuous function. Employing energy estimates again, the authors [43] show that

(II.333)
$$\begin{aligned} &\max_{[0,T)} \int_{-\infty}^{\infty} [U_t^2(x, t) + U_x^2(x, t)] \, dx \\ &\quad \leqq T e^{\mu T} \max_{[0,T)} \int_{-\infty}^{\infty} [V_t^2(x, t) + V_x^2(x, t)] \, dx, \end{aligned}$$

where μ depends only on M, $\max_{[-M,M]} |\sigma'(\cdot)|$, $\max_{[-M,M]} |\sigma''(\cdot)|$, c_0, $k(0)$, and $\int_0^\infty |k'(t)| \, dt$. With M, T chosen on the one hand so as to satisfy (II.329), and simultaneously so as to satisfy $T e^{\mu T} \leqq \frac{1}{4}$, we obtain from (II.333) and the definition of ρ,

(II.334)
$$\rho(Sv, S\bar{v}) \leqq \tfrac{1}{2}\rho(v, \bar{v}) \quad \forall v, \bar{v} \in \chi(M, T).$$

Thus $S : \chi(M, T) \to \chi(M, T)$ is actually a contraction mapping on $\chi(M, T)$ and by the Banach fixed point theorem there exists a unique fixed point $\hat{u}(x, t)$ of the map S in $\chi(M, T)$; this unique fixed point in $\chi(M, T)$ is then a solution of (II.327), and hence a solution of (II.311) on $(-\infty, \infty) \times [0, T)$; if T_0, the maximal length of the interval of existence of $\hat{u}(x, t)$ is finite and (II.323) is not satisfied, it can be shown that $\hat{u}(x, t)$ may be extended as a solution of (II.327) on a small interval $[T_0, T_0 + \varepsilon]$ beyond the maximal interval $[0, T_0]$. This completes the sketch of the proof concerning the existence of a maximally defined unique local solution to the initial-value problem (II.311).

In order to show that the initial-value problem (II.311) admits a unique, globally defined solution which decays to zero as $t \to +\infty$, when the initial data and the forcing term are small, in an appropriate sense, Dafermos and Nohel [43] proceed as follows: they consider a local solution $u(x, t)$ of (II.311) in the sense just described, which satisfies, for some T, $0 < T < \infty$, and a small $\mu > 0$, $\mu < c_0$,[4]

(II.335) $$|u_x(x,t)|, \quad |u_{xt}(x,t)|, \quad |u_{xx}(x,t)| \leqq \mu, \qquad 0 \leqq t < T,$$

for $-\infty < x < \infty$. (Note that $u(x, t)$ satisfies (II.325) as $u \in \chi(M, T)$.) By deriving energy estimates obtained by employing the evolution equation (II.311), and the equivalent equation

(II.336) $$\begin{aligned} &u_{tt}(x,t) + (r * u_{tt})(x,t) \\ &= \sigma(u_x(x,t))_x + \beta \int_0^t \sigma(u_x(x,\tau))_x \, d\tau + \tilde{\Gamma}(x,t), \qquad 0 \leqq t < \infty, \quad -\infty < x < \infty \end{aligned}$$

where

(II.337) $$\begin{aligned} \tilde{\Gamma}(x,t) &= \mathscr{F}(x,t) + (r * \mathscr{F})(x,t), \\ r(t) &= \beta + k(t) + \beta \int_0^t k(\tau)\, d\tau, \end{aligned}$$

$\beta > 0$ a suitably chosen constant, they show that the initial data and forcing term $\mathscr{F}$ may be chosen so small, so that as long as (II.335) is satisfied,

(II.338) $$\begin{aligned} &\int_{-\infty}^{\infty} [u_t^2(x,\tau) + u_x^2(x,\tau) + u_{tt}^2(x,\tau) + u_{xt}^2(x,\tau) + u_{xx}^2(x,\tau) \\ &\qquad + u_{ttt}^2(x,\tau) + u_{ttx}^2(x,\tau) + u_{txx}^2(x,\tau) + u_{xxx}^2(x,\tau)]\, dx \\ &\qquad + \int_0^s \int_{-\infty}^{\infty} [u_{tt}^2 + u_{xt}^2 + u_{xx}^2 + u_{ttt}^2 + u_{ttx}^2 + u_{txx}^2 + u_{xxx}^2]\, dx\, dt \leqq \mu^2 \end{aligned}$$

for $0 \leqq s < T$; on the other hand, it follows at once that (II.338) implies (II.335). It now follows from the proof of the local existence theorem (in particular, from (II.323)) that the maximal interval of existence of $u(x, t)$ must be $[0, \infty)$ and that (II.338) is then valid for $0 \leqq s < \infty$. From (II.338) it then follows that we have a unique global solution $u(x, t) \in C^2([-\infty, \infty) \times [0, \infty))$ of (II.311) such that

(II.339) $$u_t, u_x, u_{tt}, u_{xt}, u_{xx}, u_{ttt}, u_{ttx}, u_{txx}, u_{xxx} \in L^1([0, \infty); L^2(-\infty, \infty)),$$

---

[4] For the global existence result it is only required that $\sigma'(\zeta) \geqq p_0 > 0$ with $\zeta \in [-c_0, c_0]$ for some constant $c_0 > 0$.

(II.340) $\quad u_{tt}, u_{xt}, u_{xx}, u_{ttt}, u_{ttx}, u_{txx}, u_{xxx} \in L^2([0,\infty); L^2(-\infty,\infty)),$

from which it follows, in turn, that

(II.341) $\quad u_{tt}(\cdot, t), u_{xt}(\cdot, t), u_{xx}(\cdot, t) \to 0, \text{ as } t \to +\infty \text{ in } L^2(-\infty, \infty).$

From (II.339) and (II.341) we also obtain

$$\text{(II.342)} \qquad \begin{aligned} &u_t(x,t), u_x(x,t), u_{tt}(x,t), u_{xt}(x,t), u_{xx}(x,t) \\ &\quad \to 0 \text{ as } t \to +\infty, \quad \text{uniformly for } -\infty < x < \infty. \end{aligned}$$

For the validity of (II.338), and hence (II.339)–(II.342), we require that the data and forcing term $\mathcal{F}$ be small in the sense that the $L^1([0,\infty); L^2(-\infty,\infty))$-norms of $\mathcal{F}$, $\mathcal{F}_t$, the $L^2([0,\infty); L^2(-\infty,\infty))$-norms of $\mathcal{F}_x$, $\mathcal{F}_{tt}$, and $\mathcal{F}_{xt}$, and the $L^2(-\infty,\infty)$-norms of f', f'', f''', g, g', and g'' be sufficiently small. We will make just two additional remarks about the global existence and asymptotic decay results delineated above. First of all, in showing that (II.338) is a consequence of (II.335), for sufficiently small data, we have indicated that the authors [43] employ energy estimates based on (II.311) and (II.336). A typical estimate is obtained by multiplying (II.311) by $u_t(x,t)$, integrating over $(-\infty,\infty)\times[O,\tau)$, $0<\tau<T$, integrating by parts with respect to x, and using the property (II.317(iii)) of the resolvent kernel $k(t)$ so as to obtain

$$\text{(II.343)} \qquad \begin{aligned} &\frac{1}{2}\int_{-\infty}^{\infty} u_t^2(x,\tau)\,dx + \int_{-\infty}^{\infty} W(u_x(x,\tau))\,dx \\ &\quad \leqq \frac{1}{2}\int_{-\infty}^{\infty} u_t^2(x,0)\,dx + \int_{-\infty}^{\infty} W(u_x(x,0))\,dx + \int_0^{\tau}\int_{-\infty}^{\infty} \Gamma u_t\,dx\,dt, \end{aligned}$$

where $W(\zeta) \equiv \int_0^{\zeta} \sigma(\lambda)\,d\lambda$ satisfies

$$\text{(II.344)} \qquad W(\zeta) \geqq \tfrac{1}{2} p_0 \zeta^2, \qquad \zeta \in [-c_0, c_0]$$

in view of our hypothesis that $\sigma'(\zeta) \geqq p_0 > 0$ for $\zeta \in [-c_0, c_0]$. Using the estimate (III.344) and the fact that

$$\text{(II.345)} \qquad \int_0^{\tau}\int_{-\infty}^{\infty} \Gamma u_t\,dx\,dt \leqq \frac{1}{4}\max_{[0,\tau)}\int_{-\infty}^{\infty} u_t^2(x,\tau)\,dx + \left(\int_0^{\infty}\left(\int_{-\infty}^{\infty}\Gamma^2\,dx\right)^{1/2} dt\right)^2$$

we obtain from (II.343) the estimate

$$\text{(II.346)} \qquad \begin{aligned} &\frac{1}{2}\int_{-\infty}^{\infty} u_t^2(x,\tau)\,dx + p_0\int_{-\infty}^{\infty} u_x^2(x,\tau)\,dx \\ &\quad \leqq \int_{-\infty}^{\infty} g^2(x)\,dx + 2\int_{-\infty}^{\infty} W(f'(x))\,dx \\ &\qquad + 2\left(\int_0^{\infty}\left(\int_{-\infty}^{\infty}\Gamma^2\,dx\right)^{1/2} dt\right)^2, \qquad 0 \leqq \tau \leqq T. \end{aligned}$$

In view of the definition of $\Gamma(x, t)$, i.e., (II.312), it follows that all the terms on the right-hand side of the estimate (II.346) are "controllably small", i.e., can be made as small as one wishes by requiring that the data f, g, and $\mathcal{F}$ be small in the obvious way. Thus, energy estimates lead one to the conclusion that the terms $\int_{-\infty}^{\infty} u_t^2(x, \tau)\, dx$ and $\int_{-\infty}^{\infty} u_x^2(x, \tau)\, d\tau$ are also "controllably small". A long succession of similar energy estimates based on (II.311) and (II.336) yield the conclusion that each of the expressions on the left-hand side of the estimate (II.338) is "controllably small" and hence serves to establish this latter estimate for sufficiently small data. As for (II.336) itself, this evolution equation is obtained by noting that the solution $y(t)$ of the Volterra equation (II.313) satisfies

$$\text{(II.347)} \qquad y(t)+\beta \int_0^t y(\tau)\, d\tau = \phi(t)+(r * \phi)(t), \qquad 0 \leqq t < \infty,$$

for any $\beta > 0$, where $r(t)$ is given by (II.337). The result, i.e., (II.336) now follows from (II.229) by applying (II.347) with $y(t) = \sigma(u_x(x, t))_x$ and $\phi(t) = u_{tt}(x, t) - \mathcal{F}(x, t)$. We note that the kernel $r(\cdot)$ in (II.336) satisfies

(II.348) (i) $r(t) \in C^2[0, \infty)$, $r(t)$, $r'(t)$, $r''(t)$ all bounded on $[0, \infty)$,

(ii) $r(t) = r_\infty + R(t)$, $r_\infty = \dfrac{\beta}{g_\infty}$, $R^{(m)}(t) \in L^1(0, \infty)$ for $m = 0, 1, 2$,

(iii) For any $T > 0$ there exist constants γ, $q > 0$ with $\beta q < 1$ such that $\forall v(\cdot) \in L^2(0, T)$

$$q \int_0^T v(t) \frac{d}{dt}(r * v)(t)\, dt - \int_0^T v(t)(R * v)(t)\, dt$$
$$\geqq (1+\gamma) \int_0^T v^2(t)\, dt,$$

and that this last inequality is crucial for the derivation of those energy estimates in the global existence proof which are based on the evolution equation in the form (II.336). We will conclude our discussion of the contribution represented by [43] by simply noting that the global existence and asymptotic decay results presented there for the pure initial-value problem associated with (II.229) are easily extended to both homogeneous Neumann and Dirichlet type initial-boundary value problems, i.e., for $x \in [0, 1]$, to problems where, respectively, we have $u_x(0, t) = u_x(1, t) = 0$, $0 < t < \infty$, and $u(0, t) = u(1, t) = 0$, $0 < t < \infty$; in the latter case we must also have $\mathcal{F}(0, t) = \mathcal{F}(1, t) = 0$, $0 < t < \infty$. In the case of homogeneous Dirichlet boundary data the authors [43] note that the stronger conclusion $u(x, t) \to 0$, uniformly on $[0, 1]$, as $t \to +\infty$, will also follow from the analysis; for the details we refer the interested reader to the statement of the relevant results in [43, § 6].

Chapter III

Ill-Posed Problems for Some Partial-Integrodifferential Equations of Electromagnetic Theory

In Chapter II we presented various stability and growth estimates for ill-posed initial-history boundary value problems arising in theories of linear and nonlinear mechanical viscoelastic response; our estimates were obtained through the use of the so-called logarithmic convexity and concavity arguments that were introduced in Chapter I. In this chapter we will turn our attention to ill-posed initial-history boundary value problems which are associated with some partial-integrodifferential equations of electromagnetic theory; more specifically, these equations arise in various theories of nonconducting rigid material dielectrics with memory. All of our problems shall be posed in such a manner as to permit abstract reformulations using the basic Hilbert space structure of the last chapter. We begin by offering a derivation of the basic integrodifferential equations and the associated boundary data.

1. Integrodifferential evolution equations for theories of nonconducting material dielectrics.

Let (x^i, t), $i = 1, 2, 3$ be a Lorentz reference frame where the (x^i) represent rectangular Cartesian coordinates and t is the time parameter; in this reference frame the local forms of Maxwell's equations are

$$\frac{\partial \mathbf{B}}{\partial t} + \operatorname{curl} \mathbf{E} = \mathbf{0}, \qquad \operatorname{div} \mathbf{B} = 0,$$
(III.1)
$$\operatorname{curl} \mathbf{H} - \frac{\partial \mathbf{D}}{\partial t} = \mathbf{0}, \qquad \operatorname{div} \mathbf{D} = 0,$$

provided that the density of free current, the magnetization, and the density of free charge all vanish in the open bounded domain $\Omega \subseteq R^3$ where (III.1) apply. In (III.1) the vector $\mathbf{B}$ represents the magnetic flux density while $\mathbf{E}$, $\mathbf{H}$, and $\mathbf{D}$ represent the electric field, magnetic intensity, and electric displacement (or induction) vectors, respectively. The field $\mathbf{D}$, which is introduced so as to simplify the form of Maxwell's equations, is defined in terms of the electric field

E and the polarization field **P** via

$$\mathbf{D} = \varepsilon_0 \mathbf{E} + \mathbf{P}, \qquad \varepsilon_0 > 0 \qquad \text{(physical constant)}. \tag{III.2}$$

In order to obtain a determinate system of equations for the fields appearing in (III.1) it is necessary to append certain constitutive relations, the form of these relations being dependent upon the nature of the material in which the electric and magnetic fields occur. The simplest dielectric is a vacuum in which we have zero polarization and the classical constitutive equations

$$\mathbf{D} = \varepsilon_0 \mathbf{E}, \qquad \mathbf{H} = \mu_0^{-1} \mathbf{B} \qquad (\varepsilon_0 \mu_0 = c^{-2}), \tag{III.3}$$

where c is the speed of light in a vacuum. The next simplest kind of dielectric media in which the electromagnetic field (**E**, **B**) can occur would seem to be the rigid, linear stationary nonconducting dielectric whose constitutive relations, namely,

$$\mathbf{D} = \boldsymbol{\varepsilon} \cdot \mathbf{E}, \qquad \mathbf{B} = \boldsymbol{\mu} \cdot \mathbf{H}, \tag{III.4}$$

were given by Maxwell [112] in 1873; in (III.4) $\boldsymbol{\varepsilon}$, $\boldsymbol{\mu}$ are constant second order tensors which are proportional to the identity tensor if the material is isotropic. As was pointed out by Toupin and Rivlin [143] the constitutive relations (III.4) can not be used to account for the observed phenomena of absorption and dispersion of electromagnetic waves in nonconductors.

In 1877 Hopkinson [68], following a suggestion of Maxwell, (and in connection with his studies on the residual charge of the Leyden jar) proposed a constitutive equation for the electric displacement field in a rigid nonconducting material dielectric of the form

$$\mathbf{D}(\mathbf{x}, t) = \varepsilon \mathbf{E}(\mathbf{x}, t) + \int_{-\infty}^{t} \phi(t-\tau) \mathbf{E}(\mathbf{x}, \tau)\, d\tau, \qquad \mathbf{x} \in \Omega, \tag{III.5}$$

where $\varepsilon > 0$ and $\phi(t)$, $t \geqq 0$, is a monotonically decreasing function of t which is continuous for $0 \leqq t < \infty$. Using the constitutive relation (III.5), Hopkinson was able to correlate his data on the residual charge of Leyden jars by making suitable choices of the memory functions $\phi(t)$; he points out in [68], for instance, that a suitable memory function for glasseous materials would be a linear combination of decreasing exponentials with the coefficients in the linear combination being dependent on the silica composition of the material. To the system consisting of Maxwell's equations and (III.5) we will append the constitutive relation

$$\mathbf{H}(\mathbf{x}, t) = \mu^{-1} \mathbf{B}(\mathbf{x}, t), \qquad \mu > 0, \quad \mathbf{x} \in \Omega. \tag{III.6}$$

The material whose response is governed by the system of constitutive relations (III.5), (III.6) will then be referred to as a Maxwell-Hopkinson Dielectric. Before proceeding with our survey of constitutive relations for nonconducting material dielectrics, let us indicate how the system of relations (III.5), (III.6) leads one to a system of partial-integrodifferential equations for the components of the electric displacement field in a Maxwell-Hopkinson dielectric. We have the following:

LEMMA (Davis [45]). *In any nonconducting material dielectric governed by the constitutive relations* (III.5), (III.6), *with* $\mathbf{E}(\tau)=\mathbf{0}$, $-\infty<\tau<0$, *the components of the electric displacement field* $\mathbf{D}(\mathbf{x},t)$ *must satisfy the system of evolution equations* $(i=1,2,3)$

(III.7)
$$\varepsilon\mu\frac{\partial^2 D_i}{\partial t^2}=\nabla^2 D_i(\mathbf{x},t)+\int_0^t \Phi(t-\tau)\nabla^2 D_i(\mathbf{x},\tau)\,d\tau$$

on $\Omega\times[0,T)$, $T>0$ *where* $\Phi(t)$ *is given by*

(III.8)
$$\Phi(t)=\sum_{n=1}^{\infty}(-1)^n\phi^n(t),$$
$$\phi^1(t)=\varepsilon^{-1}\phi(t),$$
$$\phi^n(t)=\int_0^t \phi^1(t-\tau)\phi^{n-1}(\tau)\,d\tau,\qquad n\geqq 2.$$

Remarks. It will be clear from the proof given below that the Lemma remains valid, with the obvious changes, if the past history of the electric field is of the form

(III.9)
$$\mathbf{E}(\mathbf{x},\tau)=\begin{cases}\mathbf{0}, & -\infty<\tau<-\tau_\infty,\\ \boldsymbol{E}_{\tau_\infty}(\mathbf{x},\tau), & -\tau_\infty\leqq\tau<0,\end{cases}$$

for $\mathbf{x}\in\Omega$, where $\tau_\infty>0$; the validity of the evolution equation clearly depends on the convergence of the series $\sum_{n=1}^{\infty}(-1)^n\phi^n(t)$ which serves to define $\Phi(t)$ and such convergence will be discussed in § 2.

Proof (Lemma). We begin by inverting the constitutive relation (III.5) via the usual method of successive approximations so as to obtain

(III.10)
$$\mathbf{E}(\mathbf{x},t)=\varepsilon^{-1}\mathbf{D}(\mathbf{x},t)+\varepsilon^{-1}\int_0^t \Phi(t-\tau)\mathbf{D}(\mathbf{x},\tau)\,d\tau,$$

$(\mathbf{x},t)\in\Omega\times[0,T)$, where $\Phi(t)$ is given by (III.8); the relation (III.10) is valid under the conditions indicated in the above remarks. Because of the assumed continuity of ϕ, Φ will be continuous on $[0,T)$ if the series in (III.8) converges uniformly. We now recall the vector identity

(III.11)
$$\mathbf{\Delta A}=\text{grad}\,(\text{div}\,\mathbf{A})-\text{curl curl}\,\mathbf{A},$$

which is valid for any vector field $\mathbf{A}$ which is sufficiently smooth on Ω; here $\mathbf{\Delta}$ denotes the vector Laplacian, i.e., $(\mathbf{\Delta A})_i(\mathbf{x})=\nabla^2 A_i(\mathbf{x})$, $\mathbf{x}\in\Omega$. If we apply (III.11) to the electric field $\mathbf{E}$ and note that (by (III.10) and the fact that $\text{div}\,\mathbf{D}=0$) we have $\text{div}\,\mathbf{E}=0$ on Ω, then

(III.12)
$$\mathbf{\Delta E}(\mathbf{x},t)=-\text{curl curl}\,\mathbf{E}(\mathbf{x},t),\qquad \mathbf{x}\in\Omega.$$

However, by Maxwell's equations, (III.1),

$$\text{curl}\,\mathbf{E}(\mathbf{x}, t) = -\frac{\partial}{\partial t}\mathbf{B}(\mathbf{x}, t) = -\mu\frac{\partial}{\partial t}\mathbf{H}(\mathbf{x}, t), \tag{III.13}$$

where we have also used the constitutive relation (III.6). Therefore,

$$\begin{aligned} \mathbf{\Delta E}(\mathbf{x}, t) &= \mu\,\text{curl}\left[\frac{\partial}{\partial t}\mathbf{H}(\mathbf{x}, t)\right] \\ &= \mu\frac{\partial}{\partial t}[\text{curl}\,\mathbf{H}(\mathbf{x}, t)] \\ &= \mu\frac{\partial^2}{\partial t^2}\mathbf{D}(\mathbf{x}, t), \end{aligned} \tag{III.14}$$

again by Maxwell's equations. The result now follows directly if we substitute for $E(x, t)$ in (III.14) from (III.10) and then simplify. Q.E.D.

To the system of partial-integrodifferential equations (III.7) we will want to append initial and boundary data in Ω and on $\partial\Omega$, respectively; the specification of initial data offers no problem but the specification of boundary conditions can become quite involved (i.e., see [49]) unless one is careful to pose the physical problem in a convenient manner. In essence we would like to have

$$D_i(\mathbf{x}, t) = 0, \qquad (\mathbf{x}, t) \in \partial\Omega \times [0, T), \tag{III.15}$$

so that in the abstract reformulation of the initial-boundary value problem, as an initial value problem in Hilbert space, we can use the same spaces H, H_+, H_- as were used for the three-dimensional linear viscoelasticity problem in Chapter II. To this end we consider regions $\tilde{\Omega} \subseteq R^3$, $\hat{\Omega} \subseteq R^3$ with $\hat{\Omega} \subset \Omega \subset \tilde{\Omega}$ (the situation is depicted below) and assume that (i) the Maxwell-Hopkinson constitutive relations hold in $\hat{\Omega} \times [0, T)$ and (ii) the region $\tilde{\Omega}/\hat{\Omega}$ is filled with a perfect conductor so that, in effect, $\mathbf{D}(\mathbf{x}, t) = \mathbf{0}$, $(x, t) \in \tilde{\Omega}/\hat{\Omega} \times [0, T)$. At the interface $\partial\hat{\Omega}$ between the perfect conductor and the Maxwell-Hopkinson dielectric, it is well known that $\mathbf{D} \cdot \hat{\mathbf{n}} = \sigma$, where $\hat{\mathbf{n}}$ is the unit outward normal to $\partial\hat{\Omega}$ and σ is the surface charge density on $\partial\hat{\Omega}$. As $\partial\Omega$ is embedded in the perfect conductor, $\mathbf{D}(\mathbf{x}, t) = 0$, $(\mathbf{x}, t) \in \partial\Omega \times [0, T)$. However, as $\mathbf{D} \equiv \mathbf{0}$ in $\Omega/\hat{\Omega}$, the system of evolution equations (III.7) will be identically satisfied by $\mathbf{D}(\mathbf{x}, t)$ for $(\mathbf{x}, t) \in \Omega/\hat{\Omega} \times [0, T)$; i.e., the system of equations will hold not only in $\hat{\Omega}$ but in all of Ω. We then have the equations (III.7) for $D_i(\mathbf{x}, t)$ in $\Omega \times [0, T)$ and the boundary conditions (III.15) on $\partial\Omega \times [0, T)$. For initial data we have

$$D_i(\mathbf{x}, 0) = f_i(x), \qquad \frac{\partial D_i}{\partial t}(\mathbf{x}, 0) = g_i(\mathbf{x}), \qquad \mathbf{x} \in \Omega, \tag{III.16}$$

where $f_i(\mathbf{x}) = g_i(\mathbf{x}) \equiv 0$, $\mathbf{x} \in \Omega/\hat{\Omega}$.

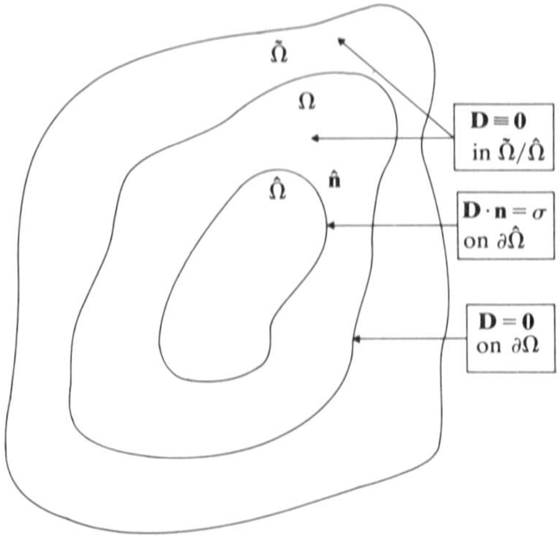

FIG. 1. *A physical situation such as the one depicted above would be realized in a capacitor.*

The constitutive relations (III.5), (III.6) which serve to define the response of a Maxwell-Hopkinson dielectric, embody three basic simplifying assumptions: they are linear, they effect an a priori separation of electric and magnetic effects, and they do not allow for magnetic memory effects. As early as 1912 Volterra [145] proposed extending the Maxwell-Hopkinson theory to treat the situation where the dielectric is anisotropic, nonlinear, and magnetized; his constitutive relations were of the form

$$\text{(III.17)} \qquad \begin{aligned} \mathbf{D}(\mathbf{x}, t) &= \boldsymbol{\varepsilon}\cdot\mathbf{E}(\mathbf{x}, t) + \underset{-\infty}{\overset{t}{\mathscr{D}}}(\mathbf{E}(\mathbf{x}, \tau)), \\ \mathbf{B}(\mathbf{x}, t) &= \boldsymbol{\mu}\cdot\mathbf{H}(\mathbf{x}, t) + \underset{-\infty}{\overset{t}{\mathscr{B}}}(\mathbf{H}(\mathbf{x}, \tau)), \end{aligned}$$

for $(\mathbf{x}, t)\in\Omega\times[0, T)$, and it can be shown that (III.17$_1$) reduces to (III.5) if the functional $\mathscr{D}$ is both linear and isotropic and the body satisfies various restrictions which follow from considerations of material symmetry. Of course, (III.17) still effects an a priori separation of electric and magnetic effects and, as pointed out by Toupin and Rivlin [143], such a separation is inadequate with respect to predicting such phenomena as the Faraday effect in dielectrics. In [143], Toupin and Rivlin thus postulated constitutive equations of the form

$$\text{(III.18a)} \qquad \begin{aligned} \mathbf{D}(\mathbf{x}, t) = {} & \sum_{\nu=0}^{n} \mathbf{a}_\nu\cdot\mathbf{E}^{(\nu)}(\mathbf{x}, t) + \sum_{\nu=0}^{n} \mathbf{c}_\nu\cdot\mathbf{B}^{(\nu)}(\mathbf{x}, t) \\ & + \int_{-\infty}^{t} \boldsymbol{\phi}_1(t, \tau)\cdot\mathbf{E}(\mathbf{x}, \tau)\, d\tau + \int_{-\infty}^{t} \boldsymbol{\phi}_2(t, \tau)\cdot\mathbf{B}(\mathbf{x}, \tau)\, d\tau \end{aligned}$$

and

$$\text{(III.18b)}\qquad \begin{aligned}\mathbf{H}(\mathbf{x}, t) = {} & \sum_{\nu=0}^{n} \mathbf{d}_\nu \cdot \mathbf{E}^{(\nu)}(\mathbf{x}, t) + \sum_{\nu=0}^{n} \mathbf{b}_\nu \cdot \mathbf{B}^{(\nu)}(\mathbf{x}, t) \\ & + \int_{-\infty}^{t} \boldsymbol{\psi}_2(t, \tau) \cdot \mathbf{E}(\mathbf{x}, \tau)\, d\tau + \int_{-\infty}^{t} \boldsymbol{\psi}_2(t, \tau) \cdot \mathbf{B}(\mathbf{x}, \tau)\, d\tau,\end{aligned}$$

for $(\mathbf{x}, t) \in \Omega \times [0, T)$, where $\mathbf{E}^{(\nu)}(\mathbf{x}, t) = \partial^\nu E(x, t)/\partial t$ and $\mathbf{a}_\nu, \cdots, \mathbf{d}_\nu$ are constant tensors; the kernels $\boldsymbol{\phi}_1, \cdots, \boldsymbol{\psi}_2$ are taken to be continuous tensor functions of t and τ which satisfy growth conditions of the form

$$\text{(III.19)}\qquad \boldsymbol{\phi}_1(t, \tau) < \frac{\mathbf{c}}{(t-\tau)^{1+\lambda}}, \qquad \lambda < 0.$$

Toupin and Rivlin [143] also assumed that the dielectric does not exhibit aging and show that as a consequence $\mathbf{D}(\cdot, t)$, $\mathbf{H}(\cdot, t)$ are periodic in t whenever $\mathbf{E}(\cdot, t)$ and $\mathbf{B}(\cdot, t)$ are; this latter result, when combined with the hypothesized growth estimates on the kernels $\boldsymbol{\phi}_1$, etc., and some early results of Volterra [145] on the theory of functionals, yields the conclusion that $\boldsymbol{\phi}_1, \cdots, \boldsymbol{\psi}_2$ depend on t and τ only through the difference $t-\tau$ (the converse of this result is also valid). It is then proven in [143] that if the dielectric exhibits holohedral isotropy, i.e., if it admits the full orthogonal group as its group of material symmetries, $\mathbf{E}(\mathbf{x}, t)$ may be eliminated from (III.18b) and $\mathbf{B}(\mathbf{x}, t)$ may be eliminated from (III.18a); for a holohedral isotropic dielectric the constitutive relations (III.18) thus reduce to

$$\text{(III.19a)}\qquad \mathbf{D}(\mathbf{x}, t) = \sum_{\nu=0}^{n} a_\nu \mathbf{E}^{(\nu)}(\mathbf{x}, t) + \int_{-\infty}^{t} \phi(t-\tau)\mathbf{E}(\mathbf{x}, \tau)\, d\tau,$$

$$\text{(III.19b)}\qquad \mathbf{H}(\mathbf{x}, t) = \sum_{\nu=0}^{n} b_\nu \mathbf{B}^{(\nu)}(\mathbf{x}, t) + \int_{-\infty}^{t} \psi(t-\tau)\mathbf{B}(\mathbf{x}, \tau)\, d\tau,$$

for $(\mathbf{x}, t) \in \Omega \times [0, T)$, where the a_ν, b_ν are constants and $\phi(t)$, $\psi(t)$ are scalar-valued functions. If the material defined by (III.18a), (III.18b) exhibits hemihedral isotropy, i.e., admits the proper orthogonal group as its group of material symmetries, then Toupin and Rivlin [143] show that the constitutive relations reduce to

$$\text{(III.20a)}\qquad \begin{aligned}\mathbf{D}(\mathbf{x}, t) = {} & \sum_{\nu=0}^{n} a_\nu \mathbf{E}^{(\nu)}(\mathbf{x}, t) + \sum_{\nu=0}^{n} c_\nu \mathbf{B}^{(\nu)}(\mathbf{x}, t) \\ & + \int_{-\infty}^{t} [\phi_1(t-\tau)\mathbf{E}(\mathbf{x}, t) + \phi_2(t-\tau)\mathbf{B}(\mathbf{x}, \tau)\mathbf{B}(\mathbf{x}, \tau)]\, d\tau\end{aligned}$$

and

$$\text{(III.20b)}\qquad \begin{aligned}\mathbf{H}(\mathbf{x}, t) = {} & \sum_{\nu=0}^{n} d_\nu \mathbf{E}^{(\nu)}(x, t) + \sum_{\nu=0}^{n} b_\nu \mathbf{B}^{(\nu)}(\mathbf{x}, t) \\ & + \int_{-\infty}^{t} [\psi_1(t-\tau)\mathbf{E}(\mathbf{x}, \tau) + \psi_2(t-\tau)\mathbf{B}(\mathbf{x}; \tau)]\, d\tau.\end{aligned}$$

The dielectric material whose behavior is defined by the constitutive hypotheses (III.20a), (III.20b) is shown, in [143], to possess the requisite characteristics needed to explain the phenomena of absorption and dispersion of electromagnetic waves. As far as this treatise is concerned, we will restrict ourselves to a special case of the constitutive relations (III.19a), (III.19b) which define the rigid holohedral isotropic dielectric, namely, the case where $a_\nu = b_\nu = 0$, $\nu \geqq 1$ so that

$$\mathbf{D}(\mathbf{x}, t) = a_0\mathbf{E}(\mathbf{x}, t) + \int_{-\infty}^{t} \phi(t-\tau)\mathbf{E}(\mathbf{x}, \tau)\, d\tau, \tag{III.21a}$$

$$\mathbf{H}(\mathbf{x}, t) = b_0\mathbf{B}(\mathbf{x}, t) + \int_{-\infty}^{t} \psi(t-\tau)\mathbf{B}(\mathbf{x}, \tau)\, d\tau \tag{III.21b}$$

for $(\mathbf{x}, t) \in \Omega \times [0, T)$. From [14] we then have the following:

LEMMA (Bloom [14]). *If* $\mathbf{E}(\mathbf{x}, t) = \mathbf{B}(\mathbf{x}, \tau) = \mathbf{0}$, $\mathbf{x} \in \Omega$, $-\infty < \tau < 0$, *then the evolution of the electric displacement field in any material dielectric which conforms to the constitutive hypotheses* (III.21) *is governed by the system of partial-integrodifferential equations*

$$\begin{aligned} &\frac{\partial^2 D_i(x, t)}{\partial t^2} + \Psi(0)\frac{\partial D_i}{\partial t}(\mathbf{x}, t) - b_0\dot{\Psi}(0)[c_0\nabla^2 D_i(\mathbf{x}, t) - D_i(\mathbf{x}, t)] \\ &\quad + b_0 \int_0^t [\ddot{\Psi}(t-\tau)D_i(\mathbf{x}, \tau) - \Phi_0(t-\tau)\nabla^2 D_i(\mathbf{x}, \tau)]\, d\tau \\ &= b_0\dot{\Psi}(t)D_i(\mathbf{x}, 0), \end{aligned} \tag{III22}$$

where $c_0 = 1/a_0\dot{\Psi}(0)$, $\Phi_0(t) = \Phi(t)/a_0$, *with* $\Phi(t)$ *defined in terms of* $\phi(t)$ *by* (III.8) *and* $\Psi(t)$ *defined in an analogous manner in terms of* $\psi(t)$.

Proof. We invert both (III.21a), (III.21b) via successive approximations so as to obtain the equations

$$\mathbf{E}(\mathbf{x}, t) = a_0^{-1}\mathbf{D}(\mathbf{x}, t) + a_0^{-1}\int_0^t \Phi(t-\tau)\mathbf{D}(\mathbf{x}, \tau)\, d\tau, \tag{III.23a}$$

$$\mathbf{B}(\mathbf{x}, t) = b_0^{-1}\mathbf{H}(\mathbf{x}, t) + b_0^{-1}\int_0^t \Psi(t-\tau)\mathbf{H}(\mathbf{x}, \tau)\, d\tau, \tag{III.23b}$$

which are valid provided the respective series defining the kernel functions Φ, Ψ converge; in obtaining (III.23a, b) we have used the assumption of zero past histories for **E** and **B**. By (III.23a) and Maxwell's equations, div $\mathbf{E} = 0$ so the vector identity (III.11), applied to **E**, again yields (III.12). Now, however,

$$\begin{aligned} \operatorname{curl} \mathbf{E}(\mathbf{x}, t) &= -\frac{\partial}{\partial t}\mathbf{B}(\mathbf{x}, t) \\ &= -b_0^{-1}\frac{\partial}{\partial t}\mathbf{H}(\mathbf{x}, t) - b_0^{-1}\Psi(0)\mathbf{H}(\mathbf{x}, t) \\ &\quad -\int_0^t \Psi_t(t-\tau)\mathbf{H}(\mathbf{x}, \tau)\, d\tau, \end{aligned} \tag{III.24}$$

where we have used Maxwell's equations and (III.23b). Therefore

$$
\begin{aligned}
-\operatorname{curl}\operatorname{curl}\mathbf{E}(\mathbf{x}, t) &= b_0^{-1}\frac{\partial}{\partial t}(\operatorname{curl}\mathbf{H}(\mathbf{x}, t)) \\
&\quad + b_0^{-1}\Psi(0)\operatorname{curl}\mathbf{H}(\mathbf{x}, t) + \int_0^t \Psi_t(t-\tau)\operatorname{curl}\mathbf{H}(\mathbf{x}, \tau)\, d\tau \\
&= b_0^{-1}\frac{\partial^2}{\partial t^2}\mathbf{D}(\mathbf{x}, t) + b_0^{-1}\Psi(0)\frac{\partial}{\partial t}\mathbf{D}(\mathbf{x}, t) \\
&\quad + \int_0^t \Psi_t(t-\tau)\frac{\partial}{\partial \tau}\mathbf{D}(\mathbf{x}, \tau)\, d\tau,
\end{aligned}
\tag{III.25}
$$

where we have once again employed the Maxwell equations. Combining (II.25) with (III.12) and using the inverted constitutive relation (II.23a) we obtain

$$
\begin{aligned}
&\frac{\partial^2}{\partial t^2}\mathbf{D}(\mathbf{x}, t) + \Psi(0)\frac{\partial}{\partial t}\mathbf{D}(x, t) + b_0^{-1}\int_0^t \Psi_t(t-\tau)\frac{\partial}{\partial \tau}\mathbf{D}(\mathbf{x}, \tau)\, d\tau \\
&\quad = b_0 a_0^{-1}\Delta\mathbf{D}(\mathbf{x}, t) + b_0 a_0^{-1}\int_0^t \Phi(t-\tau)\Delta\mathbf{D}(\mathbf{x}, \tau)\, d\tau.
\end{aligned}
\tag{III.26}
$$

However,

$$
\begin{aligned}
&\int_0^t \Psi_t(t-\tau)\frac{\partial}{\partial \tau}\mathbf{D}(\mathbf{x}, \tau)\, d\tau \\
&\quad = \dot{\Psi}(0)\mathbf{D}(\mathbf{x}, t) - \dot{\Psi}(t)\mathbf{D}(\mathbf{x}, 0) + \int_0^t \Psi_{\tau\tau}(t-\tau)\mathbf{D}(\mathbf{x}, \tau)\, d\tau \\
&\quad = \dot{\Psi}(0)\mathbf{D}(\mathbf{x}, t) - \dot{\Psi}(t)\mathbf{D}(\mathbf{x}, 0) + \int_0^t \ddot{\Psi}(t-\tau)\mathbf{D}(\mathbf{x}, \tau)\, d\tau,
\end{aligned}
\tag{III.27}
$$

and substitution into (III.26) now produces the stated result, i.e., (III.22). The derivation depends not only on the existence of the kernels Φ, Ψ in the inverted constitutive relations but on their being sufficiently differentiable as well. Q.E.D.

Remark. If we drop the assumption of zero past histories for **E**, **B** and assume, instead, past histories of the form

$$
\begin{aligned}
\mathbf{E}(\mathbf{x}, \tau) &= \begin{cases} \mathbf{0}, & -\infty < \tau < -\tau_\infty, \\ \boldsymbol{E}_{\tau_\infty}(\mathbf{x}, \tau), & -\tau_\infty \leqq \tau < 0, \end{cases} \\
\mathbf{B}(\mathbf{x}, \tau) &= \begin{cases} \mathbf{0}, & -\infty < \tau < -\tau_\infty, \\ \mathbf{B}_{\tau_\infty}(\mathbf{x}, \tau), & -\tau_\infty \leqq \tau < 0, \end{cases}
\end{aligned}
\tag{III.28}
$$

for some $\tau_\infty > 0$, then it is not too difficult to show that (III.22) must be replaced by

$$\frac{\partial^2 D_i(\mathbf{x},t)}{\partial t^2}+\Psi(0)\frac{\partial D_i}{\partial t}(\mathbf{x},t)+\dot{\Psi}(0)[D_i(\mathbf{x},t)-c_0\nabla^2 D_i(\mathbf{x},\tau)]$$

(III.29)

$$+\int_{-\tau_\infty}^{t}\left[\Psi(t-\tau)D_i(\mathbf{x},\tau)-\left(\frac{b_0}{a_0}\right)\Phi(t-\tau)\nabla^2 D_i(\mathbf{x},\tau)\right]d\tau=0,$$

provided the past history $\mathbf{D}_{\tau_\infty}$ satisfies $\mathbf{D}_{\tau_\infty}(\cdot, -\tau_\infty)=\mathbf{0}$ uniformly on Ω (as a consequence of the constitutive hypotheses and our assumptions (III.28), the past history of $\mathbf{D}$ will have the same form as those of $\mathbf{E}$ and $\mathbf{B}$). In (III.29), $\Phi(t)$ is given by

$$\Phi(t)=\sum_{n=1}^{\infty}(-1)^n\phi^n(t),$$

(III.30)

$$\phi^1(t)=a_0^{-1}\phi(t),$$

$$\phi^n(t)=\int_{-\tau_\infty}^{t}\phi^1(t-\tau)\phi^{n-1}(\tau)\,d\tau,\qquad n\geqq 2,$$

with a similar definition, of course, for $\Psi(t)$ in terms of $\psi(t)$. More general assumptions can be made about the past histories of $\mathbf{E}$, $\mathbf{B}$ than that represented by (II.28) but, as for the viscoelasticity problems of Chapter II, this would necessitate the introduction, say, of certain fading memory spaces and would involve us in analysis considerably beyond that needed to deal with the elementary spaces H_+, H, and H_- introduced in the previous chapter. Note that with either of the systems of evolution equations (III.26), (III.29) we may associate initial and boundary data of the form (III.15), (III.16) by embedding the holohedral isotropic dielectric, which is governed by the constitutive hypotheses (III.21a), (III.21b), in a perfect conductor.

2. Stability and growth estimates for Maxwell-Hopkinson dielectrics: Logarithmic convexity arguments.

In the previous section, we proved that the components of the electric displacement field, in a material dielectric which is defined by the system of constitutive relations (III.5), (III.6), satisfy the system of partial-integrodifferential equations (III.7) whenever the past history of the electric field vanishes identically; for past histories of the form (III.9) we may simply replace the integral expression on the right-hand side of (III.7) by $\int_{-\tau_\infty}^{t}\Phi(t-\tau)\nabla^2 D_i(\mathbf{x},\tau)\,d\tau$ and then make the obvious changes in what follows below. We want to recast the initial-boundary value problem (III.7), (III.15), (III.16) as an initial-value problem for an integrodifferential equation in

Hilbert space. Because of the physical setting of the problem, and the resulting boundary data, it is very easy to see that the abstract initial-value problem will be of the form (II.23), (II.24), with $\mathbf{U}(\tau)=\mathbf{0}$, $-\infty<\tau<0$, in (II.25). The spaces H_+, H, H_- are also the same as in Chapter II, i.e., $(H_0^1(\Omega))^3$, $(L_2(\Omega))^3$, and $(H^{-1}(\Omega))^3$, respectively, while the operators $\mathbf{N}, \mathbf{K}(t)$ which appear in the abstract integrodifferential equation are now defined by

(III.31)
$$(\mathbf{N}\mathbf{w})_i(\mathbf{x}) \equiv \frac{1}{\varepsilon\mu}\frac{\partial^2 w_i}{\partial x_k\,\partial x_k}(\mathbf{x}), \qquad \mathbf{x}\in\Omega \quad \forall \mathbf{w}\in(H_0^1(\Omega))^3,$$
$$(\mathbf{K}(t)\mathbf{w})_i(\mathbf{x}) \equiv \Phi(t)(\mathbf{N}\mathbf{w})_i(\mathbf{x}), \qquad \begin{cases}\mathbf{x}\in\Omega, & t\in(-\infty,\infty),\\ \forall\mathbf{w}\in(H_0^1(\Omega))^3,\end{cases}$$

where the second order spatial derivatives are taken in the sense of distributions. It is very easy to see that $\mathbf{N}\in\mathscr{L}_s(H_+,H_-)$, $\mathbf{K}(\cdot)\in L^2((-\infty,\infty);\mathscr{L}_s(H_+,H_-))$ and, in fact, for any $\mathbf{v}\in H_+$

(III.32)
$$\begin{aligned}\|\mathbf{K}(t)\mathbf{v}\|_- &= \sup_{\mathbf{w}\in H_+}\frac{|\langle\mathbf{K}(t)\mathbf{v},\mathbf{w}\rangle|}{\|\mathbf{w}\|_+}\\
&= \sup_{\mathbf{w}\in H_+}\frac{\left|\int_\Omega \frac{1}{\varepsilon\mu}\Phi(t)\frac{\partial^2 v_i}{\partial x_k\,\partial x_k}w_i\,d\mathbf{x}\right|}{\|\mathbf{w}\|_+}\\
&= \sup_{\mathbf{w}\in H_+}\left(\frac{|\Phi(t)|}{\varepsilon\mu}\right)\frac{\left|\int_\Omega \frac{\partial^2 v_i}{\partial x_k\,\partial x_k}\cdot w_i\,d\mathbf{x}\right|}{\|\mathbf{w}\|_+}\\
&= \frac{|\Phi(t)|}{\varepsilon\mu}\sup_{\mathbf{w}\in H_+}\frac{\left|\int_\Omega \frac{\partial v_i}{\partial x_k}\frac{\partial w_i}{\partial x_k}\,d\mathbf{x}\right|}{\|\mathbf{w}\|_+}\\
&\leqq \frac{|\Phi(t)|}{\varepsilon\mu}\sup_{\mathbf{w}\in H_+}\frac{\left(\int_\Omega \frac{\partial v_i}{\partial x_k}\frac{\partial v_i}{\partial x_k}\,d\mathbf{x}\right)^{1/2}\left(\int_\Omega \frac{\partial w_i}{\partial x_k}\frac{\partial w_i}{\partial x_k}\,d\mathbf{x}\right)^{1/2}}{\|\mathbf{w}\|_+}\\
&= \frac{|\Phi(t)|}{\varepsilon\mu}\|\mathbf{v}\|_+.\end{aligned}$$

Therefore, for all $\mathbf{v}\in H_+$
$$\frac{\|\mathbf{K}(t)\mathbf{v}\|_-}{\|\mathbf{v}\|_+}\leqq\frac{|\Phi(t)|}{\varepsilon\mu},$$
which implies that for any $t\in(-\infty,\infty)$

(III.33)
$$\|\mathbf{K}(t)\|_{\mathscr{L}_s(H_+,H_-)}\equiv\sup_{\mathbf{v}\in H_+}\frac{\|\mathbf{K}(t)\mathbf{v}\|_-}{\|\mathbf{v}\|_+}\leqq\frac{|\Phi(t)|}{\varepsilon\mu}.$$

However, as

$$\text{(III.34)} \qquad \frac{\left|\int_\Omega \frac{\partial v_i}{\partial x_k}\frac{\partial w_i}{\partial x_k}\,d\mathbf{x}\right|}{\|\mathbf{w}\|_+} = \|\mathbf{v}\|_+, \quad \text{for} \quad \mathbf{w}=\mathbf{v},$$

the computation in (III.32) actually yields

$$\text{(III.35)} \qquad \|\mathbf{K}(t)\|_{\mathscr{L}_s(H_+,H_-)} = \frac{1}{\varepsilon\mu}|\Phi(t)|, \qquad t\in(-\infty,\infty),$$

and, in a completely analogous fashion,

$$\text{(III.36)} \qquad \|\mathbf{K}_t(t)\|_{\mathscr{L}_s(H_+,H_-)} = \frac{1}{\varepsilon\mu}|\dot\Phi(t)|, \qquad t\in(-\infty,\infty).$$

Now, for the initial-value problem (II.23), (II.24), with $\mathbf{U}(\tau)=\mathbf{0}$, $-\infty<\tau<0$, and $\mathbf{N}$, $\mathbf{K}(t)$ defined by (III.31), the growth estimates of Chapter II will be valid for solutions $\mathbf{u}\in\mathcal{N}$, where $\mathcal{N}$ is the class of bounded functions defined by (II.53), provided the basic hypotheses (II.55) are satisfied. In order to satisfy the first inequality in (II.55) we clearly must have, by virtue of (III.31),

$$\text{(III.37)} \qquad -\int_\Omega v_i[\Phi(0)(\mathbf{N}\mathbf{w})_i]\,d\mathbf{x} \geqq \kappa\int_\Omega \frac{\partial v_i}{\partial x_j}\frac{\partial v_i}{\partial x_j}\,d\mathbf{x} \quad \forall \mathbf{v}\in H_+,$$

or

$$\text{(III.38)} \qquad -\frac{\Phi(0)}{\varepsilon\mu}\int_\Omega v_i\frac{\partial^2 v_i}{\partial x_k\,\partial x_k}\,d\mathbf{x} \geqq \kappa\int_\Omega \frac{\partial v_i}{\partial x_j}\frac{\partial v_i}{\partial x_j}\,d\mathbf{x} \quad \forall \mathbf{v}\in H_+,$$

which, is easily seen to be equivalent to

$$\text{(III.39)} \qquad \Phi(0) \leqq -\kappa\,\varepsilon\mu$$

since $\mathbf{v}\in H_+$ implies that $\mathbf{v}$ vanishes on $\partial\Omega$. In view of (III.36), the second inequality in (II.55) will be satisfied provided

$$\text{(III.40)} \qquad \kappa \geqq \frac{\gamma T}{\varepsilon\mu}\sup_{[0,T)}|\dot\Phi(t)|$$

and, thus, if we combine (III.39) and (III.40) we find that $\mathbf{K}(t)$, as defined by (III.31), will satisfy the hypotheses (II.55) if $\Phi(t)$ satisfies

$$\text{(III.41)} \qquad \Phi(0) \leqq -\gamma T\sup_{[0,T)}|\dot\Phi(t)|.$$

Before we proceed with a brief analysis of the important inequality (III.41), let us note that, for the problem at hand, $\mathscr{F}(t)$, as given by (II.57), (II.77) will reduce to

$$\text{(III.42)} \qquad \hat{\mathscr{F}}(t) = \mathscr{E}(t) + \frac{\theta}{2}\sup_{[0,T)}|\Phi(t)|,$$

$\theta \equiv \gamma TN^4/\varepsilon\mu$, by virtue of (III.35) and our assumption of zero past history, and $F(t;\beta,t_0)$, as defined by (II.54) will then satisfy

$$FF'' - F'^2 \geqq -2F(2\hat{\mathcal{F}}(0)+\beta), \qquad 0 \leqq t < T, \tag{III.43}$$

whenever the condition (III.41) relative to $\Phi(t)$ is satisfied. From the differential inequality (III.43) various stability and growth estimates, analogous to those obtained in Chapter II for the viscoelastic displacement vector, will follow for the $\|(\cdot)\|$ norm of the electric displacement vector, once specific hypotheses are laid down concerning the relation of the initial energy $\mathscr{E}(0)$ and $\sup_{[0,T)}|\Phi(t)|$.

In view of (III.41) and (III.42) it is clearly important to have some information about the magnitudes of $\sup_{[0,T)}|\Phi(t)|$ and $\sup_{[0,T)}|\dot{\Phi}(t)|$ where $\Phi(t)$ is defined by (III.8). To this end we have the following:

LEMMA (Bloom [20]). *Let $\phi(\cdot)\in C^1[0,T)$ and assume that* (III.8), *and the derived series which is obtained by term-by-term differentiation of* (III.8), *are both uniformly convergent for $0 \leqq t < T$. Then, if $\sup_{[0,T)}|\phi(t)| < \varepsilon/T$,*

$$\sup_{[0,T)}|\Phi(t)| \leqq \alpha(T), \tag{III.44}$$

$$\sup_{[0,T)}|\dot{\Phi}(t)| \leqq \frac{\alpha(T)}{T}\left[1 + T\left(\frac{\sup_{[0,T)}|\dot{\phi}(t)|}{\sup_{[0,T)}|\phi(t)|}\right)\right], \tag{III.45}$$

where

$$\alpha(T) \equiv \frac{\sup_{[0,T)}|\phi(t)|}{\varepsilon - T\sup_{[0,T)}|\phi(t)|}. \tag{III.46}$$

Proof. By (III.8),

$$|\Phi(t)| \leqq \sum_{n=1}^{\infty}|\phi^n(t)|, \qquad 0 \leqq t < T. \tag{III.47}$$

But, for $n \geqq 2$

$$\phi^n(t) = \frac{1}{\varepsilon}\int_0^t \phi(t-\tau)\phi^{n-1}(\tau)\,d\tau, \qquad 0 \leqq t < T, \tag{III.48}$$

so

$$|\phi^n(t)| \leqq \frac{T}{\varepsilon}\sup_{[0,T)}|\phi(\tau)|\sup_{[0,T)}|\phi^{n-1}(\tau)|, \qquad 0 \leqq t < T, \tag{III.49}$$

from which it is immediate that for $n \geqq 2$

$$\sup_{[0,T)}|\phi^n(t)| \leqq \frac{1}{\varepsilon}\sup_{[0,T)}|\phi(t)|\sup_{[0,T)}|\phi^{n-1}(t)|. \tag{III.50}$$

Successive application of the recursive estimates (III.50) then yields

(III.51)
$$\begin{aligned}\sup_{[0,T)} |\phi^n(t)| &\leqq \left(\frac{T}{\varepsilon} \sup_{[0,T)} |\phi(t)|\right)^{n-1} \sup_{[0,T)} |\phi^1(t)| \\ &= \frac{1}{\varepsilon}\left(\frac{T}{\varepsilon} \sup_{[0,T)} |\phi(t)|\right)^{n-1} \sup_{[0,T)} |\phi(t)| \\ &= \frac{T^{n-1}}{\varepsilon^n} (\sup_{[0,T)} |\phi(t)|)^n.\end{aligned}$$

Therefore, from (III.47) we have, for $0 \leqq t < T$

(III.52)
$$\begin{aligned}|\Phi(t)| &\leqq \sum_{n=1}^{\infty} \frac{T^{n-1}}{\varepsilon^n} (\sup_{[0,T)} |\phi(\tau)|)^n \\ &\leqq \frac{1}{T} \sum_{n=1}^{\infty} \left(\frac{T}{\varepsilon} \sup_{[0,T)} |\phi(\tau)|\right)^n.\end{aligned}$$

By our assumption that $\sup_{[0,T)}|\phi(t)| \leqq \varepsilon/T$, it follows that the geometric series in (III.52) converges and, in fact, we have

(III.53)
$$|\Phi(t)| \leqq \frac{1}{T}\left(\frac{T \sup_{[0,T)} |\phi(t)|}{\varepsilon - T \sup_{[0,T)} |\phi(t)|}\right) \equiv \alpha(T),$$

from which (III.44) follows immediately. To prove (III.45) we begin with the obvious observation that

(III.54)
$$|\dot{\Phi}(t)| \leqq \sum_{n=1}^{\infty} |\dot{\phi}^n(t)|, \qquad 0 \leqq t < T,$$

by virtue of our hypothesis concerning the uniform convergence of the derived series for (III.8). However,

(III.55)
$$\begin{aligned}\dot{\phi}^n(t) &= \frac{1}{\varepsilon} \frac{d}{dt} \int_0^t \phi(t-\tau)\phi^{n-1}(\tau)\, d\tau \qquad (n \geqq 2) \\ &= \frac{\phi(0)}{\varepsilon} \cdot \phi^{n-1}(t) + \frac{1}{\varepsilon} \int_0^t \phi_t(t-\tau)\phi^{n-1}(\tau)\, d\tau\end{aligned}$$

for $0 \leqq t < T$. Therefore,

(III.56)
$$\begin{aligned}|\dot{\phi}^n(t)| &\leqq \frac{1}{\varepsilon} |\phi(0)|\, |\phi^{n-1}(t)| + \frac{T}{\varepsilon} \sup_{[0,T)} |\dot{\phi}(\tau)| \sup_{[0,T)} |\phi^{n-1}(\tau)| \\ &\leqq \frac{1}{\varepsilon} (|\phi(0)| + T \sup_{[0,T)} |\dot{\phi}(\tau)|) \sup_{[0,T)} |\phi^{n-1}(\tau)|.\end{aligned}$$

If we now apply the recursion formula (III.51) we have

(III.57)
$$\sup_{[0,T)} |\phi^{n-1}(\tau)| \leqq \frac{T^{n-2}}{\varepsilon^{n-1}} (\sup_{[0,T)} |\phi(\tau)|)^{n-1}$$

and, therefore, (III.56) yields

(III.58)
$$|\dot{\phi}^n(t)| \leqq (|\phi(0)| + T \sup_{[0,T)} |\dot{\phi}(\tau)|) \frac{T^{n-2}}{\varepsilon^n} (\sup_{[0,T)} |\phi(\tau)|)^{n-1}.$$

By substituting from (III.58) in (III.54) we obviously obtain

$$|\Phi(t)| \leqq (|\phi(0)| + T \sup_{[0,T)} |\dot{\phi}(\tau)|) \sum_{n=1}^{\infty} \frac{T^{n-2}}{\varepsilon^n} (\sup_{[0,T)} |\phi(\tau)|)^{n-1}$$

(III.59)
$$\leqq \frac{(\sup_{[0,T)} |\phi(\tau)| + T \sup_{[0,T)} |\dot{\phi}(\tau)|)}{T^2 \sup_{[0,T)} |\phi(\tau)|} \sum_{n=1}^{\infty} \left(\frac{T}{\varepsilon} \sup_{[0,T)} |\phi(\tau)|\right)^n$$

(III.60)
$$= \frac{1}{T^2} \left(1 + T \frac{\sup_{[0,T)} |\dot{\phi}(\tau)|}{\sup_{[0,T)} |\phi(\tau)|} \right) \cdot \frac{T \sup_{[0,T)} |\phi(\tau)|}{(\varepsilon - T \sup_{[0,T)} |\phi(\tau)|)}.$$

Therefore,

(III.61)
$$|\dot{\Phi}(t)| \leqq \frac{\alpha(T)}{T} \left(1 + T \frac{\sup_{[0,T)} |\dot{\phi}(\tau)|}{\sup_{[0,T)} |\phi(\tau)|} \right), \qquad 0 \leqq t < T,$$

from which the estimate (III.45) is immediate. Q.E.D.

Remark. There is an alternative means of deriving the estimates stated in the Lemma above which yields further useful information concerning the relationship between the kernel functions $\phi(t)$ and $\Phi(t)$. If we multiply (III.8_3) through by $(-1)^n$ and sum over n, $2 \leqq n < \infty$, we obtain

(III.62)
$$\sum_{n=2}^{\infty} (-1)^n \phi^n(t) = \int_0^t \phi^1(t-\tau) \left[\sum_{n=2}^{\infty} (-1)^n \phi^{n-1}(\tau) \right] d\tau,$$

where we have used the assumed uniform convergence of the series to interchange the integration and summation operations. However, by (III.8_1), (III.8_2),

(III.63)
$$\sum_{n=2}^{\infty} (-1)^n \phi^n(t) = \Phi(t) + \frac{1}{\varepsilon} \phi(t), \qquad 0 \leqq t < T,$$

so we may rewrite (III.62) in the form

$$\text{(III.64)}\qquad \begin{aligned}\Phi(t)+\frac{1}{\varepsilon}\phi(t) &= -\frac{1}{\varepsilon}\int_0^t \phi(t-\tau)\left[\sum_{n=2}^{\infty}(-1)^{n-1}\phi^{n-1}(\tau)\right]d\tau\\ &= -\frac{1}{\varepsilon}\int_0^t \phi(t-\tau)\Phi(\tau)\,d\tau.\end{aligned}$$

As a direct consequence of (III.64) we have

$$\text{(III.65)}\qquad \Phi(0) = -\frac{1}{\varepsilon}\phi(0).$$

If we now differentiate (III.64) through with respect to t we obtain, for $0 \leqq t < T$

$$\text{(III.66)}\qquad \varepsilon\dot{\Phi}(t) = -\dot{\phi}(t)-\phi(0)\Phi(t)-\int_0^t \phi_t(t-\tau)\Phi(\tau)\,d\tau.$$

In conjunction with (III.65), (III.66) implies that

$$\text{(III.67)}\qquad \dot{\Phi}(0) = -\frac{1}{\varepsilon}\dot{\phi}(0)+\frac{1}{\varepsilon^2}\phi^2(0).$$

In order to establish (III.44) we rewrite (III.64) in the form

$$\text{(III.68)}\qquad \Phi(t) = -\frac{1}{\varepsilon}\phi(t)-\frac{1}{\varepsilon}\int_0^t \phi(t-\tau)\Phi(\tau)\,d\tau, \qquad 0 \leqq t < T,$$

from which it follows that for $0 \leqq t < T$

$$\text{(III.69)}\qquad \begin{aligned}|\Phi(t)| &\leqq \frac{1}{\varepsilon}|\phi(t)|+\frac{1}{\varepsilon}\sup_{[0,T)}|\phi(\tau)|\int_0^t |\Phi(\tau)|\,d\tau\\ &\leqq \frac{1}{\varepsilon}\sup_{[0,T)}|\phi(\tau)|(1+T\sup_{[0,T)}|\Phi(\tau)|).\end{aligned}$$

This last estimate clearly implies that

$$\text{(III.70)}\qquad \varepsilon\sup_{[0,T)}|\Phi(t)| \leqq \sup_{[0,T)}|\phi(t)|(1+T\sup_{[0,T)}|\Phi(t)|);$$

which under the hypotheses of the above lemma is easily seen to be equivalent to (III.44); the estimate (III.45) may be established in an analogous fashion.

In addition to the upper bounds given by the last lemma we also have a set of lower bounds which are delineated in the following:

LEMMA (Bloom [20]). *Under the conditions which prevail in the previous lemma,*

$$\text{(III.71)}\qquad \sup_{[0,T)}|\Phi(t)| \geqq \frac{\sup_{[0,T)}|\phi(t)|}{(\varepsilon+T\sup_{[0,T)}|\phi(t)|)}$$

and, if $\sup_{[0,T)} |\dot{\phi}(t)| \geqq |\phi(0)|^2/(\varepsilon - T|\phi(0)|)$, *then*

$$\sup_{[0,T)} |\dot{\phi}(t)| \geqq \frac{\sup_{[0,T)} |\dot{\phi}(t)|(\varepsilon - T|\phi(0)|) - |\phi(0)|^2}{2\varepsilon^2 + \varepsilon T^2 \sup_{[0,T)} |\dot{\phi}(t)|}. \tag{III.72}$$

Proof. By (III.64) we have

$$\phi(t) = -\varepsilon\Phi(t) - \int_0^t \phi(t-\tau)\Phi(\tau)\, d\tau, \qquad 0 \leqq t < T, \tag{III.73}$$

so that

$$\begin{aligned} |\phi(t)| &\leqq \varepsilon|\Phi(t)| + \sup_{[0,T)} |\phi(\tau)| \int_0^t |\Phi(\lambda)|\, d\lambda \\ &\leqq \varepsilon|\Phi(t)| + T \sup_{[0,T)} |\phi(\tau)| \sup_{[0,T)} |\Phi(\tau)| \\ &\leqq (\varepsilon + T \sup_{[0,T)} |\phi(\tau)|) \sup_{[0,T)} |\Phi(\tau)| \end{aligned} \tag{III.74}$$

for all t, $0 \leqq t < T$. The estimate (III.71) follows from (III.74$_3$). We now turn our attention to the estimate (III.72) under the one additional hypothesis on $\phi(t)$ which is stated above. By (III.66)

$$\dot{\phi}(t) = -\varepsilon\dot{\Phi}(t) - \phi(0)\Phi(t) - \int_0^t \phi_t(t-\tau)\Phi(\tau)\, d\tau \tag{III.75}$$

for $0 \leqq t < T$. Therefore,

$$|\dot{\phi}(t)| \leqq \varepsilon|\dot{\Phi}(t)| + (|\phi(0)| + T \sup_{[0,T)} |\dot{\phi}(\tau)|) \sup_{[0,T)} |\Phi(\tau)|. \tag{III.76}$$

However, $\Phi(t) = \int_0^t \dot{\Phi}(\tau)\, d\tau + \Phi(0)$, so

$$|\Phi(t)| \leqq T \sup_{[0,T)} |\dot{\Phi}(\tau)| + \frac{1}{\varepsilon}|\phi(0)|, \qquad 0 \leqq t < T, \tag{III.77}$$

where we have used (III.65). Taking the supremum over $[0, T)$ in (III.77) and using the resulting estimate in (III.76) yields

$$\begin{aligned} |\dot{\phi}(t)| &\leqq \varepsilon|\dot{\Phi}(t)| + (|\phi(0)| + T \sup_{[0,T)} |\dot{\phi}(\tau)|)\left(T \sup_{[0,T)} |\dot{\Phi}(\tau)| + \frac{1}{\varepsilon}|\phi(0)|\right) \\ &\leqq \sup_{[0,T)} |\dot{\Phi}(t)|(\varepsilon + T[|\phi(0)| + T \sup_{[0,T)} |\dot{\phi}(\tau)|]) \\ &\quad + \frac{1}{\varepsilon}|\phi(0)|(|\phi(0)| + T \sup_{[0,T)} |\dot{\phi}(\tau)|) \end{aligned} \tag{III.78}$$

for $0 \leqq t < T$. Taking the supremum over $[0, T)$ in (III.78), and rearranging terms, we get

(III.79) $$\sup_{[0,T)} |\dot{\Phi}(t)| \geqq \frac{\sup_{[0,T)} |\dot{\phi}(\tau)|(1-(T/\varepsilon)|\phi(0)|) - (1/\varepsilon)|\phi(0)|^2}{\varepsilon + T(|\phi(0)| + T \sup_{[0,T)} |\dot{\phi}(\tau)|)},$$

where the numerator on the right-hand side of the estimate is positive by virtue of the hypotheses of the lemma. Thus,

(III.80) $$\begin{aligned}\sup_{[0,T)} |\dot{\Phi}(t)| &\geqq \frac{\sup_{[0,T)} |\dot{\phi}(t)|(\varepsilon - T|\phi(0)|) - |\phi(0)|^2}{\varepsilon^2 + \varepsilon T|\phi(0)| + \varepsilon T^2 \sup_{[0,T)} |\dot{\phi}(\tau)|} \\ &\geqq \frac{\sup_{[0,T)} |\dot{\phi}(t)|(\varepsilon - T|\phi(0)|) - |\phi(0)|^2}{2\varepsilon^2 + \varepsilon T^2 \sup_{[0,T)} |\dot{\phi}(\tau)|}\end{aligned}$$

as $\varepsilon T|\phi(0)| \leqq \varepsilon T \sup_{[0,T)} |\phi(t)| \leqq \varepsilon T \cdot \varepsilon/T = \varepsilon^2$; this establishes the lower bound (III.72). Q.E.D.

The importance of the previous lemmas lies with the fact that they allow us to replace the terms $\Phi(0)$, $\sup_{[0,T)} |\Phi(t)|$, $\sup_{[0,T)} |\dot{\Phi}(t)|$, which appear in (III.41) and (III.42), with expressions which depend on the basic memory function in the constitutive theory that defines the Maxwell-Hopkinson dielectric, i.e., $\phi(t)$. In fact by (III.65), (III.41) is equivalent to

(III.81) $$\phi(0) \geqq \varepsilon\gamma T \sup_{[0,T)} |\dot{\Phi}(t)|$$

and, in view of (III.45), if $\sup_{[0,T)} |\phi(t)| < \varepsilon/T$ this estimate will certainly be satisfied if

(III.82) $$\phi(0) \geqq \varepsilon\gamma\alpha(T)\left(1 + T\frac{\sup_{[0,T)} |\dot{\phi}(t)|}{\sup_{[0,T)} |\phi(t)|}\right),$$

where $\alpha(t)$ is given by (III.46). If $\phi(t)$ satisfies $\sup_{[0,T)} |\phi(t)| < \varepsilon/T$ and (III.82), then $\mathbf{K}(t)$, as defined by (III.31) will satisfy the basic hypotheses (II.55). The stability and growth estimates which follow from the differential inequality (III.43), where $\hat{\mathscr{F}}(t)$ is given by (III.42), may then be applied to study the growth behavior of the norm of the electric displacement field in the Maxwell-Hopkinson dielectric which is governed by the system of partial-integrodifferential equations (III.7).

Example. Let $\phi(t) = ke^{-t}$ with $k > 0$. Then $\sup_{[0,T)} |\phi(t)| = \sup_{[0,T)} |\dot{\phi}(t)| = k$. Thus, $\sup_{[0,T)} |\phi(t)| < \varepsilon/T$ if and only if $k < \varepsilon/T$. If $k < \varepsilon/T$ then, by (III.46),

$\alpha(T)=k/(\varepsilon-kT)$ and (III.82) will be satisfied provided

$$k>\left(\frac{\varepsilon\gamma k}{\varepsilon-kT}\right)[1+T]\Leftrightarrow k<\frac{\varepsilon}{T}[1-\gamma(1+T)]. \tag{III.83}$$

Therefore, if $\gamma<1/(1+T)$ and k satisfies (III.83) (which automatically implies that $k<\varepsilon/T$) the inequality (II.82) will be satisfied for $\phi(t)=k\,e^{-t}$ and the corresponding $\mathbf{K}(t)$ will satisfy the hypotheses (II.55). Another way to view the condition expressed by (III.83) is as follows: Suppose that $\gamma<1, k>0$ are given; then (III.83), and, as a consequence the condition $k<\varepsilon/T$, will be satisfied if $T>0$ is chosen so that

$$T<\frac{\varepsilon(1-\gamma)}{(\gamma\varepsilon+k)}. \tag{III.84}$$

Remark. For $\phi(t)$ a decreasing exponential, say, $\phi(t)=e^{-t}$, the corresponding kernel $\Phi(t)$, which enters directly into the basic condition (III.41), may be computed in a straightforward manner by using (III.8). Suppose we denote $\Phi(t)$ corresponding to $\phi(t)=e^{-t}$ by $\Phi(\tau;e^{-t})$; then by the definition of $\phi^n(\tau)$, $n\geqq 2$, and the fact that $\phi^1(\tau)=(1/\varepsilon)e^{-\tau}$ we have

$$\phi^n(\tau)=\frac{1}{\varepsilon^{n-1}}e^{-\tau}\frac{\tau^{n-1}}{(n-1)!},\qquad n\geqq 2. \tag{III.85}$$

Therefore,

$$\begin{aligned}\Phi(\tau;e^{-t})&=\sum_{n=1}^{\infty}(-1)^n\phi^n(\tau)\\&=-\frac{1}{\varepsilon}e^{-\tau}+e^{-\tau}\sum_{n=2}^{\infty}(-1)^n\frac{1}{\varepsilon^{n-1}}\frac{\tau^{n-1}}{(n-1)!}\\&=-\frac{1}{\varepsilon}e^{-\tau}+e^{-\tau}\sum_{n=1}^{\infty}(-1)^{n+1}\frac{\tau^n}{\varepsilon^n n!}\\&=-e^{-\tau}\left(\frac{1}{\varepsilon}+\sum_{n=1}^{\infty}\frac{\left(\frac{-\tau}{\varepsilon}\right)^n}{n!}\right)\\&=-e^{-\tau}\left(\frac{1}{\varepsilon}+[e^{-\tau/\varepsilon}-1]\right).\end{aligned} \tag{III.86}$$

Since $k=1$ (see the example above), we have chosen $0\leqq\tau<T<\varepsilon$ so that $|\tau/\varepsilon|<1$ in the geometric series in (III.86). Our result can readily be put into the form

$$\Phi(\tau;e^{-t})=\left(1-\frac{1}{\varepsilon}\right)e^{-\tau}-e^{-\alpha\tau},\qquad \alpha=\frac{1+\varepsilon}{\varepsilon}. \tag{III.87}$$

Elementary analysis then establishes the following:

$$\text{(III.88)}\qquad \begin{aligned}&\text{(i)}\quad \Phi(0;e^{-t})=-\frac{1}{\varepsilon}<0,\qquad \dot{\Phi}(0;e^{-t})=\frac{2}{\varepsilon}>0,\\ &\text{(ii)}\quad \Phi(\tau;e^{-t})>0\Leftrightarrow e^{\tau/\varepsilon}>\varepsilon/(\varepsilon-1),\\ &\text{(iii)}\quad \dot{\Phi}(\tau;e^{-t})>0\Leftrightarrow \frac{\varepsilon+1}{\varepsilon-1}>e^{\tau/\varepsilon}.\end{aligned}$$

In other words $\Phi(\tau; e^{-t})$ is initially negative and increasing and continues to increase on the interval $[0, \varepsilon\cdot\ln[(\varepsilon+1)/(\varepsilon-1)])$ becoming positive for $\tau>\varepsilon\ln[\varepsilon/(\varepsilon-1)]$. It is demonstrated in [20] that for $T<\varepsilon\ln[(\varepsilon+1)/(\varepsilon-1)]$,

$$\text{(III.89)}\qquad \sup_{[0,T)}|\Phi(\tau;e^{-t})|=\frac{1}{\varepsilon}.$$

Using (III.87) one may compute, directly, the consequences of the condition (III.41); we will not, however, pursue that direction here for the exponentially decreasing memory function considered above as it is simpler in this case to work with the consequences of (III.82) (e.g., the above example) which, in turn, implies that (III.41) is satisfied.

With $\phi(t)$ chosen so as to satisfy $\sup_{[0,T)}|\phi(t)|<\varepsilon/T$, and (III.82), it follows that $\mathbf{K}(t)$ as defined by (III.31) will satisfy the hypothesis (II.55) of Chapter II. In view of our assumption of zero past history for the electric field, the differential inequality (III.43) will then be applicable where $\hat{\mathscr{F}}$ is given by (III.42) and

$$\text{(III.90)}\qquad F(t;\beta,t_0)=\int_\Omega D_i(\mathbf{x},t)D_i(\mathbf{x},t)\,d\mathbf{x}+\beta(t+t_0)^2.$$

In view of (III.31)

$$\text{(III.91)}\qquad \mathscr{E}(0)=\frac{1}{2}\int_\Omega g_i(\mathbf{x})g_i(\mathbf{x})\,d\mathbf{x}+\frac{1}{2\varepsilon\mu}\int_\Omega \frac{\partial f_i}{\partial x_j}(\mathbf{x})\frac{\partial f_i}{\partial x_j}(\mathbf{x})\,d\mathbf{x}>0,$$

so that $\hat{\mathscr{F}}(0)>0$. If

$$\text{(III.92)}\qquad \int_\Omega f_i(\mathbf{x})g_i(\mathbf{x})\,d\mathbf{x}\geqq(2\hat{\mathscr{F}}(0))^{1/2}\int_\Omega f_i(\mathbf{x})f_i(\mathbf{x})\,d\mathbf{x},$$

then the estimates of case III, Chapter II, will apply with $\beta=0$; i.e., with the strict inequality in (III.92) we have the growth estimate

$$\text{(III.93)}\qquad \begin{aligned}\int_\Omega D_i(\mathbf{x},t)D_i(\mathbf{x},t)\,d\mathbf{x}\geqq&\left[\int_\Omega f_i(\mathbf{x})f_i(\mathbf{x})\,d\mathbf{x}+\frac{4\hat{\mathscr{F}}(0)}{\hat{\Gamma}^2}\right]\cosh\hat{\Gamma}t\\ &+\left[\frac{2\int_\Omega f_i(\mathbf{x})g_i(\mathbf{x})\,d\mathbf{x}}{\hat{\Gamma}}\right]\sinh\hat{\Gamma}t-\frac{4\hat{\mathscr{F}}(0)}{\hat{\Gamma}^2}\end{aligned}$$

for $0 \leqq t < T$, where

$$\hat{\Gamma}^2 = 4\left[\frac{\int_\Omega f_i(\mathbf{x})g_i(\mathbf{x})\,d\mathbf{x}}{\int_\Omega f_i(\mathbf{x})f_i(\mathbf{x})\,d\mathbf{x}}\right]^2 - \frac{8\hat{\mathcal{F}}(0)}{\int_\Omega f_i(\mathbf{x})f_i(\mathbf{x})\,d\mathbf{x}} > 0, \tag{III.94}$$

while for the equality sign in (III.92) we have

$$\begin{aligned}&\int_\Omega D_i(\mathbf{x}, t)D_i(\mathbf{x}, t)\,d\mathbf{x}\\ &\quad \geqq \int_\Omega f_i(\mathbf{x})f_i(\mathbf{x})\,d\mathbf{x} + 2^{3/2}\left(\int_\Omega f_i(\mathbf{x})f_i(\mathbf{x})\,d\mathbf{x}\right)^{1/2}(\hat{\mathcal{F}}(0))^{1/2}t + 2\hat{\mathcal{F}}(0)t^2\end{aligned} \tag{III.95}$$

on $[0, T)$. Other growth and stability estimates follow directly from our work in Chapter II, but we will not pursue this matter further here as the carryover from Chapter II to the present situation is now fairly straightforward; in particular, as $\mathscr{E}(0) > 0$ the exponential growth estimate (II.167) can also be applied to bound $\int_\Omega D_iD_i\,d\mathbf{x}$ from below whenever the memory function $\phi(t)$, in the Maxwell-Hopkinson theory, satisfies $\sup_{[0,T)} |\phi(t)| < \varepsilon/T$ and the estimate (III.82).

Remarks. Once we have established stability and growth estimates for the electric displacement field **D**, it is a fairly simple matter to use those estimates, in conjunction with the Maxwell-Hopkinson constitutive equations, to derive the corresponding stability and growth estimates for the electric field **E**; for a derivation of such estimates for the electric field in a Maxwell-Hopkinson dielectric we refer the reader to §§ 4, 5 of [20]. We note in passing that it is shown in [20] that the electric field **E**, in a Maxwell-Hopkinson dielectric, satisfies the partial-integrodifferential equation

$$\begin{aligned}&\frac{\partial^2}{\partial t^2}\mathbf{E}(\mathbf{x}, t) - \frac{1}{\varepsilon}\left(\frac{1}{\mu}\mathbf{\Delta} + \mathbf{I}\right)\mathbf{E}(\mathbf{x}, t)\\ &\qquad + \frac{1}{\varepsilon}\int_0^t \phi_{tt}(t-\tau)\mathbf{E}(\mathbf{x}, \tau)\,d\tau = -\frac{\phi(0)}{\varepsilon}\frac{\partial}{\partial t}\mathbf{E}(\mathbf{x}, t),\end{aligned} \tag{III.96}$$

provided $\mathbf{E}(\cdot, \tau) = \mathbf{0}$ on Ω for $-\infty < \tau < 0$. It is easy to see that (III.96) can be put in the form (II.23), with $\mathbf{U}(\tau) = \mathbf{0}$ for $\tau < 0$, via the introduction of suitably defined operators $\mathbf{N} \in \mathscr{L}_s(H_+, H_-)$, $\mathbf{K}(\cdot) \in L^2((-\infty, \infty); \mathscr{L}_s(H_+, H_-))$, if and only if $\phi(0) = 0$; for many physically reasonable $\phi(\cdot)$ associated with Maxwell-Hopkinson dielectrics, however, the assumption that $\phi(0) = 0$ is not a valid one (e.g., the linear combination of decreasing exponentials considered by Hopkinson in [68]). Formally, the system of partial-integrodifferential equations defined by (III.96) is of the same basic form as the system (III.29), for the components of the electric displacement field in a holohedral isotropic material dielectric, when $\tau_\infty = 0$; considered as abstract integrodifferential equations in Hilbert space, in fact, these two systems will assume the same form with, of course, different definitions for the basic operators $\mathbf{N}$, $\mathbf{K}(t)$. The reformulation of the initial-history boundary value problems for holohedral isotropic dielectrics as initial-history value problems in Hilbert space, and the subsequent derivation of stability and growth estimates for electric displacement fields in such materials, by means of a logarithmic convexity argument, is the subject of the next section.

3. Stability and growth estimates for holohedral isotropic dielectrics: Logarithmic convexity arguments.

We consider in this section the isotropic holohedral material dielectric defined by the constitutive relations (III.21a, b) on $\Omega\times[0, T)$; for our work in this section we will assume that $\mathbf{E}(\cdot, \tau)=\mathbf{B}(\cdot, \tau)=\mathbf{0}$, $-\infty<\tau<0$, on Ω, so that, by the Lemma of the previous section, the governing system of partial-integrodifferential equations for the components of the electric displacement field in $\Omega\times[0, T)$ is given by (III.22). We again assume initial and boundary data of the form (III.15), (III.16).

Remark. In the following section we will assume that the past histories of $\mathbf{E}, \mathbf{B}$ are of the form (III.28) so that the governing system of evolution equations will be (III.29), provided the past history $\mathbf{D}_{\tau_\infty}(\cdot, -\tau_\infty)=\mathbf{0}$, uniformly on Ω. The problem of deriving growth estimates for solutions of initial-boundary value problems associated with the system (III.29) will be approached in § 4 from a different viewpoint and via a concavity argument.

We now want to recast the initial-boundary value problem (III.22), (III.15), (III.16) as an initial-value problem for an integrodifferential equation in Hilbert space; to this end we again employ the spaces $H_+=(H_0^1(\Omega))^3$, $H=(L_2(\Omega))^3$, and $H_-=(H^{-1}(\Omega))^3$ and define operators $\mathbf{N}\in\mathscr{L}_s(H_+, H_-)$, $\mathbf{K}(\cdot)\in L^2((-\infty, \infty); \mathscr{L}_s(H_+, H_-))$ via

$$(\text{III.97})\qquad (\mathbf{N}\mathbf{v})_i(\mathbf{x})\equiv b_0\dot{\Psi}(0)[c_0^2 v_i(\mathbf{x})-v_i(\mathbf{x})],\qquad \mathbf{x}\in\Omega,\qquad \mathbf{v}\in H_+,$$

and

$$(\text{III.98})\qquad (\mathbf{K}(t)\mathbf{v})_i(\mathbf{x})\equiv b_0[\ddot{\Psi}(t)\mathbf{v}_i(\mathbf{x})-\Phi_0(t)\nabla^2 v_i(\mathbf{x})],\qquad \mathbf{v}\in H_+,$$

for $(\mathbf{x}, t)\in\Omega\times(-\infty, \infty)$. With these definitions of $\mathbf{N}, \mathbf{K}(\cdot)$, the problem of finding a solution of the initial-boundary value problem (III.22), (III.15), (III.16) becomes formally equivalent to finding a solution $\mathbf{u}\in C^2([0, T); H_+)$ of the initial-value problem

$$(\text{III.99})\qquad \mathbf{u}_{tt}-\alpha\mathbf{u}_t-\mathbf{N}\mathbf{u}+\int_0^t \mathbf{K}(t-\tau)\mathbf{u}(\tau)\,d\tau=\zeta(t)\mathbf{f},$$

for $0\leqq t<T$, with

$$(\text{III.100})\qquad \mathbf{u}(0)=\mathbf{f},\quad \mathbf{u}_t(0)=\mathbf{g},\quad \mathbf{f}, \mathbf{g}\in H_+,$$

where $\alpha = -\Psi(0)$, $\zeta(t) = b_0\dot{\Psi}(t)$, $0 \leqq t < T$; if the initial value $\mathbf{f} = \mathbf{0}$ then (III.99) reduces to a homogeneous integrodifferential equation and if $\Psi(0) > 0$ the equation possesses strong damping.

For the initial-value problem (III.99), (III.100), stability and growth estimates for $\mathbf{u}(t)$ in the $\|(\cdot)\|$-norm may be obtained via a logarithmic convexity argument analogous to the one employed in Chapter II. In fact, we have the following:

THEOREM (Bloom [14]). *Let $\mathbf{u} \in \mathcal{N}$ be any solution of* (III.100) *and suppose that $\mathbf{K}(\cdot)$, as defined by* (III.98), *satisfies the hypotheses* (II.55). *Then $\exists \mu > 0$ such that for all t, $0 \leqq t < T$, the real-valued function*

$$F(t; \beta, t_0) = \|\mathbf{u}(t)\|^2 + \beta(t + t_0)^2, \qquad \beta, t_0 \geqq 0,$$

satisfies the differential inequality

$$\begin{aligned} FF'' - F'^2 \geqq& -2F(2\mathscr{E}(0) + \mu) + \alpha FF' \\ & -2\alpha F\left(\beta(t + t_0) + 4\int_0^t K(t)\, d\tau\right) \\ & + 2F\left(2\int_0^t \dot{\zeta}(\tau)\langle \mathbf{u}, \mathbf{f}\rangle\, d\tau - \zeta(t)\langle \mathbf{u}, \mathbf{f}\rangle\right) \\ & + 4F\zeta(0)\|\mathbf{f}\|^2, \end{aligned} \tag{III.101}$$

where $\mathscr{K}(t)$ denotes the kinetic energy $\frac{1}{2}\|\mathbf{u}_t\|^2$.

Proof. From the definition of $F(t; \beta, t_0)$ and the evolution equation (III.99) we have

$$F'(t; \beta, t_0) = 2\langle \mathbf{u}, \mathbf{u}_t\rangle + \beta(t + t_0),$$

$$\begin{aligned} F''(t; \beta, t_0) =& 2\|\mathbf{u}_t\|^2 + 2\alpha\langle \mathbf{u}, \mathbf{u}_t\rangle + 2\langle \mathbf{u}, \mathbf{N}\mathbf{u}\rangle \\ & -2\left\langle \mathbf{u}, \int_0^t \mathbf{K}(t-\tau)\mathbf{u}(\tau)\, d\tau\right\rangle + 2\zeta(t)\langle \mathbf{u}, \mathbf{f}\rangle + 2\beta. \end{aligned} \tag{III.102}$$

By employing the definitions of the total and kinetic energies $\mathscr{E}(t)$ and $\mathscr{K}(t)$, respectively, we may rewrite (III.102) in the form

$$\begin{aligned} F''(t; \beta, t_0) =& 2\alpha\langle \mathbf{u}, \mathbf{u}_t\rangle + 2\zeta(t)\langle \mathbf{u}, \mathbf{f}\rangle \\ & -2\left\langle \mathbf{u}, \int_0^t \mathbf{K}(t-\tau)\mathbf{u}(\tau)\, d\tau\right\rangle \\ & + 4(2\mathscr{K}(t) + \beta) - 2(2\mathscr{E}(0) + \beta) - 4(\mathscr{E}(t) - \mathscr{E}(0)). \end{aligned} \tag{III.103}$$

However, for any τ, $0 \leqq \tau < T$

$$\begin{aligned}\mathscr{E}'(\tau) &= \langle \mathbf{u}_\tau, \mathbf{u}_{\tau\tau} \rangle - \langle \mathbf{u}_\tau, \mathbf{N}\mathbf{u} \rangle \\ &= \alpha \|\mathbf{u}_\tau\|^2 + \zeta(\tau)\langle \mathbf{u}_\tau, \mathbf{f} \rangle \\ &\quad - \left\langle \mathbf{u}_\tau, \int_0^\tau \mathbf{K}(\tau-\sigma)\mathbf{u}(\sigma)\, d\sigma \right\rangle,\end{aligned} \tag{III.104}$$

and thus

$$\begin{aligned}\mathscr{E}'(\tau) &= 2\alpha \mathscr{K}(\tau) + \zeta(\tau)\langle \mathbf{u}_\tau, \mathbf{f} \rangle \\ &\quad - \frac{d}{d\tau}\left\langle \mathbf{u}(\tau), \int_0^t \mathbf{K}(\tau-\sigma)\mathbf{u}(\sigma)\, d\sigma \right\rangle \\ &\quad + \left\langle \mathbf{u}(\tau), \int_0^\tau \mathbf{K}_\tau(\tau-\sigma)\mathbf{u}(\sigma)\, d\sigma \right\rangle + \langle \mathbf{u}(\tau), \mathbf{K}(0)\mathbf{u}(\tau) \rangle.\end{aligned} \tag{III.105}$$

By integrating this last result over $[0, T)$ and then substituting for $\mathscr{E}(t) - \mathscr{E}(0)$ in (III.103) we obtain

$$\begin{aligned}F''(t; \beta, t_0) &= 2\alpha \langle \mathbf{u}, \mathbf{u}_t \rangle + 2\zeta(t)\langle \mathbf{u}, \mathbf{f} \rangle \\ &\quad + 2\left\langle \mathbf{u}, \int_0^t \mathbf{K}(t-\tau)\mathbf{u}(\tau)\, d\tau \right\rangle \\ &\quad + 2(2\mathscr{E}(t) + \beta) - 2(2\mathscr{E}(0) + \beta) - 8\alpha \int_0^t \mathscr{K}(\tau)\, d\tau \\ &\quad - 4\int_0^t \zeta(\tau)\langle \mathbf{u}_\tau, \mathbf{f} \rangle\, d\tau - 4\int_0^t \langle \mathbf{u}(\tau), \mathbf{K}(0)\mathbf{u}(\tau) \rangle\, d\tau \\ &\quad - 4\int_0^t \left\langle \mathbf{u}(\tau), \int_0^\tau \mathbf{K}_\tau(\tau-\sigma)\mathbf{u}(\sigma)\, d\sigma \right\rangle d\tau.\end{aligned} \tag{III.106}$$

By combining (III.105) with (III.102), and making use of the Schwarz inequality, we obtain, in a strightforward manner, the differential inequality

$$\begin{aligned}FF'' - F'^2 &\geqq -2F(2\mathscr{E}(0) + \beta) + \alpha F\left(\frac{d}{dt}\|\mathbf{u}\|^2 - 8\int_0^t \mathscr{K}(\tau)\, d\tau\right) \\ &\quad + 2F\left(2\int_0^t \dot{\zeta}(\tau)\langle \mathbf{u}, \mathbf{f} \rangle\, d\tau - \zeta(t)\langle \mathbf{u}, \mathbf{f} \rangle\right) \\ &\quad + 4F\zeta(0)\|\mathbf{f}\|^2 + 2F\left\langle \mathbf{u}, \int_0^t \mathbf{K}(t-\tau)\mathbf{u}(\tau)\, d\tau \right\rangle \\ &\quad - 4F\int_0^t \left\langle \mathbf{u}(\tau), \int_0^\tau \mathbf{K}_\tau(\tau-\sigma)\mathbf{u}(\sigma)\, d\sigma \right\rangle d\tau \\ &\quad - 4F\int_0^t \langle \mathbf{u}(\tau), \mathbf{K}(0)\mathbf{u}(\tau) \rangle\, d\tau,\end{aligned} \tag{III.107}$$

or, as $(d/dt)\|\mathbf{u}\|^2 = F'(t;\beta,t_0) - 2\beta(t+t_0)$,

$$
\begin{aligned}
FF''-F'^2 &\geqq -2F(2\mathscr{E}(0)+\beta) \geqq \alpha FF' \\
&\quad -2\alpha F\Big(\beta(t+t_0)+4\int_0^t \mathscr{K}(\tau)\,d\tau\Big) \\
&\quad +2F\Big(2\int_0^t \dot{\zeta}(\tau)\langle \mathbf{u},\mathbf{f}\rangle\,d\tau - \zeta(t)\langle \mathbf{u},\mathbf{f}\rangle\Big) \\
&\quad +4F\zeta(0)\|\mathbf{f}\|^2 + 2F\Big\langle \mathbf{u}, \int_0^t \mathbf{K}(t-\tau)\mathbf{u}(\tau)\,d\tau\Big\rangle \\
&\quad -4F\int_0^t \Big\langle \mathbf{u}(\tau), \int_0^t \mathbf{K}_\tau(\tau-\sigma)\mathbf{u}(\sigma)\,d\sigma\Big\rangle\,d\tau \\
&\quad -4F\int_0^t \langle \mathbf{u}(\tau), \mathbf{K}(0)\mathbf{u}(\tau)\rangle\,d\tau.
\end{aligned}
\tag{III.108}
$$

By using a procedure completely analogous to that employed in Chapter II it is easy to see that, in view of our hypothesis that $\mathbf{u}\in\mathcal{N}$, and (II.55), we may bound the last three expressions in (III.108) from below by

$$2F\Big\langle \mathbf{u}, \int_0^t \mathbf{K}(t-\tau)\mathbf{u}(\tau)\,d\tau\Big\rangle \geqq -2\gamma N^2 T \sup_{[0,T)} \|\mathbf{K}(t)\|_{\mathscr{L}_s(H_+,H_-)}F, \tag{III.109}$$

$$-4F\int_0^t \langle \mathbf{u}(\tau), \mathbf{K}(0)\mathbf{u}(\tau)\rangle\,d\tau \geqq 4\kappa F\int_0^t \|\mathbf{u}(\tau)\|_+^2\,d\tau, \tag{III.110}$$

and

$$
\begin{aligned}
&-4F\int_0^t \Big\langle \mathbf{u}(\tau), \int_0^T \mathbf{K}_\tau(\tau-\sigma)\mathbf{u}(\sigma)\,d\sigma\Big\rangle\,d\tau \\
&\qquad \geqq -4\gamma T \sup_{[0,T)} \|\mathbf{K}_t\|_{\mathscr{L}_s(H_+,H_-)} F\int_0^t \|\mathbf{u}(\tau)\|_+^2\,d\tau,
\end{aligned}
\tag{III.111}
$$

respectively. By combining the differential inequality (III.108) with the estimates (III.109)–(III.110) we obtain (III.101) with

$$\mu \equiv \beta + \gamma N^2 T \sup_{[0,T)} \|\mathbf{K}(t)\|_{\mathscr{L}_s(H_+,H_-)}. \tag{III.112}$$

Q.E.D.

Before proceeding to delineate some of the stability and growth estimates which result fom the differential inequality (III.101) in various special cases, we want to examine the consequences of requiring that the fundamental hypotheses (II.55) hold for the operator $\mathbf{K}(t)$ as defined by (III.98). Directly from (III.98) we have for any $\mathbf{v} \in H_+$

$$
\begin{aligned}
\langle \mathbf{v}, \mathbf{K}(0)\mathbf{v}\rangle &= \int_\Omega \mathbf{K}(0)\mathbf{v})_i(\mathbf{x}) v_i(\mathbf{x})\, d\mathbf{x} \\
&= b_0\ddot{\Psi}(0) \int_\Omega v_i(\mathbf{x}) v_i(\mathbf{x})\, d\mathbf{x} \\
&\quad - \frac{b_0}{a_0}\Phi(0) \int_\Omega v_i(\mathbf{x}) \nabla^2 v_i(\mathbf{x})\, d\mathbf{x} \\
&= b_0\ddot{\Psi}(0)\|\mathbf{v}\|^2 + \frac{b_0}{a_0}\Phi(0)\|\mathbf{v}\|_+^2,
\end{aligned} \tag{III.113}
$$

so that the first condition in (II.55) will be satisfied if and only if

$$
-b_0\ddot{\Psi}(0)\|\mathbf{v}\|^2 - \frac{b_0}{a_0}\Phi(0)\|\mathbf{v}\|_+^2 \geqq \kappa\|\mathbf{v}\|_+^2. \tag{III.114}
$$

However, $\|\mathbf{v}\|^2 \leqq \gamma^2\|\mathbf{v}\|_+^2$ and therefore

$$
\begin{aligned}
-b_0\ddot{\Psi}(0)\|\mathbf{v}\|^2 - \frac{b_0}{a_0}\Phi(0)\|\mathbf{v}\|_+^2 &\geqq -b_0|\ddot{\Psi}(0)|\,\|\mathbf{v}\|^2 - \frac{b_0}{a_0}\Phi(0)\|\mathbf{v}\|_+^2 \\
&\geqq -b_0(\gamma^2|\ddot{\Psi}(0)| + \frac{1}{a_0}\Phi(0))\|\mathbf{v}\|_+^2,
\end{aligned} \tag{III.115}
$$

which implies that (III.114) will be satisfied if

$$
-b_0\left(\gamma^2|\ddot{\Psi}(0)| + \frac{1}{a_0}\Phi(0)\right) \geqq \kappa. \tag{III.116}
$$

The second condition in (II.55) requires that

$$
\kappa \geqq \gamma T \sup_{[0,T)} \|\mathbf{K}_t\|_{\mathscr{L}_s(H_+,H_-)}.
$$

However, for any $\mathbf{v} \in H_+$, it follows from (III.98) that

$$\|\mathbf{K}_t\mathbf{v}\|_- = \sup_{\mathbf{w}\in H_+} \left(\frac{|\langle \mathbf{K}_t\mathbf{v}, \mathbf{w}\rangle|}{\|\mathbf{w}\|_+}\right)$$

$$= \sup_{\mathbf{w}\in H_+} \left(\frac{1}{\|\mathbf{w}\|_+}\left|\int_\Omega (\mathbf{K}_t\mathbf{v})_i(\mathbf{x}) w_i(\mathbf{x})\, d\mathbf{x}\right|\right)$$

$$= \sup_{\mathbf{w}\in H_+} \frac{1}{\|\mathbf{w}\|_+}\left(\left|\int_\Omega b_0\Psi^{(3)}(t) v_i(\mathbf{x}) w_i(\mathbf{x})\, d\mathbf{x}\right.\right.$$

$$\left.\left. - \int_\Omega \left(\frac{b_0}{a_0}\right)\dot{\Phi}(t)\nabla^2 v_i(\mathbf{x}) w_i(\mathbf{x})\, d\mathbf{x}\right|\right)$$

$$\text{(III.117)} \qquad = \sup_{\mathbf{w}\in H_+} \frac{1}{\|\mathbf{w}\|_+}\left(\left| b_0\Psi^{(3)}(t)\int_\Omega v_i(\mathbf{x}) w_i(\mathbf{x})\, d\mathbf{x}\right.\right.$$

$$\left.\left. + \left(\frac{b_0}{a_0}\right)\dot{\Phi}(t)\int_\Omega \frac{\partial v_i}{\partial x_j}(\mathbf{x})\frac{\partial w_i}{\partial x_j}(\mathbf{x})\, d\mathbf{x}\right|\right)$$

$$\leqq \sup_{\mathbf{w}\in H_+} \frac{1}{\|\mathbf{w}\|_+}\left(b_0|\Psi^{(3)}(t)|\left|\int_\Omega v_i(\mathbf{x}) w_i(\mathbf{x})\, d\mathbf{x}\right|\right.$$

$$\left. + \left(\frac{b_0}{a_0}\right)|\dot{\Phi}(t)|\left|\int_\Omega \frac{\partial v_i}{\partial x_j}(\mathbf{x})\frac{\partial w_i}{\partial x_j}(\mathbf{x})\, d\mathbf{x}\right|\right)$$

$$\leqq \sup_{\mathbf{w}\in H_+} \frac{1}{\|\mathbf{w}\|_+}\left(b_0|\Psi^{(3)}(t)|\gamma^2 + \left(\frac{b_0}{a_0}\right)|\dot{\Phi}(t)|\right)\|\mathbf{v}\|_+\|\mathbf{w}\|_+$$

$$= \left(b_0\gamma^2|\Psi^{(3)}(t)| + \left(\frac{b_0}{a_0}\right)|\dot{\Phi}(t)|\right)\|\mathbf{v}\|_+,$$

and therefore

$$\text{(III.118)} \qquad \begin{aligned}\|\mathbf{K}_t\|_{\mathscr{L}_s(H_+,H_-)} &= \sup_{\mathbf{v}\in H_+}\frac{\|\mathbf{K}_t\mathbf{v}\|_-}{\|\mathbf{v}\|_+}\\ &\leqq b_0\left[\gamma^2|\Psi^{(3)}(t)| + \frac{1}{a_0}|\dot{\Phi}(t)|\right].\end{aligned}$$

It follows that the second condition in (II.55) will be satisfied if

$$\text{(III.119)} \qquad \kappa \geqq \gamma b_0 T\left(\gamma^2 \sup_{[0,T)} |\Psi^{(3)}(t)| + \frac{1}{a_0}\sup_{[0,T)}|\dot{\Phi}(t)|\right).$$

Combining (III.116), (III.119) gives that $\mathbf{K}(t)$, as defined by (III.98), will satisfy (II.55) if

$$\text{(III.120)} \qquad -\left(\gamma^2|\ddot{\Psi}(0)| + \frac{1}{a_0}\Phi(0)\right) \geqq \gamma T\left(\gamma^2 \sup_{[0,T)} |\Psi^{(3)}(t)| + \frac{1}{a_0}\sup_{[0,T)}|\dot{\Phi}(t)|\right).$$

It is easy to see that (III.120) will be satisfied if and only if $\Phi(0)<0$ with $|\Phi(0)|$ sufficiently large, specifically, with

(III.121) $$|\Phi(0)| \geqq a_0\gamma\Big(T\Big(\gamma^2 \sup_{[0,T)} |\Psi^{(3)}(t)| + \frac{1}{a_0} \sup_{[0,T)} |\dot{\Phi}(t)|\Big) + \gamma|\ddot{\Psi}(0)|\Big).$$

Remark. By using the definitions of the kernel functions $\Phi(t)$, $\Psi(t)$, respectively, in terms of the basic memory functions $\phi(t)$, $\psi(t)$ which serve to define the response of the holohedral isotropic material dielectric governed by (III.21a, b), it is demonstrated in [14] that (III.121), and hence the hypotheses (II.55) relative to the operator $\mathbf{K}(t)$ given by (III.98), will be satisfied if

(III.122) $$\frac{1}{a_0^2}\phi(0) - \frac{\gamma^2}{b_0}\left|\frac{1}{b_0^2}\psi^3(0) - \frac{2}{b_0}\psi(0)\dot{\psi}(0) + \ddot{\psi}(0)\right| \geqq \mathscr{D},$$

where

$$\mathscr{D} = \mathscr{D}(\gamma, T, a_0, b_0, |\psi^{(i)}(0)|, \sup_{[0,T)} |\phi^{(j)}(t)|, \sup_{[0,T)} |\psi^{(k)}(t)|),$$

for $i=0,1,2,\ j=0,1$, and $k=0,1,2,3$, is computable. The condition represented by (III.122) requires that $\phi(0)>0$ be sufficiently large (as far as the satisfaction of (III.21) is concerned, the only troublesome term which appears in $\mathscr{D}$, namely, $\sup_{[0,T)}|\phi(t)|$ occurs solely within the denominator of a rational expression). In fact, it is easily shown that (III.122) will be satisfied for $\phi(t)$ of the form

(III.123) $$\phi(t) = k\exp(-t/K), \qquad k>0, \quad K>0,$$

for k, K both sufficiently large, if $\psi(t)$ is independent of k. For further details we refer the reader to § 4 of [14].

We now proceed to delineate the stability and growth estimates which follow from the differential inequality (III.101) in two special cases.

Case 1. $\mathbf{f}=\mathbf{0}$ and $\alpha=-\Psi(0)<0$.

In this case $\mathscr{E}(0)=\frac{1}{2}\|\mathbf{g}\|^2$ and the expression $-2\alpha F(\beta(t+t_0)+4\int_0^t \mathscr{K}(\tau)\,d\tau)$, on the right-hand side of (III.101) is nonnegative for all β, $t_0 \geqq 0$. Thus, (III.101) implies that

(III.124) $$FF'' - F'^2 \geqq -2F(\|\mathbf{g}\|^2 + \mu) - |\alpha|FF', \qquad 0 \leqq t < T,$$

where μ is given by (III.112). By a computation analogous to the one which led to (III.18), and (III.112)

(III.125) $$\mu \leqq \beta + \gamma N^2 T\Big(b_0\gamma^2 \sup_{[0,T)} |\ddot{\Psi}(t)| + \Big(\frac{b_0}{a_0}\Big) \sup_{[0,T)} |\Phi(t)|\Big).$$

For arbitrary β, t_0 we have for any $\lambda>0$

(III.126) $$\lambda\beta t_0^2 \leqq \lambda\|\mathbf{u}(t)\|^2 + \lambda\beta(t+t_0)^2 = \lambda F(t;\beta,t_0).$$

If, in particular, we choose

(III.127) $$\lambda = \lambda(\beta; t_0) = \frac{2(\|\mathbf{g}\|^2 + \mu)}{\beta t_0^2}$$

then for all t, $0 \leqq t < T$, and all β, $t_0 \geqq 0$

(III.128) $$2(\|\mathbf{g}\|^2 + \mu) \leqq \lambda(\beta; t_0) F(t; \beta, t_0)$$

and (III.124) then implies that

(III.129) $$FF'' - F'^2 \geqq -\lambda(\beta; t_0) F^2 - |\alpha| FF', \qquad 0 \leqq t < T.$$

However, the differential inequality (III.129) is equivalent to the statement that

(III.130) $$\frac{d^2}{d\sigma^2} \ln\left(\hat{F}(\sigma; \beta, t_0)_0 \sigma^{-\lambda(\beta, t_0)/\alpha^2}\right) \geqq 0, \qquad \sigma_1 < \sigma < \sigma_2,$$

where $\sigma = \exp(-|\alpha| t)$, $\sigma_j = \exp(-|\alpha| t_j)$, $j = 1, 2$, where $[t_1, t_2] \subseteq [0, T)$ is any interval such that $F(t; \beta, t_0) > 0$ for $t_1 < t < t_2$, and

(III.131) $$\hat{F}(\sigma; \beta, t_0) = F\left(\frac{1}{|\alpha|} \ln \sigma^{-1}; \beta, t_0\right).$$

If we integrate (III.130) over $[\sigma_1, \sigma_2]$, and then substitute for σ in terms of t, we find the estimate

(III.132) $$F(t; \beta, t_0) \leqq e^{-\lambda t/|\alpha|} (F(t_1; \beta, t_0)\, e^{\lambda t_1/|\alpha|})^{\delta(t)} (F(t_2; \beta, t_0)\, e^{\lambda t_2 |\alpha|})^{1-\delta(t)}$$

valid for $t_1 < t < t_2$, where

(III.133) $$\delta(t) = \frac{(e^{-|\alpha| t} - e^{-|\alpha| t_2})}{(e^{-|\alpha| t_1} - e^{-|\alpha| t_2})}, \qquad t_1 < t < t_2.$$

In particular, if $F(t; \beta, t_0) > 0$ on $[0, T)$ then (II.132) yields

(III.134) $$F(t; \beta, t_0) \leqq e^{-\lambda t/|\alpha|} (\beta t_0^2)^{\bar{\delta}(t)} (F(T; \beta, t_0)^{\lambda T/|\alpha|})^{1-\bar{\delta}(t)},$$

$0 \leqq t < T$, where $\bar{\delta}(t) = (e^{-|\alpha| t} - e^{-|\alpha| T})/(1 - e^{-|\alpha| T})$, and we have used the hypothesis that $\mathbf{f} = \mathbf{0}$. We now choose $\beta = 1/t_0^2$ and then take the limit in (III.134) as $t_0 \to +\infty$. As

(III.135)
$$F\left(t; \frac{1}{t_0^2}, t_0\right) = \|\mathbf{u}(t)\|^2 + \left(\frac{t}{t_0} + 1\right)^2,$$
$$\lim_{t_0 \to \infty} F\left(t; \frac{1}{t_0^2}, t_0\right) = \|\mathbf{u}(t)\|^2 + 1, \qquad 0 \leqq t < T.$$

Also, in view of our hypothesis that $\mathbf{u} \in \mathcal{N}$

(III.136) $$\lim_{t_0 \to \infty} F\left(T; \frac{1}{t_0^2}, t_0\right) \leqq \gamma^2 N^2 + 1,$$

and finally, by virtue of (III.127) and (III.112),

(III.137) $$\lim_{t_0\to\infty} \lambda\left(\frac{1}{t_0^2}; t_0\right) = 2(\|\mathbf{g}\|^2 + \gamma N^2 T \sup_{[0,T)} \|\mathbf{K}\|_{\mathscr{L}_s(H_+,H_-)}) \equiv \bar{\lambda}.$$

Therefore, as $t_0 \to \infty$ in (III.134), with $\beta = 1/t_0^2$, we find that

(III.138) $$\|\mathbf{u}(t)\|^2 \leqq e^{-\bar{\lambda}t/|\alpha|}((\gamma^2 N^2+1)\, e^{\bar{\lambda}T/|\alpha|})^{1-\bar{\delta}(t)}, \qquad 0 \leqq t < T,$$

or, if we choose $M > 0$ so large that

$$\gamma^2 N^2 + 1 < M \exp\left(\frac{-\bar{\lambda}}{|\alpha|}\right) T,$$

(III.139) $$\|\mathbf{u}(t)\|^2 \leqq M^{1-\bar{\delta}(t)}\, e^{-\bar{\lambda}t/|\alpha|}, \qquad 0 \leqq t < T,$$

so that $\|\mathbf{u}\|^2$ is bounded from above by an exponentially decreasing function of t on $[0, T)$. Lower bounds in the present situation may be obtained as follows: we begin by integrating (III.130) according to the "tangent property" of convex functions, assuming that $F(t; \beta, t_0) > 0$ on $[0, T)$, and we obtain the estimate

(III.140) $$F(t;\beta,t_0) \geqq F(0;\beta,t_0) \times \exp\left[\left\{\frac{F'(0;\beta,t_0) + (\lambda/|\alpha|)F(0;\beta,t_0)}{|\alpha| F(0;\beta,t_0)}\right\}(1-e^{-|\alpha|t}) - \frac{\lambda}{|\alpha|} t\right]$$

for $0 \leqq t < T$. Setting $\beta = 1/t_0^2$ in (III.140) we get

(III.141) $$\|\mathbf{u}(t)\|^2 + \left(\frac{t}{t_0}+1\right)^2 \geqq \exp\left[\chi(t;t_0)\right], \qquad 0 \leqq t < T,$$

where

(III.142) $$\chi(t;t_0) = \frac{1}{|\alpha|}\left[\left(\frac{2}{t_0} + \frac{\lambda(1/t_0^2;t_0)}{|\alpha|}\right)(1-e^{-|\alpha|t}) - \lambda\left(\frac{1}{t_0^2}; t_0\right)t\right],$$

so that $\chi(0; t_0) = 0$ for all $t_0 \geqq 0$. For the sake of convenience we now set

(III.143) $$\eta(t_0) = \frac{2}{t_0} + \frac{1}{|\alpha|}\left(\lambda\left(\frac{1}{t_0^2}; t_0\right)\right)$$

and note that

(III.144) $$\chi'(t;t_0) = \eta(t_0)\, e^{-|\alpha|t} - \lambda\left(\frac{1}{t_0^2}; t_0\right),$$

so that $\chi'(t; t_0) > 0$, for $0 < t < (1/|\alpha|) \ln(\eta(t_0)/\lambda(1/t_0^2; t_0))$, if $\eta(t_0) >$

$\lambda(1/t_0^2; t_0)$. We now take the limit in (III.141) as $t_0 \to +\infty$ and obtain

(III.145)
$$\|\mathbf{u}(t)\|^2 + 1 \geqq \lim_{t_0 \to \infty} \exp[\chi(t; t_0)] = \exp[\lim_{t_0 \to \infty} \chi(t; t_0)] = \exp(\bar{\chi}(t)),$$

for $0 \leqq t < T$, where

(III.146)
$$\bar{\chi}(t) = \frac{\bar{\lambda}}{|\alpha|^2}(1 - e^{-|\alpha|t}) - \bar{\lambda}t$$

with $\bar{\lambda}$ given by (III.137). Clearly

(III.147)
$$\bar{\chi}'(t) = \bar{\lambda}\left(\frac{e^{-|\alpha|t}}{|\alpha|} - 1\right) > 0, \qquad \begin{cases} 0 \leqq t < \dfrac{1}{|\alpha|} \ln\left(\dfrac{1}{|\alpha|}\right), \\ |\alpha| < 1, \end{cases}$$

so that for $0 \leqq t < (1/|\alpha|) \ln(1/|\alpha|)$, $\|\mathbf{u}(t)\|^2$ is bounded from below by an exponentially increasing function if $|\alpha| < 1$.

Case 2. $\mathbf{f} \neq 0$, $\zeta(t) \not\equiv 0$, $\alpha < 0$, and $\zeta(0) > 0$.

In this case it is easily seen that (III.101) implies that

(III.148)
$$\begin{aligned} FF'' - F'^2 &\geqq -2F(2\mathscr{E}(0) + \mu) - |\alpha| FF' \\ &\quad + 2F\left(2\int_0^t \dot{\zeta}(\tau)\langle \mathbf{u}, \mathbf{f}\rangle\, d\tau - \zeta(t)\langle \mathbf{u}, \mathbf{f}\rangle\right) + 4F\zeta(0)\|\mathbf{f}\|^2 \\ &= 2F(2\mathscr{E}(0) - 2\zeta(0)\|\mathbf{f}\|^2 + \mu) - |\alpha| FF' \\ &\quad + 2F\left(2\int_0^t \dot{\zeta}(\tau)\langle \mathbf{u}, \mathbf{f}\rangle\, d\tau - \zeta(t)\langle \mathbf{u}, \mathbf{f}\rangle\right). \end{aligned}$$

In order to proceed further, in this case, we first note the following:

LEMMA. *If $\dot{\zeta}(t)$ is bounded on $[0, T)$ then $\exists C > 0$ such that*

(III.149)
$$2\int_0^t \dot{\zeta}(\tau)\langle \mathbf{u}, \mathbf{f}\rangle\, d\tau - \zeta(t)\langle \mathbf{u}, \mathbf{f}\rangle \geqq -C\|\mathbf{f}\|, \qquad 0 \leqq t < T.$$

Proof. We set $\lambda_0 = \sup_{[0,T)} |\dot{\zeta}(t)| < \infty$. Then

(III.150)
$$\begin{aligned} \left|\int_0^t \dot{\zeta}(\tau)\langle \mathbf{u}(\tau), \mathbf{f}\rangle\, d\tau\right| &= \left|\left\langle \int_0^t \dot{\zeta}(\tau)\mathbf{u}(\tau)\, d\tau, \mathbf{f}\right\rangle\right| \\ &\leqq \left[\int_0^t |\dot{\zeta}(\tau)|\|\mathbf{u}(\tau)\|\, d\tau\right]\|\mathbf{f}\| \\ &\leqq \rho_0\left(\int_0^T \|\mathbf{u}(\tau)\|\, d\tau\right)\|\mathbf{f}\| \\ &\leqq \rho_0 \gamma N T \|\mathbf{f}\|, \end{aligned}$$

so that

$$\int_0^t \dot{\zeta}(\tau)\langle \mathbf{u}, \mathbf{f}\rangle\, d\tau \geqq -\rho_0\gamma NT\|\mathbf{f}\|, \qquad 0\leqq t<T. \tag{III.151}$$

Also,

$$\begin{aligned} |\zeta(t)\langle \mathbf{u}, \mathbf{f}\rangle| &\leqq |\zeta(t)||\langle \mathbf{u}, \mathbf{f}\rangle| \leqq \gamma N|\zeta(t)|\|\mathbf{f}\| \\ &\leqq \left(\gamma N\left|\int_0^t \dot{\zeta}(\tau)\, d\tau + \zeta(0)\right|\right)\|\mathbf{f}\| \\ &\leqq \gamma N(\rho_0 T + \zeta(0))\|\mathbf{f}\|, \end{aligned} \tag{III.152}$$

so

$$-\zeta(t)\langle \mathbf{u}, \mathbf{f}\rangle \geqq -\gamma N(\rho_0 T + \zeta(0))\|\mathbf{f}\|, \qquad 0\leqq t<T. \tag{III.153}$$

Combining (III.151) with (III.153) we obtain (III.149) with

$$C = \gamma N(3\rho_0 T + \zeta(0)) > 0. \tag{III.154}$$

We now return to the differential inequality (III.148) and combine this estimate with the result of the above lemma so as to obtain

$$FF'' - F'^2 \geqq -2F(\|\mathbf{g}\|^2 + \Sigma(\mathbf{f}) + \mu) - |\alpha|FF', \tag{III.155}$$

for $0\leqq t<T$, where $\Sigma: H_+ \to R^+$ is given by

$$\Sigma(\mathbf{w}) = 2\zeta(0)\|\mathbf{w}\|\left[\frac{C}{2\zeta(0)} - \|\mathbf{w}\|\right] - \langle \mathbf{w}, \mathbf{N}\mathbf{w}\rangle \quad \forall \mathbf{w}\in H_+. \tag{III.156}$$

If we set $\beta = 0$, then (III.155) reduces to

$$FF'' - F'^2 \geqq -2F(\|\mathbf{g}\|^2 + \Sigma(\mathbf{f}) + \bar{\mu}) - |\alpha|FF', \tag{III.157}$$

for $0\leqq t<T$ where, in (III.157), $F(t) = \|\mathbf{u}(t)\|^2$ and, by (III.112),

$$\begin{aligned} \bar{\mu} &= \gamma N^2 T \sup_{[0,T)} \|\mathbf{K}(t)\|_{\mathscr{L}_s(H_+,H_-)} \\ &\leqq \gamma N^2 T\left[b_0\gamma^2 \sup_{[0,T)} |\ddot{\Psi}(t)| + \left(\frac{b_0}{a_0}\right) \sup_{[0,T)} \Phi(t)|\right]. \end{aligned} \tag{III.158}$$

Thus, (III.157) is valid under the hypotheses of Case 2 above and the additional assumption that $\dot{\zeta}(t)$ is bounded on $[0, T)$. Various stability and growth estimates can be easily obtained from the differential inequality (III.157), given specific assumptions relative to $\Sigma(\mathbf{f})$; i.e., suppose that

$$\|\mathbf{g}\|^2 + \Sigma(\mathbf{f}) \leqq -\bar{\mu}. \tag{III.159}$$

Then (III.157) clearly implies that $F(t) = \|\mathbf{u}(t)\|^2$ satisfies

$$FF'' - F'^2 \geqq -|\alpha|FF', \qquad 0\leqq t<T, \tag{III.160}$$

from which it easily follows that

$$\text{(III.161)}\qquad \|\mathbf{u}(t)\|^2 \geqq \|\mathbf{f}\|^2 \exp\left[\frac{2\langle \mathbf{f}, \mathbf{g}\rangle}{|\alpha|\,\|\mathbf{f}\|^2}(1-e^{-|\alpha|t})\right], \qquad 0 \leqq t < T.$$

For a further discussion of other cases which are possible relative to the differential inequality (III.101), and a derivation of the associated stability and growth estimates for $\|\mathbf{u}(t)\|^2$, we refer the reader to § 3 of [14].

4. Growth estimates for nonconducting material dielectrics with memory: Concavity arguments.

In the last two sections we have derived various stability and growth estimates for solutions of initial-boundary value problems associated with systems of partial-integrodifferential equations that arise in two theories of nonconducting material dielectrics with memory, the Maxwell-Hopkinson dielectric, as defined by (III.5), (III.6), and the special case of the holohedral isotropic dielectric which is defined by (III.21a, b). Our estimates were obtained by means of logarithmic convexity arguments, which required the a priori restriction of the solutions to classes of bounded functions of the form $\mathcal{N}$ as given by (II.53). In this section we will derive some new growth estimates for Maxwell-Hopkinson and isotropic holohedral dielectrics by means of a concavity argument; our analysis will allow us to drop the a priori restriction that solutions $\mathbf{u} \in \mathcal{N}$ (for some N) and will also enable us to weaken the hypotheses (II.55) relative to the appropriate operators $\mathbf{K}(t)$. We will do the complete computation for the system of evolution equations associated with that special case of the isotropic holohedral material, which is defined by (III.21a, b), with past histories of the form (III.28); for the Maxwell-Hopkinson dielectric which is defined by (III.5), (III.6), with past history of the electric field $\mathbf{E}$ of the form (III.9), we will then state the appropriate corresponding results and give suitable references.

As indicated in § 1, for the constitutive theory defined by (III.21a, b), with past histories prescribed to be of the form (III.28), the system of partial-integrodifferential equations which governs the evolution of the components of the electric displacement field $\mathbf{D}$, on $\Omega \times [0, T)$, will be of the form (III.29). Letting H, H_+, H_- denote $(L_2(\Omega))^3$, $(H_0^1(\Omega))^3$, and $(H^{-1}(\Omega))^3$ again and defining operators $\mathbf{N} \in \mathcal{L}_s(H_+, H_-)$ and $\mathbf{K}(\cdot) \in L^2((-\infty, \infty); \mathcal{L}_s(H_+, H_-))$ via

$$\text{(III.162)}\qquad \begin{aligned} (\mathbf{N}\mathbf{v})_i(\mathbf{x}) &\equiv \dot{\Psi}(0)[c_0\nabla^2 v_i(x) - v_i(\mathbf{x})], && \mathbf{x} \in \Omega,\\ \mathbf{K}(t)\mathbf{v})_i(\mathbf{x}) &\equiv \ddot{\Psi}(t)v_i(\mathbf{x}) - \left(\frac{b_0}{a_0}\right)\Phi(t)\nabla^2 v_i(\mathbf{x}), && \mathbf{x} \in \Omega, \end{aligned}$$

where $c_0 = b_0/a_0\dot{\Psi}(0)$, it is easy to see that the initial-history boundary value for $\mathbf{D}$, i.e., (III.29), (III.15), (III.16) and

(III.163) $$\mathbf{D}(\mathbf{x}, \tau) = \begin{cases} \mathbf{0}, & -\infty < \tau < -\tau_\infty, \\ \mathbf{D}_{\tau_\infty}, & -\tau_\infty \leqq \tau < 0 \end{cases}$$

(where $\mathbf{D}_{\tau_\infty}(\cdot, -\tau_\infty) = \mathbf{0}$, uniformly on Ω), assumes the abstract form

(III.164) $$\mathbf{u}_{tt} + \Psi(0)\mathbf{u}_t - \mathbf{N}\mathbf{u} + \int_{-\infty}^{t} \mathbf{K}(t-\tau)\mathbf{u}(\tau)\, d\tau = \mathbf{0}, \qquad 0 \leqq t < T,$$

(III.165) $$\mathbf{u}(0) = \mathbf{f}, \qquad \mathbf{u}_t(0) = \mathbf{g}(\mathbf{f}, \mathbf{g} \in H_+),$$

(III.166) $$\mathbf{u}(\tau) = \mathbf{U}(\tau), \qquad -\infty < \tau < 0,$$

where

$$\mathbf{U}(\tau) = \begin{cases} \mathbf{0}, & -\infty < \tau < -\tau_\infty, \\ \mathbf{U}_{\tau_\infty}, & -\tau_\infty \leqq \tau < 0. \end{cases}$$

We assume that $\mathbf{U}_{\tau_\infty} : [-\tau_\infty, 0) \to H_+$, and satisfies

(III.167) $$\int_{-\tau_\infty}^{0} \|\mathbf{U}_{\tau_\infty}(\tau)\|_+ \, d\tau = \int_{-\infty}^{0} \|\mathbf{U}(\tau)\|_+ \, d\tau < \infty.$$

For the rest of our work in this section we will assume that $\Psi(0) > 0$. As for the conditions on the operator $\mathbf{K}(\cdot)$ which is defined by (III.162$_2$) we will weaken the hypotheses (II.55) and assume only that

(III.168) $$-\langle \mathbf{v}, \mathbf{K}(0)\mathbf{v}\rangle \geqq 0 \quad \forall \mathbf{v} \in H_+.$$

We require in addition that

(III.169) $$\int_0^\infty \|\mathbf{K}(\tau)\|_{\mathscr{L}_s(H_+, H_-)} \, d\tau < \infty, \qquad \int_0^\infty \|\mathbf{K}_\tau(\tau)\|_{\mathscr{L}_s(H_+, H_-)} \, d\tau < \infty.$$

The consequences of requiring that (III.168), (III.169) hold for the operator $\mathbf{K}(\cdot)$, which is defined by (III.162$_2$), will be examined after we have stated and proved our basic growth estimates; these estimates will require that we place the following restrictions on the initial data and the past history:

(III.170)
(i) $\langle \mathbf{f}, \mathbf{g}\rangle > 0$,
(ii) $\langle \mathbf{f}, \mathbf{N}\mathbf{f}\rangle > 0$,
(iii) $$\left\langle \mathbf{f}, \int_{-\infty}^{0} \mathbf{K}(-\tau)\mathbf{U}(\tau)\, d\tau \right\rangle = \left\langle \mathbf{f}, \int_{-\tau_\infty}^{0} \mathbf{K}(-\tau)\mathbf{U}_{\tau_\infty}(\tau)\, d\tau \right\rangle < 0.$$

Having presented our basic assumptions relative to $\mathbf{K}(\cdot)$ and the data $\mathbf{f}$, $\mathbf{g}$, and $\mathbf{U}$ we can now state and prove the following:

THEOREM (Bloom [19]). *Let* $\mathbf{u}^\delta \in C^2([0, T); H_+)$ *denote any solution of the initial-history value problem*

$$\mathbf{u}^\delta_{tt} + \Psi(0)\mathbf{u}^\delta_t - \mathbf{N}\mathbf{u}^\delta + \int_{-\infty}^{t} \mathbf{K}(t-\tau)\mathbf{u}^\delta(\tau)\,d\tau = \mathbf{0}, \qquad 0 \leqq t < T,$$

(III.171)
$$\mathbf{u}^\delta(0) = \delta\mathbf{f}, \qquad \mathbf{u}^\delta_t(0) = \mathbf{g}, \qquad \delta > 0,$$

$$\mathbf{u}^\delta(\tau) = \mathbf{U}(\tau), \qquad -\infty < \tau < 0,$$

where $\mathbf{K}(\cdot)$ *satisfies* (III.168), (III.169) *and the data satisfy* (III.170). *If*

(III.172)
$$\text{(i)}\quad \|\mathbf{f}\|^2 \leqq \frac{2}{\Psi(0)}\langle \mathbf{f}, \mathbf{g}\rangle,$$

$$\text{(ii)}\quad T > \frac{1}{\Psi(0)} \ln\left(\frac{2\langle \mathbf{f}, \mathbf{g}\rangle}{2\langle \mathbf{f}, \mathbf{g}\rangle - \Psi(0)\|\mathbf{f}\|^2}\right),$$

then for each $\delta \geqq \|\mathbf{g}\|/\langle \mathbf{f}, \mathbf{N}\mathbf{f}\rangle^{1/2}$

(III.173)
$$\sup_{-\infty<t<T} \|\mathbf{u}^\delta\|_+ \geqq \left[\frac{|\langle \mathbf{f}, \int_{-\infty}^0 \mathbf{K}(-\tau)\mathbf{U}(\tau)\,d\tau\rangle|}{\gamma\pi_T}\right]^{1/2}\sqrt{\delta},$$

where

(III.174)
$$\pi_T = \tfrac{1}{2}\|\mathbf{N}\|_{\mathcal{L}_s(H_+,H_-)} + \int_0^\infty \|\mathbf{K}(\tau)\|_{\mathcal{L}_s(H_+,H_-)}\,d\tau + T\int_0^\infty \|\mathbf{K}_\tau(\tau)\|_{\mathcal{L}_s(H_+,H_-)}\,d\tau.$$

Remarks. The estimate (III.173) provides a lower bound for $\sup_{-\infty<t<T} \|\mathbf{u}^\delta\|_+$ in terms of δ, γ, the data $\mathbf{f}$, $\mathbf{g}$, and $\mathbf{U}$, and the operator norms of $\mathbf{N}$, $\mathbf{K}$, and $\mathbf{K}_t$; note that the initial-history value problem (III.171) results from effecting a one-parameter variation of the initial datum $\mathbf{u}(0)$ in (III.164)–(III.166). We will demonstrate below that estimates of the form (III.173), which also hold under appropriate conditions for solutions of initial-history value problems associated with Maxwell-Hopkinson materials, can be used to obtain bounds for constitutive parameters which occur in the definition of the corresponding $\mathbf{K}(\cdot)$.

Proof. We wish to establish the estimate (III.173) under the conditions given above. Suppose that for some $\delta = \bar{\delta}$, with $\bar{\delta} > \|\mathbf{g}\|/\langle \mathbf{f}, \mathbf{N}\mathbf{f}\rangle^{1/2}$, (III.173) is not satisfied when T, $\Psi(0)$, $\mathbf{f}$, and $\mathbf{g}$ satisfy (III.172), i.e., suppose instead that

(III.175)
$$\sup_{-\infty<t<T} \|\mathbf{u}^{\bar{\delta}}\|_+ < \left[\frac{|\langle \mathbf{f}, \int_{-\infty}^0 \mathbf{K}(-\tau)\mathbf{U}(\tau)\,d\tau\rangle|}{\gamma\pi_T}\right]^{1/2}\sqrt{\bar{\delta}}.$$

If we define $F_\delta(t)=\langle \mathbf{u}^\delta(t), \mathbf{u}^\delta(t)\rangle$, $0\leqq t<T$, then a direct computation yields

$$F_\delta F_\delta'' - (\delta+1)F_\delta'^2 = 4(\delta+1)S_\delta'^2 + 2F_\delta\{\langle \mathbf{u}^\delta, \mathbf{u}_{tt}^\delta\rangle - (2\delta+1)\|\mathbf{u}_t^\delta\|^2\} \tag{III.176}$$

for any $\delta>0$, where

$$S_\delta^2(t) = \|\mathbf{u}^\delta\|^2\|\mathbf{u}_t^\delta\|^2 - \langle \mathbf{u}^\delta, \mathbf{u}_t^\delta\rangle^2 \geqq 0, \qquad t\geqq 0, \tag{III.177}$$

by virtue of the Schwarz inequality. Combining (III.176), (III.177), and the evolution equation (III.171) we obtain the differential inequality

$$F_\delta F_\delta'' - (\delta+1)F_\delta'^2 \geqq 2F_\delta J_\delta, \qquad 0\leqq t<T, \tag{III.178}$$

where

$$J_\delta(t) = \langle \mathbf{u}^\delta, \mathbf{N}\mathbf{u}^\delta\rangle - \left\langle \mathbf{u}^\delta, \int_{-\infty}^t \mathbf{K}(t-\tau)\mathbf{u}^\delta(\tau)\,d\tau\right\rangle - \Psi(0)\langle \mathbf{u}^\delta, \mathbf{u}_t^\delta\rangle - (2\delta+1)\|\mathbf{u}_t^\delta\|^2. \tag{III.179}$$

We will show that $J_\delta(t)\geqq -(\Psi(0)/2)F_\delta'(t)$ for $t\in[0, T)$; this, in turn, will lead to the conclusion that $[e^{\Psi(0)t}(F_\delta^{-\delta}(t))']'\leqq 0$, for $0\leqq t<T$, which will produce, upon integration, a contradiction to (III.175). Now, directly from (III.179) we have

$$\begin{aligned} J_\delta'(t) &= 2\langle \mathbf{u}_t^\delta, \mathbf{N}\mathbf{u}^\delta\rangle - \frac{d}{dt}\left\langle \mathbf{u}^\delta, \int_{-\infty}^t \mathbf{K}(t-\tau)\mathbf{u}^\delta(\tau)\,d\tau\right\rangle \\ &\quad - \Psi(0)\|\mathbf{u}_t^\delta\|^2 - \Psi(0)\langle \mathbf{u}^\delta, \mathbf{u}_{tt}^\delta\rangle - 2(2\delta+1)\langle \mathbf{u}_t^\delta, \mathbf{u}_{tt}^\delta\rangle \\ &= -4\delta\langle \mathbf{u}_t^\delta, \mathbf{N}\mathbf{u}^\delta\rangle - \frac{d}{dt}\left\langle \mathbf{u}^\delta, \int_{-\infty}^t \mathbf{K}(t-\tau)\mathbf{u}^\delta(\tau)\,d\tau\right\rangle \\ &\quad - \Psi(0)\|\mathbf{u}_t^\delta\|^2 - \Psi(0)\langle \mathbf{u}^\delta, \mathbf{u}_{tt}^\delta\rangle + 2\Psi(0)(2\delta+1)\|\mathbf{u}_t^\delta\|^2 \\ &\quad + 2(2\delta+1)\left\langle \mathbf{u}_t^\delta, \int_{-\infty}^t \mathbf{K}(t-\tau)\mathbf{u}^\delta(\tau)\,d\tau\right\rangle. \end{aligned} \tag{III.180}$$

Integrating this last result over $[0, t)$, using the fact that

$$J_\delta(0) = \delta^2\langle \mathbf{f}, \mathbf{N}\mathbf{f}\rangle - \delta\left\langle \mathbf{f}, \int_{-\infty}^0 \mathbf{K}(-\tau)\mathbf{U}(\tau)\,d\tau\right\rangle - \Psi(0)\delta\langle \mathbf{f}, \mathbf{g}\rangle - (2\delta+1)\|\mathbf{g}\|^2, \tag{III.181}$$

and then dropping a term proportional to $\|\mathbf{u}_t^\delta\|^2$, we obtain

$$\begin{aligned}
J_\delta(t) &\geqq J_\delta(0) - 2\delta(\langle \mathbf{u}^\delta, \mathbf{N}\mathbf{u}^\delta\rangle - \delta^2\langle \mathbf{f}, \mathbf{N}\mathbf{f}\rangle) \\
&\quad - \left\langle \mathbf{u}^\delta, \int_{-\infty}^{t} \mathbf{K}(t-\tau)\mathbf{u}^\delta(\tau)\, d\tau \right\rangle \\
&\quad + \delta\left\langle \mathbf{f}, \int_{-\infty}^{0} \mathbf{K}(-\tau)\mathbf{U}(\tau)\, d\tau \right\rangle - \Psi(0)(\langle \mathbf{u}^\delta, \mathbf{u}_t^\delta\rangle - \delta\langle \mathbf{f}, \mathbf{g}\rangle) \\
&\quad + 2(2\delta+1)\int_0^t \left\langle \mathbf{u}_\tau^\delta, \int_{-\infty}^{\tau} \mathbf{K}(\tau-\lambda)\mathbf{u}^\delta(\lambda)\, d\lambda \right\rangle d\tau \\
&= (2\delta+1)[\delta^2\langle \mathbf{f}, \mathbf{N}\mathbf{f}\rangle - \|\mathbf{g}\|^2] - 2\delta\langle \mathbf{u}^\delta, \mathbf{N}\mathbf{u}^\delta\rangle \\
&\quad + 2(2\delta+1)\int_0^t \left\langle \mathbf{u}_\tau^\delta, \int_{-\infty}^{\tau} \mathbf{K}(\tau-\lambda)\mathbf{u}^\delta(\lambda)\, d\lambda \right\rangle d\tau \\
&\quad - \left\langle \mathbf{u}^\delta, \int_{-\infty}^{t} \mathbf{K}(t-\tau)\mathbf{u}^\delta(\tau)\, d\tau \right\rangle - \Psi(0)\langle \mathbf{u}^\delta, \mathbf{u}_t^\delta\rangle.
\end{aligned} \tag{III.182}$$

However, in view of our hypothesis (III.168) relative to $\mathbf{K}(0)$,

$$\begin{aligned}
&\int_0^t \left\langle \mathbf{u}_\tau^\delta, \int_{-\infty}^{\tau} \mathbf{K}(\tau-\lambda)\mathbf{u}^\delta(\lambda)\, d\lambda \right\rangle d\tau \\
&\quad \geqq \int_0^t \frac{d}{d\tau}\left\langle \mathbf{u}^\delta, \int_{-\infty}^{t} \mathbf{K}(\tau-\lambda)\mathbf{u}^\delta(\lambda)\, d\lambda \right\rangle d\tau \\
&\quad\quad - \int_0^t \left\langle \mathbf{u}^\delta(\tau), \int_{-\infty}^{\tau} \mathbf{K}_\tau(\tau-\lambda)\mathbf{u}^\delta(\lambda)\, d\lambda \right\rangle d\tau,
\end{aligned} \tag{III.183}$$

and therefore

$$\begin{aligned}
J_\delta(t) &\geqq (2\delta+1)\left[\delta^2\langle \mathbf{f}, \mathbf{N}\mathbf{f}\rangle - \|\mathbf{g}\|^2 + 2\delta\left|\left\langle \mathbf{f}, \int_{-\infty}^{0} \mathbf{K}(-\tau)\mathbf{U}(\tau)\, d\tau \right\rangle\right|\right] \\
&\quad - 2\delta\langle \mathbf{u}^\delta, \mathbf{N}\mathbf{u}^\delta\rangle + (4\delta+1)\left\langle \mathbf{u}^\delta, \int_{-\infty}^{t} \mathbf{K}(t-\tau)\mathbf{u}^\delta(\tau)\, d\tau \right\rangle \\
&\quad - 2(2\delta+1)\int_0^t \left\langle \mathbf{u}^\delta, \int_{-\infty}^{\tau} \mathbf{K}_\tau(\tau-\lambda)\mathbf{u}^\delta(\lambda)\, d\lambda \right\rangle d\tau - \Psi(0)\langle \mathbf{u}^\delta, \mathbf{u}_t^\delta\rangle,
\end{aligned} \tag{III.184}$$

where we have used hypothesis (III.170, (iii)). For the sake of convenience, we now define

$$M_{T,\delta} \equiv \left[\frac{|\langle \mathbf{f}, \int_{-\infty}^{0} \mathbf{K}(-\tau)\mathbf{U}(\tau)\, d\tau\rangle|}{\gamma\pi_T}\right]^{1/2}\sqrt{\delta}, \tag{III.185}$$

so that (III.175) is equivalent to

(III.175′) $$\sup_{-\infty<\tau<T} \|\mathbf{u}^{\delta}\|_{+} < M_{T,\bar{\delta}}.$$

Routine estimates employing the Schwarz inequality and (III.175′) then yield the lower bounds

(III.186) $$\left\langle \mathbf{u}^{\bar{\delta}}, \int_{-\infty}^{t} \mathbf{K}(t-\tau)\mathbf{u}^{\bar{\delta}}(\tau)\, d\tau \right\rangle \geqq -\gamma M_{T,\bar{\delta}}^{2} \int_{0}^{\infty} \|\mathbf{K}(\rho)\|_{\mathscr{L}_s(H_+,H_-)}\, d\rho,$$

(III.187) $$\begin{aligned} -\int_0^t \left\langle \mathbf{u}^{\bar{\delta}}, \int_{-\infty}^{\tau} \mathbf{K}_{\tau}(\tau-\lambda)\mathbf{u}^{\bar{\delta}}(\lambda)\, d\lambda \right\rangle d\tau & \\ \geqq -\gamma M_{T,\bar{\delta}}^{2} \int_0^T \int_{-\infty}^{t} \|\mathbf{K}_t(t-\tau)\|_{\mathscr{L}_s(H_+,H_-)}\, d\tau\, dt & \\ = -\gamma T M_{T,\bar{\delta}}^{2} \int_0^{\infty} \|\mathbf{K}_{\rho}(\rho)\|_{\mathscr{L}_s(H_+,H_-)}\, d\rho & \end{aligned}$$

and

(III.188) $$-\langle \mathbf{u}^{\bar{\delta}}, \mathbf{N}\mathbf{u}^{\bar{\delta}} \rangle \geqq -\gamma M_{T,\bar{\delta}}^{2} \|\mathbf{N}\|_{\mathscr{L}_s(H_+,H_-)}.$$

Now, in view of our assumption that $\bar{\delta} \geqq \|\mathbf{g}\|/\langle \mathbf{f}, \mathbf{N}\mathbf{f}\rangle^{1/2}$, (III.185) and (III.184), with $\delta = \bar{\delta}$, imply that

(III.189) $$\begin{aligned} J_{\bar{\delta}}(t) \geqq\ & 2(2\bar{\delta}+1)\gamma\pi_T M_{T,\bar{\delta}}^{2} - 2\bar{\delta}\langle \mathbf{u}^{\bar{\delta}}, \mathbf{N}\mathbf{u}^{\bar{\delta}}\rangle \\ & + (4\bar{\delta}+1)\left\langle \mathbf{u}^{\bar{\delta}}, \int_{-\infty}^{t} \mathbf{K}(t-\tau)\mathbf{u}^{\bar{\delta}}(\tau)\, d\tau \right\rangle \\ & - 2(2\bar{\delta}+1) \int_0^t \left\langle \mathbf{u}^{\bar{\delta}}, \int_{-\infty}^{\tau} \mathbf{K}_{\tau}(\tau-\lambda)\mathbf{u}^{\bar{\delta}}(\lambda)\, d\lambda \right\rangle d\tau \\ & - \Psi(0)\langle \mathbf{u}^{\bar{\delta}}, \mathbf{u}_t^{\bar{\delta}}\rangle. \end{aligned}$$

If we now combine the estimate (III.189) with (III.186)–(III.188), the definition of π_T, and the fact that $F'_{\delta} = 2\langle \mathbf{u}^{\delta}, \mathbf{u}_t^{\delta}\rangle$, we have

(III.190) $$J_{\bar{\delta}}(t) \geqq -\frac{\Psi(0)}{2} F'_{\bar{\delta}}(t), \qquad 0 \leqq t < T.$$

Therefore, in view of (III.178), we obtain the differential inequality

(III.191) $$F_{\bar{\delta}}F''_{\bar{\delta}} - (\bar{\delta}+1)F'^{2}_{\bar{\delta}} \geqq -\Psi(0)F_{\bar{\delta}}F'_{\bar{\delta}}, \qquad 0 \leqq t < T,$$

which is equivalent to

(III.192) $$\frac{d}{dt}\left(e^{\Psi(0)t} \frac{d}{dt}\left(F_{\bar{\delta}}^{-\bar{\delta}}(t) \right) \right) \geqq 0, \qquad 0 \leqq t < T.$$

Two successive integrations of (III.192) readily yield the lower bound

$$F_{\bar{\delta}}^{\bar{\delta}}(t) \geqq F_{\bar{\delta}}^{\bar{\delta}}(0)\left(1-(1-e^{-\Psi(0)t})\bar{\delta}\frac{F'_{\bar{\delta}}(0)}{\Psi(0)F_{\bar{\delta}}(0)}\right)^{-1} \tag{III.193}$$

for $0 \leqq t < T$, where $T > 0$ is assumed to satisfy (III.172(ii)). The bracketed expression in (III.193) will clearly vanish at

$$t_\infty = \frac{1}{\Psi(0)} \ln\left[\frac{2\langle \mathbf{f}, \mathbf{g}\rangle}{2\langle \mathbf{f}, \mathbf{g}\rangle - \Psi(0)\|\mathbf{f}\|^2}\right] < T \tag{III.194}$$

(note that t_∞ exists by virtue of (III.172(i)). It then follows that

$$+\infty = \sup_{-\infty<t<T} \|\mathbf{u}^{\bar{\delta}}(t)\| \leqq \gamma \sup_{-\infty<t<T} \|\mathbf{u}^{\bar{\delta}}(t)\|_+, \tag{III.195}$$

contradicting (III.175) and thus establishing the estimate (III.173) for any solution $\mathbf{u}^\delta$ of (III.171); note that (III.173) is valid even if $\sup_{-\infty<t<T} \|\mathbf{u}^\delta\|_+ = +\infty$ but is not, of course, of any value in this case. Q.E.D.

As a companion to the above result, we have the following theorem, which applies to the situation where we introduce a one-parameter family of past histories into the initial-history value problem (III.164)–(III.166):

THEOREM (Bloom [19]). *Let* $\mathbf{u}^\varepsilon \in C^2((0, T); H_+)$ *denote any solution of the initial-history value problem*

$$\begin{aligned} &\mathbf{u}_{tt}^\varepsilon + \Psi(0)\mathbf{u}_t^\varepsilon - \mathbf{N}\mathbf{u}^\varepsilon + \int_{-\infty}^t \mathbf{K}(t-\tau)\mathbf{u}^\varepsilon(\tau)\,d\tau = \mathbf{0}, \qquad 0 \leqq t < T, \\ &\mathbf{u}^\varepsilon(0) = \mathbf{f}, \qquad \mathbf{u}_t^\varepsilon(0) = \mathbf{g}, \\ &\mathbf{u}^\varepsilon(\tau) = \varepsilon\mathbf{U}(\tau), \qquad -\infty < \tau < 0, \quad \varepsilon > 0, \end{aligned} \tag{III.196}$$

where $\mathbf{K}(\cdot)$ satisfies (III.168), (III.169) and the data satisfy (III.170). If, in addition,

$$\text{(i)} \qquad \langle \mathbf{f}, \mathbf{N}\mathbf{f}\rangle \geqq \|\mathbf{g}\|^2$$

and

$$\text{(ii)} \qquad T > \frac{1}{\Psi(0)} \ln\left(\frac{2\bar{\omega}\langle \mathbf{f}, \mathbf{g}\rangle}{2\bar{\omega}\langle \mathbf{f}, \mathbf{g}\rangle - \Psi(0)\|\mathbf{f}\|^2}\right) \tag{III.197}$$

for some

$$\bar{\omega} > \frac{\Psi(0)}{2}\left(\frac{\|\mathbf{f}\|^2}{\langle \mathbf{f}, \mathbf{g}\rangle}\right),$$

then for all ε, $0 < \varepsilon < \infty$,

$$\sup_{-\infty<t<T} \|\mathbf{u}^\varepsilon\|_+ \geqq M_{T,\varepsilon}, \tag{III.198}$$

where $M_{T,\varepsilon}$ is given by (III.185) with $\delta \to \varepsilon$.

Proof. Suppose that for some $\varepsilon=\bar{\varepsilon}$, $0<\bar{\varepsilon}<\infty$, (III.198) is not satisfied when the conditions in (III.197) are met, i.e.,

$$\sup_{-\infty<t<T}\|\mathbf{u}^{\bar{\varepsilon}}\|_{+}<M_{T,\bar{\varepsilon}}. \tag{III.199}$$

Let $\omega>0$ be arbitrary and set $F_\varepsilon(t)=\|\mathbf{u}^\varepsilon(t)\|^2$. Then we compute that

$$F_\varepsilon F_\varepsilon''-(\omega+1)F_\varepsilon'^2\geqq 2F_\varepsilon H_{\omega,\varepsilon},\qquad 0\leqq t<T, \tag{III.200}$$

where

$$\begin{aligned}H_{\omega,\varepsilon}(t)=\langle\mathbf{u}^\varepsilon,\mathbf{N}\mathbf{u}^\varepsilon\rangle-\left\langle\mathbf{u}^\varepsilon,\int_{-\infty}^{t}\mathbf{K}(t-\tau)\mathbf{u}^\varepsilon(\tau)\,d\tau\right\rangle\\ -\Psi(0)\langle\mathbf{u}_t^\varepsilon,\mathbf{u}^\varepsilon\rangle-(2\omega+1)\|\mathbf{u}_t^\varepsilon\|^2.\end{aligned} \tag{III.201}$$

Employing a computation directly analogous to the one used in the previous theorem, we find that

$$\begin{aligned}H_{\omega,\varepsilon}(t)\geqq&(2\omega+1)[\langle\mathbf{f},\mathbf{N}\mathbf{f}\rangle-\|\mathbf{g}\|^2]-\Psi(0)\langle\mathbf{u}^\varepsilon,\mathbf{u}_t^\varepsilon\rangle\\ &+2\varepsilon(2\omega+1)\left|\left\langle\mathbf{f},\int_{-\infty}^{0}\mathbf{K}(-\tau)\mathbf{U}(\tau)\,d\tau\right\rangle\right|\\ &-2\omega\langle\mathbf{u}^\varepsilon,\mathbf{N}\mathbf{u}^\varepsilon\rangle+(4\omega+1)\left\langle\mathbf{u}^\varepsilon,\int_{-\infty}^{t}\mathbf{K}(t-\tau)\mathbf{u}^\varepsilon(\tau)\,d\tau\right\rangle\\ &-2(2\omega+1)\int_0^t\left\langle\mathbf{u}^\varepsilon,\int_{-\infty}^{\tau}\mathbf{K}_\tau(\tau-\lambda)\mathbf{u}^\varepsilon(\lambda)\,d\lambda\right\rangle d\tau.\end{aligned} \tag{III.202}$$

By using (III.199) so as to bound, from below, the last three expressions on the right-hand side of (III.202), and then the hypothesis (III.197, (i)) in order to drop the bracketed expression, we easily find that (III.202) implies that for any $\omega>0$

$$\begin{aligned}H_{\omega,\bar{\varepsilon}}(t)\geqq&-\frac{\Psi(0)}{2}F_{\bar{\varepsilon}}'(t)+2(2\omega+1)\bar{\varepsilon}\left|\left\langle\mathbf{f},\int_{-\infty}^{0}\mathbf{K}(-\tau)\mathbf{U}(\tau)\,d\tau\right\rangle\right|\\ &-(2\omega+1)\gamma M_{T,\bar{\varepsilon}}^2\left[\left(\frac{2\omega}{2\omega+1}\right)\|\mathbf{N}\|_{\mathscr{L}_s(H_+,H_-)}\right.\\ &\qquad+\left(\frac{4\omega+1}{2\omega+1}\right)\int_0^\infty\|\mathbf{K}(\rho)\|_{\mathscr{L}_s(H_+,H_-)}\,d\rho\\ &\qquad\left.+2T\int_0^\infty\|\mathbf{K}_\rho(\rho)\|_{\mathscr{L}_s(H_+,H_-)}\,d\rho\right]\\ \geqq&-\Psi(0)F_{\bar{\varepsilon}}'(t)+2(2\omega+1)\\ &\times\left[\bar{\varepsilon}\left|\left\langle\mathbf{f},\int_{-\infty}^{0}\mathbf{K}(-\tau)\mathbf{U}(\tau)\,d\tau\right\rangle\right|\right.\\ &\qquad-\gamma M_{T,\bar{\varepsilon}}^2\left(\frac{1}{2}\|\mathbf{N}\|_{\mathscr{L}_s(H_+,H_-)}+\int_0^\infty\|\mathbf{K}(\rho)\|_{\mathscr{L}_s(H_+,H_-)}\,d\rho\right.\\ &\qquad\left.\left.+T\int_0^\infty\|\mathbf{K}_\rho(\rho)\|_{\mathscr{L}_s(H_+,H_-)}\,d\rho\right)\right],\end{aligned} \tag{III.203}$$

or

$$H_{\omega,\bar{\varepsilon}}(t) \geqq -\Psi(0)F'_{\bar{\varepsilon}}(t), \qquad 0 \leqq t < T, \quad \omega > 0, \tag{III.204}$$

in view of the definitions of π_T and $M_{T,\varepsilon}$. Combining the differential inequality (III.200) with the estimate (III.204), we immediately obtain, for any $\omega > 0$, the inequality

$$F_{\bar{\varepsilon}}F''_{\bar{\varepsilon}} - (\omega+1)F'^2_{\bar{\varepsilon}} \geqq -\Psi(0)F_{\bar{\varepsilon}}F'_{\bar{\varepsilon}}, \qquad 0 \leqq t < T, \tag{III.205}$$

which may be integrated so as to yield

$$F^{\omega}_{\bar{\varepsilon}}(t) \geqq F^{\omega}_{\bar{\varepsilon}}(0)\left[1-(1-e^{-\Psi(0)t})\omega\frac{F'_{\bar{\varepsilon}}(0)}{\Psi(0)F_{\bar{\varepsilon}}(0)}\right]^{-1}. \tag{III.206}$$

However, (III.206) implies that $F_{\bar{\varepsilon}}(t)$ tends to $+\infty$ as t tends to

$$t_{\infty} = \frac{1}{\Psi(0)}\ln\left(\frac{2\omega\langle \mathbf{f}, \mathbf{g}\rangle}{2\omega\langle \mathbf{f}, \mathbf{g}\rangle - \Psi(0)\|\mathbf{f}\|^2}\right). \tag{III.207}$$

If we now choose $\omega = \bar{\omega}$, where $\bar{\omega} > (\Psi(0)/2)(\|\mathbf{f}\|^2/\langle \mathbf{f}, \mathbf{g}\rangle)$ then, as per our hypothesis (III.197, (ii)), $t_{\infty} < T$ and $\sup_{-\infty<t<T} \|\mathbf{u}^{\varepsilon}\|_{+} = +\infty$; this contradicts (III.109) and serves to establish the estimate (III.198) for all $\varepsilon > 0$. Q.E.D.

Before proceeding with a statement of the analogous results for Maxwell-Hopkinson dielectrics, we want to delineate the restrictions on the kernel functions $\Phi(t)$, $\Psi(t)$, in the system of partial-integrodifferential equations (III.29) for the components of the electric displacement field, which are implied by the requirement that $\mathbf{N}$, as defined by (III.162$_1$) and $\mathbf{K}(\cdot)$, as defined by (III.162$_2$), satisfy the hypotheses (III.168), (III.169) and (III.170). From (III.162$_1$) we compute that

$$\begin{aligned}\langle \mathbf{v}, \mathbf{N}\mathbf{v}\rangle &= \int_{\Omega} v_i(\mathbf{x})[\mathbf{N}\mathbf{v}]_i(\mathbf{x})\,d\mathbf{x} \\ &= \dot{\Psi}(0)\left(\int_{\Omega} c_0\nabla^2 v_i(\mathbf{x})v_i(\mathbf{x})\,d\mathbf{x} - \int_{\Omega} v_i(\mathbf{x})v_i(\mathbf{x})\,d\mathbf{x}\right) \\ &= -\dot{\Psi}(0)[c_0\|\mathbf{v}\|^2_{+} + \|\mathbf{v}\|^2]\end{aligned} \tag{III.208}$$

for any $\mathbf{v} \in H_{+}$, The condition (III.170(ii)) then requires that

$$\dot{\Psi}(0)[c_0\|\mathbf{f}\|^2_{+} + \|\mathbf{f}\|^2] < 0 \tag{III.209}$$

or, in view of the definition of c_0, that

$$\left(\frac{b_0}{a_0}\right)\|\mathbf{f}\|^2_{+} + \dot{\Psi}(0)\|\mathbf{f}\|^2 < 0. \tag{III.210}$$

Clearly, (III.210) will be satisfied if $\dot{\Psi}(0)<0$ with

(III.211)
$$|\dot{\Psi}(0)| > \frac{b_0\|\mathbf{f}\|_+^2}{a_0\|\mathbf{f}\|^2}.$$

Also, for any $\mathbf{v} \in H_+$

$$\begin{aligned}\langle \mathbf{v}, \mathbf{K}(0)\mathbf{v}\rangle &= \int_\Omega v_i(\mathbf{x})[\mathbf{K}(0)\mathbf{v}]_i(\mathbf{x})\,d\mathbf{x} \\ &= \ddot{\Psi}(0)\int_\Omega v_i(\mathbf{x})v_i(\mathbf{x})\,d\mathbf{x} - \left(\frac{b_0}{a_0}\right)\Phi(0)\int_\Omega \nabla^2 v_i(\mathbf{x})v_i(\mathbf{x})\,d\mathbf{x} \qquad \text{(III.212)} \\ &= \ddot{\Psi}(0)\|\mathbf{v}\|^2 + \left(\frac{b_0}{a_0}\right)\Phi(0)\|\mathbf{v}\|_+^2,\end{aligned}$$

and thus

(III.213)
$$-\langle \mathbf{v}, \mathbf{K}(0)\mathbf{v}\rangle \geqq 0 \Leftrightarrow \ddot{\Psi}(0)\|\mathbf{v}\|^2 + \left(\frac{b_0}{a_0}\right)\Phi(0)\|\mathbf{v}\|_+^2 \leqq 0.$$

If we assume that $\ddot{\Psi}(0) \geqq 0$ then, via the embedding of H_+ into H, (III.213) will be satisfied for all $\mathbf{v} \in H_+$ if

(III.214)
$$\gamma^2\ddot{\Psi}(0) + \left(\frac{b_0}{a_0}\right)\Phi(0) \leqq 0.$$

It then follows that $\mathbf{K}(0)$, as defined by (III.162$_2$), will satisfy the basic hypothesis (III.168) if

(III.215)
$$\ddot{\Psi}(0) \geqq 0 \quad \text{and} \quad \Phi(0) < 0$$

with

(III.216)
$$|\Phi(0)| > \frac{a_0\gamma^2\ddot{\Psi}(0)}{b_0}.$$

Finally, computations completely analogous to (III.17), (III.18) easily yield, for $-\infty < t < \infty$

(III.217)
$$\|\mathbf{K}(t)\|_{\mathscr{L}_s(H_+,H_-)} \leqq \gamma^2|\ddot{\Psi}(t)| + \left(\frac{b_0}{a_0}\right)|\Phi(t)|,$$

(III.218)
$$\|\mathbf{K}_t(t)\|_{\mathscr{L}_s(H_+,H_-)} \leqq \gamma^2|\Psi^{(3)}(t)| + \left(\frac{b_0}{a_0}\right)|\dot{\Phi}(t)|,$$

and, therefore, $\mathbf{K}(\cdot)$, as defined by (III.162$_2$) will satisfy the hypotheses (III.169) if

(III.219)
$$\int_0^\infty |\ddot{\Psi}(t)|\,d\tau < \infty, \qquad \int_0^\infty |\Phi(\tau)|\,d\tau < \infty,$$

(III.220)
$$\int_0^\infty |\Psi^{(3)}(\tau)|\,d\tau < \infty, \qquad \int_0^\infty |\dot{\Phi}(\tau)|\,d\tau < \infty.$$

For the Maxwell-Hopkinson dielectric defined by the constitutive relations (III.5), (III.6), with past history of the form (III.9), and associated initial and boundary data (III.15), (III.16), results such as the ones obtained above for holohedral isotropic dielectrics also exist in the literature ([17]). The initial-history boundary value problem in question is modeled by the abstract initial-history value problem

$$\begin{aligned} &\mathbf{u}_{tt}-\mathbf{N}\mathbf{u}+\int_{-\infty}^{t}\mathbf{K}(t-\tau)\mathbf{u}(\tau)\,d\tau=\mathbf{0}, \qquad 0\leqq t<T,\\ \text{(III.221)}\qquad &\mathbf{u}(0)=\mathbf{f},\quad \mathbf{u}_t(0)=\mathbf{g},\quad \mathbf{f},\mathbf{g}\in H_+,\\ &\mathbf{u}(\tau)=\mathbf{U}(\tau), \qquad -\infty<\tau<0,\end{aligned}$$

where for some $\tau_\infty>0$,

$$\mathbf{U}(\tau)=\begin{cases}\mathbf{0}, & -\infty<\tau<-\tau_\infty,\\ \mathbf{U}_{\tau_\infty}, & -\tau_\infty\leqq\tau<0,\end{cases}$$

with $\mathbf{U}_{\tau_\infty}$ satisfying (III.167). The appropriate operators $\mathbf{N}\in\mathscr{L}_s(H_+,H_-)$, and $\mathbf{K}(\cdot)\in L^2((-\infty,\infty);\mathscr{L}_s(H_+,H_-))$ in (III.221), are given by (III.31). Formally, the system (III.221) is of the same form as the system (III.164)–(III.165) appropriate to the holohedral isotropic case considered in this section with $\Psi(0)=0$. However, it is clear from the form of the hypotheses (III.172), in the first theorem of this section, that we can not simply set $\Psi(0)=0$ in those hypotheses and recover conditions which guarantee, for instance, the validity of the estimate (III.173) for solutions of the problem obtained from (III.221) by replacing $\mathbf{f}$ by $\delta\mathbf{f}$, $\delta>0$; the same is true concerning the validity of the second estimate, i.e., (III.198), if we wish to consider the problem obtained from (III.221) by replacing $\mathbf{U}$ by $\varepsilon\mathbf{U}$, $\varepsilon>0$. It turns out, however, that the estimates (III.173) and (III.198) are valid for the solutions of the initial-history value problems obtained from (III.221) by making the transformations $\mathbf{f}\to\delta\mathbf{f}$ and $\mathbf{U}\to\varepsilon\mathbf{U}$, respectively, and we delineate below the various conditions under which these estimates hold; as was the case for the isotropic holohedral material, we again assume that $\mathbf{N}$, $\mathbf{K}(\cdot)$, this time as defined by (III.31), satisfy (III.168)–(III.170).

THEOREM (Bloom [17]). *Let $\mathbf{u}^\delta\in C^2([0,T);H_+)$ denote any solution of the initial-history value problem*

$$\begin{aligned} &\mathbf{u}^\delta_{tt}-\mathbf{N}\mathbf{u}^\delta+\int_{-\infty}^{t}\mathbf{K}(t-\tau)\mathbf{u}^\delta(\tau)\,d\tau=\mathbf{0}, \qquad 0\leqq t<T,\\ \text{(III.222)}\qquad &\mathbf{u}^\delta(0)=\delta\mathbf{f},\qquad \mathbf{u}^\delta_t(0)=\mathbf{g},\qquad \delta>0,\\ &\mathbf{u}^\delta(\tau)=\mathbf{U}(\tau), \qquad -\infty<\tau<0,\end{aligned}$$

where $\mathbf{K}(\cdot)$ *satisfies* (III.168), (III.169) *and the data satisfy* (III.170). *If* $T>\|\mathbf{f}\|^2/2\langle\mathbf{f},\mathbf{g}\rangle$ *then for each* $\delta\geqq\|\mathbf{g}\|/\langle\mathbf{f},\mathbf{N}\mathbf{f}\rangle^{1/2}$, $\mathbf{u}^\delta$ *satisfies the estimate* (III.173).

THEOREM (Bloom [17]). *Let* $\mathbf{u}^\varepsilon \in C^2([0, T); H_+)$ *denote any solution of the initial-history value problem*

$$\mathbf{u}^\varepsilon_{tt} - \mathbf{N}\mathbf{u}^\varepsilon + \int_{-\infty}^{t} \mathbf{K}(t-\tau)\mathbf{u}^\varepsilon(\tau)\,d\tau = \mathbf{0}, \qquad 0 \leqq t < T,$$

(III.223) $$\mathbf{u}^\varepsilon(0) = \mathbf{f}, \quad \mathbf{u}^\varepsilon_t(0) = \mathbf{g},$$

$$\mathbf{u}^\varepsilon(\tau) = \varepsilon\,\mathbf{U}(\tau), \qquad -\infty < \tau < 0, \quad \varepsilon > 0,$$

where $\mathbf{K}(\cdot)$ *satisfies* (III.168), (III.169) *and the data satisfy* (III.170). *If* $\langle \mathbf{f}, \mathbf{N}\mathbf{f}\rangle \geqq \|\mathbf{g}\|^2$ *then for any* $\beta > 0$ *and all* $T > 0$, $\mathbf{u}^\varepsilon$ *satisfies the estimate* (III.198).

Remark. The proofs of the two theorems above are formally analogous to the proofs for the holohedral isotropic case and may be found in [17], § 2; note that neither of the parameters ε, T is restricted as far as the growth estimate for the system (III.223) is concerned.

Remark. We comment briefly here on an application of the estimate (III.198), as it applies to the system (III.223), which models the initial-history boundary value problem for the Maxwell-Hopkinson dielectric under continuous one-parameter perturbations of the past history of the electric displacement field. We rewrite (III.198), with $\mathbf{u}$ replaced by $\mathbf{D}$ in the form

(III.224a) $$\sup_{-\infty < t < T} \left(\int_\Omega \frac{\partial D_i^\varepsilon}{\partial x_j}(\mathbf{x}, t) \frac{\partial D_i^\varepsilon}{\partial x_j}(\mathbf{x}, t)\,d\mathbf{x} \right)^{1/2} \geqq M_{T,\varepsilon},$$

(III.224b) $$M_{T,\varepsilon} = \sqrt{\frac{\varepsilon}{\gamma \pi_T}} \left(\left| \int_\Omega f_i(\mathbf{x}) \int_{-T_\infty}^{0} [\mathbf{K}(-\tau)\mathbf{U}_{T_\infty}(\mathbf{x}, \tau)]_i\,d\tau\,d\mathbf{x} \right| \right)^{1/2},$$

$$\pi_T \equiv \tfrac{1}{2}\|\mathbf{N}\|_{\mathcal{L}_s(H_+, H_-)} + \int_0^\infty \|\mathbf{K}(\rho)\|_{\mathcal{L}_s(H_+, H_-)}\,d\rho$$

(III.224c) $$+ T \int_0^\infty \|\mathbf{K}_\rho(\rho)\|_{\mathcal{L}_s(H_+, H_-)}\,d\rho$$

$$= \frac{1}{2\varepsilon\mu} + \frac{1}{\varepsilon\mu} \left(\int_0^\infty |\Phi(\rho)|\,d\rho + T \int_0^\infty |\dot{\Phi}(\rho)|\,d\rho \right),$$

where we have used (III.31), (III.35), (III.36). Suppose that $\phi(\cdot)$ is such that $\Phi(\cdot)$ has the form

(III.225) $$\Phi(t) = e^{-\lambda t}, \quad \text{some } \lambda > 0,$$

where λ is an unknown constitutive parameter. By (III.224c)

$$\pi_T = \frac{1}{2\varepsilon\mu} + \frac{1}{\varepsilon\mu}\left(\frac{1}{\lambda} + T \right)$$

(III.226) $$= \frac{1}{\varepsilon\mu} \left(\frac{\lambda + 2}{2\lambda} + T \right)$$

$$\equiv \pi_\gamma(\varepsilon, \mu, T).$$

Also, by (III.224b), (III.225), and (III.31),

$$
\begin{aligned}
&\int_\Omega f_i(\mathbf{x}) \int_{-\tau_\infty}^{0} [\mathbf{K}(-\tau)\mathbf{U}_{\tau_\infty}(\mathbf{x},\tau)]_i \, d\tau \, d\mathbf{x} \\
&\qquad = \int_\Omega f_i(\mathbf{x}) \int_{-\tau_\infty}^{0} \frac{\Phi(-\tau)}{\varepsilon\mu} [\nabla^2 \mathbf{U}_{\tau_\infty}(\mathbf{x},\tau)]_i \, d\tau \, d\mathbf{x} \\
&\qquad = \frac{1}{\varepsilon\mu} \int_{-\tau_\infty}^{0} \Phi(-\tau) \left(\int_\Omega f_i(\mathbf{x}) [\nabla^2 \mathbf{U}_{\tau_\infty}(\mathbf{x},\tau)]_i \, d\mathbf{x} \right) d\tau \\
&\qquad = -\frac{1}{\varepsilon\mu} \int_{-\tau_\infty}^{0} \Phi(-\tau) \left(\int_\Omega \frac{\partial f_i(\mathbf{x})}{\partial x_j} \cdot \frac{\partial [\mathbf{U}_{\tau_\infty}(\mathbf{x},\tau)]_i}{\partial x_j} \, d\mathbf{x} \right) d\tau,
\end{aligned}
\tag{III.227}
$$

so that, in view of our hypothesis (III.170, (iii))

$$
\begin{aligned}
&\left| \int_\Omega f_i(\mathbf{x}) \int_{-\tau_\infty}^{0} [\mathbf{K}(-\tau)\mathbf{U}_{\tau_\infty}(\mathbf{x},\tau)]_i \, d\tau \, d\mathbf{x} \right| \\
&\qquad = \frac{1}{\varepsilon\mu} \left(\int_{-\tau_\infty}^{0} \Phi(-\tau) \mathcal{D}(\tau) \, d\tau \right) \\
&\qquad = \frac{1}{\varepsilon\mu} \left(\int_{-\tau_\infty}^{0} e^{\lambda\tau} \mathcal{D}(\tau) \, d\tau \right),
\end{aligned}
\tag{III.228}
$$

where

$$
\mathcal{D}(\tau) = \int_\Omega \frac{\partial f_i(\mathbf{x})}{\partial x_j} \frac{\partial [\mathbf{U}_{\tau_\infty}(\mathbf{x},\tau)]_i}{\partial x_j} \, d\mathbf{x}
\tag{III.229}
$$

is computable once the data $\mathbf{f}$, $\mathbf{U}_{\tau_\infty}$ are prescribed. Thus, the estimate (III.224a, b, c) assumes the form

$$
\begin{aligned}
&\sup_{-\infty<t<T} \left[\int_\Omega \frac{\partial D_i^\varepsilon}{\partial x_j}(\mathbf{x},t) \frac{\partial D_i^\varepsilon}{\partial x_j}(\mathbf{x},t) \, d\mathbf{x} \right]^{1/2} \\
&\qquad \geqq \sqrt{\frac{\varepsilon}{\gamma\pi_\gamma}} \cdot \frac{1}{\varepsilon\mu} \left(\int_{-\tau_\infty}^{0} e^{\lambda\tau} \mathcal{D}(\tau) \, d\tau \right) \\
&\qquad \equiv Q(\lambda) \cdot \frac{1}{\mu} \sqrt{\frac{\varepsilon}{\gamma}},
\end{aligned}
\tag{III.230}
$$

where

$$
Q(\lambda) \equiv \frac{\displaystyle\int_{-\tau_\infty}^{0} e^{\lambda\tau} \mathcal{D}(\tau) \, d\tau}{\sqrt{\pi_\gamma}}.
\tag{III.231}
$$

If we rewrite (III.230) in the equivalent form

$$\text{(III.232)} \qquad Q(\lambda) \leqq \varepsilon\mu\sqrt{\frac{\gamma}{\varepsilon}} \sup_{-\infty<t<T} \left[\int_\Omega \frac{\partial D_i^\varepsilon}{\partial x_j} \frac{\partial D_i^\varepsilon}{\partial x_j} \, d\mathbf{x} \right]^{1/2},$$

then we have an estimate which can serve as the basis for a series of laboratory experiments aimed at determining a bound for the constitutive parameter λ; in fact, by the constitutive relation (III.5), with $\mathbf{E}(\tau) = \mathbf{0}$, $\rho < -\tau_\infty$, and $\phi(t) = \phi_\lambda(t)$, to indicate the dependence of $\phi(\cdot)$ on the parameter λ,

$$\text{(III.233)} \qquad \|\mathbf{D}^\varepsilon(t)\|_+ \leqq \varepsilon \|\mathbf{E}^\varepsilon(t)\|_+ + \sup_{[-\tau_\infty, T)} |\phi_\lambda| \int_{-\tau_\infty}^{t} \|\mathbf{E}^\varepsilon(\tau)\|_+ \, d\tau$$

for, $0 \leqq t < T$; therefore

$$\sup_{-\infty<t<T} \|\mathbf{D}^\varepsilon(t)\|_+ \leqq [\varepsilon + (T + \tau_\infty) \sup_{[-\tau_\infty, T)} |\phi_\lambda|] \sup_{-\infty<t<T} \|\mathbf{E}^\varepsilon(t)\|_+.$$

If we set

$$\text{(III.234)} \qquad R(\lambda) = \frac{Q(\lambda)}{[\varepsilon + (T + \tau_\infty) \sup_{[-\tau_\infty, T)} |\phi_\lambda|]},$$

then (III.232) implies that

$$\text{(III.235)} \qquad R(\lambda) \leqq \varepsilon\mu\sqrt{\frac{\gamma}{\varepsilon}} \sup_{-\infty<t<T} \left[\int_\Omega \frac{\partial E_i^\varepsilon}{\partial x_j} \frac{\partial E_i^\varepsilon}{\partial x_j} \, d\mathbf{x} \right]^{1/2}.$$

Besides the constitutive parameter λ, $R(\lambda)$ involves quantities which are either measurable or controllable, i.e., ε, μ, T, τ_∞, $\mathbf{f}$, $\mathbf{U}_{\tau_\infty}$, while the right-hand side of (III.235) involves some of these quantities, the variable parameter ε, and the corresponding value of the electric field, which should be measurable in the laboratory sense; in fact, what is needed is just the peak value of $\|\mathbf{E}^\varepsilon(t)\|_+$ over $[-\tau_\infty, T)$, and many such experiments may be conducted as we are free to vary τ_∞, T, $\mathbf{f}$, $\mathbf{U}_{\tau_\infty}$, and ε subject to the hypotheses (III.170).

5. Asymptotic estimates for electric displacement fields in material dielectrics.

In this concluding section we again consider the isotropic holohedral material dielectric which is defined by the constitutive relations (III.21a, b) with past histories given in the form (III.28); the system of partial-integrodifferential equations for the components of the electric displacement field $\mathbf{D}$ in such a dielectric is then given by (III.29), on $\Omega \times [0, T)$, $T > 0$, and the past history of $\mathbf{D}$ will be of the form (III.163). If we associate with the system (III.29) initial and boundary data of the form (III.15), (III.16) and again introduce the operators $\mathbf{N} \in \mathscr{L}_s(H_+, H_-)$, $\mathbf{K}(\cdot) \in L^2((-\infty, \infty); \mathscr{L}_s(H_+, H_-))$, which are defined by (III.162), our initial-history boundary value problem is then equivalent to the initial-history value problem (III.164)–(III.166), where we again assume that the past history $\mathbf{U}_{\tau_\infty}$ satisfies (III.167). In this section we will be interested in the behavior as $t \to +\infty$ of solutions $\mathbf{u}(t)$ to the system (III.164)–(III.166) which lie in the (globally) bounded class of functions

$$\text{(III.236)} \qquad \mathcal{N}_\infty = \{\mathbf{v} \in C^1([-\tau_\infty; 0), H_+) \cap C^2([0, \infty); H_+) | \sup_{[-\tau_\infty, \infty)} \|\mathbf{v}(\tau)\|_+ \leqq N\}$$

for some $N > 0$. We will show that under relatively mild conditions on $\mathbf{K}(\cdot)$, namely, (III.168) and

$$\text{(III.237)} \qquad \begin{aligned} &\mathcal{K}(\cdot), \hat{\mathcal{K}}(\cdot) \in L_1[0, \infty), && \hat{\mathcal{K}}(0) = 0, \\ &\mathcal{K}(t) = \|\mathbf{K}(t)\|_{\mathcal{L}_s(H_+, H_-)}, && \hat{\mathcal{K}}(t) = \int \|\mathbf{K}_t(t)\|_{\mathcal{L}_s(H_+, H_-)}\, dt, \end{aligned}$$

$\|\mathbf{u}(t)\|$ is bounded from below as $t \to +\infty$, whenever the initial energy $\mathcal{E}(0)$ is negative and sufficiently large in magnitude; this result will be valid regardless of the magnitude of the damping factor $\Psi(0) > 0$ in (III.164).

Remark. Results concerning the asymptotic stability of solutions to the initial-history value problem (III.164)–(III.166) may be gleaned from the work of Dafermos [38]; as was the case in Chapter II, however, Dafermos' theorems require that one impose rather strong definiteness assumptions on $\mathbf{K}(t)$ and $\mathbf{K}_t(t)$ for all $t > 0$.

Remark. Many authors ([41], [133]), and the references cited therein, have studied the asymptotic behavior of solutions to initial-value problems associated with damped evolution equations of the form

$$\text{(III.238)} \qquad \mathbf{u}_{tt} + \mathbf{A}\mathbf{u}_t + \mathbf{B}\mathbf{u} = \mathbf{0},$$

where $\mathbf{u}: [0, \infty) \to \mathcal{H}$, a real Hilbert space with inner-product $(\cdot, \cdot)$ and associated natural norm $|||(\cdot)|||$; it is usually assumed that $\mathbf{B} \in \mathcal{L}_S(\mathcal{H}; \mathcal{H})$ and satisfies a coerciveness condition of the form

$$\text{(III.239)} \qquad (\mathbf{v}, \mathbf{B}\mathbf{v}) \geqq |||\mathbf{v}|||^2, \qquad \lambda > 0 \quad \forall \mathbf{v} \in \mathcal{D}(\mathbf{B}),$$

where $\overline{\mathcal{D}(\mathbf{B})} = \mathcal{H}$. If the linear operator $\mathbf{A}$ satisfies $(\mathbf{A}\mathbf{v}, \mathbf{v}) \geqq 0$ and $\mathbf{A}^{-1}$ exists (the so-called strongly damped case) then it is well-known that the total energy $\mathcal{E}(t) = \frac{1}{2}(|||\mathbf{u}_t|||^2 + (\mathbf{u}(t), \mathbf{B}\mathbf{u}(t)))$ decays at a uniform exponential rate; even when $\mathbf{A}^{-1}$ does not exist (the weakly damped case) it can often be shown that $\mathcal{E}(t) \to 0$ as $t \to +\infty$. In [15] we considered the system

$$\text{(III.240)} \qquad \begin{aligned} &\mathbf{u}^\delta_{tt} + \Gamma \mathbf{u}^\delta_t - \mathbf{N}\mathbf{u}^\delta = \mathbf{0}, && \Gamma > 0, \quad 0 \leqq t < \infty, \\ &\mathbf{u}^\delta(0) = \delta \mathbf{f}, \quad \mathbf{u}^\delta_t(0) = \mathbf{g}, && \delta > 0, \quad \mathbf{f}, \mathbf{g} \in \mathcal{D}(\mathbf{N}). \end{aligned}$$

If $(\mathbf{v}, \mathbf{N}\mathbf{v}) \leqq -\lambda |||\mathbf{v}|||^2$, for some $\lambda > 0$, and all $\mathbf{v} \in \mathcal{D}(N)$, then asymptotic stability in the energy norm follows immediately; however, it is shown in [15] that if $\mathbf{N}$ is symmetric with $(\mathbf{v}, \mathbf{N}\mathbf{v}) \geqq 0$ $\forall \mathbf{v} \in \mathcal{D}(\mathbf{N})$, and $\exists \hat{\mathbf{f}} \in \mathcal{D}(\mathbf{N})$ such that $(\hat{\mathbf{f}}, \mathbf{N}\hat{\mathbf{f}}) > 0$, then any solution $\mathbf{u}^\delta \in C^2([0, \infty); \mathcal{D}(\mathbf{N}))$ must satisfy, for $\mathbf{f} = \hat{\mathbf{f}}$, and δ sufficiently large,

(III.241) $$\lim_{t\to+\infty} \|\mathbf{u}^\delta(t)\|^2 \geqq \delta^2\|\hat{\mathbf{f}}\|^2 \exp(-\Sigma_0(\delta, \Gamma)),$$

where $\Sigma_0(\delta, \Gamma)$ satisfies $\lim_{\Gamma\to+\infty} \Sigma_0(\delta, \Gamma) = 0$ (i.e., solutions are asymptotically bounded away from zero, for $\delta > 0$ sufficiently large, no matter how strong the damping is). The asymptotic lower bound (III.241) is obtained in [15] by employing a differential inequality argument, which is neither of the pure logarithmic convexity or concavity type, to establish the estimate

(III.242) $$\|\mathbf{u}^\delta(t)\|^2 \geqq \delta^2\|\hat{\mathbf{f}}\|^2 \exp\left[\left(\frac{(\hat{\mathbf{f}}, \mathbf{g})}{\delta\Gamma\|\hat{\mathbf{f}}\|^2}\right)(1-e^{-\Gamma t})\right]$$

for all $t \geqq 0$, $\delta > \|\mathbf{g}\|/\sqrt{(\hat{\mathbf{f}}, \mathbf{N}\hat{\mathbf{f}})}$; it is not required that $\mathbf{u}^\delta$ be a priori restricted to lie in a class of bounded perturbations. The estimate (III.242) may be extended to the case where $\mathbf{N} \in \mathscr{L}_s(H_+, H_-)$, $\mathbf{u}^\delta : [0, \infty) \to H_+$ where $H_+ \subseteq H$, both algebraically and topologically, (H a real Hilbert space with inner-product $\langle\ ,\ \rangle$, e.g., $H = (L_2(\Omega))^n$, $\Omega \subseteq R^n$ an open bounded domain) and H_- is the completion of H under the norm $\|(\cdot)\|_-$ defined via

$$\|\mathbf{w}\|_- = \sup_{\mathbf{v}\in H_+} \left(\frac{|\langle \mathbf{v}, \mathbf{w}\rangle|}{\|\mathbf{v}\|_+}\right),$$

$$(\text{e.g., } H_+ = (H_0^1(\Omega))^n, \quad H_- = (H^{-1}(\Omega))^n).$$

For the system (III.164)–(III.166) we will derive asymptotic bounds of the form (III.241) without introducing a one-parameter family of initial-data functions of the form $\delta\mathbf{f}$, $\delta > 0$, *and without making any definiteness assumptions on* $\mathbf{N}$; our assumptions on $\mathbf{K}(\cdot)$ will be given by (III.168) and (III.237). We have, in fact, the following preliminary results:

THEOREM (Bloom [13]). *Let* $\mathbf{u} \in \mathcal{N}_\infty$ *be a solution of* (III.164)–(III.166) *and let* $\mathbf{K}(\cdot)$ *satisfy the hypotheses* (III.168), (III.237). *If* $\mathscr{E}(0) = \frac{1}{2}\|\mathbf{g}\|^2 - \frac{1}{2}\langle \mathbf{f}, \mathbf{N}\mathbf{f}\rangle < 0$ *with*

(III.243) $$|\mathscr{E}(0)| > \tfrac{3}{2}\gamma N^2(\|\mathscr{K}\|_{L_1[0,\infty)} + \|\hat{\mathscr{K}}\|_{L_1[0,\infty)}),$$

then for all t, $0 \leqq t < \infty$, *and any* $\beta > 0$, $F(t) = \|\mathbf{u}(t)\|^2$ *satisfies the differential inequality*

(III.244) $$FF'' - \left(\frac{\beta+1}{2\beta+1}\right)F'^2 \geqq -\Psi(0)FF', \qquad \Psi(0) > 0.$$

Proof. Directly from the definition of $F(t)$, we find, in the usual manner, that for any $\beta > 0$

(III.245) $$FF'' - (\beta+1)F'^2 \geqq 4(\beta+1)S_\beta^2 + 2F(\langle \mathbf{u}, \mathbf{u}_{tt}\rangle - (2\beta+1)\|\mathbf{u}_t\|^2),$$

where $S_\beta^2(t) = \|\mathbf{u}\|^2\|\mathbf{u}_t\|^2 - \langle \mathbf{u}, \mathbf{u}_t\rangle^2 \geqq 0$ by the Schwarz inequality. Therefore, for

$\beta>0$ we have

$$FF''-(\beta+1)F'^2 \geqq 2FG_\beta, \qquad 0 \leqq t < \infty, \tag{III.246}$$

where, by virtue of the integrodifferential equation (III.164)

$$\begin{aligned} G_\beta(t) &= \langle \mathbf{u}, \mathbf{N}\mathbf{u}\rangle - \Psi(0)\langle \mathbf{u}, \mathbf{u}_t\rangle - (2\beta+1)\|\mathbf{u}_t\|^2 \\ &\quad - \left\langle \mathbf{u}, \int_{-\infty}^{t} \mathbf{K}(t-\tau)\mathbf{u}(\tau)\, d\tau \right\rangle \\ &= -\frac{\Psi(0)}{2} F'(t) - (2\beta+1)[\|\mathbf{u}_t\|^2 - \langle \mathbf{u}, \mathbf{N}\mathbf{u}\rangle] \\ &\quad - 2\langle \mathbf{u}, \mathbf{N}\mathbf{u}\rangle - \left\langle \mathbf{u}, \int_{-\infty}^{t} \mathbf{K}(t-\tau)\mathbf{u}(\tau)\, d\tau \right\rangle \\ &= -\frac{\Psi(0)}{2} F'(t) - 2(2\beta+1)\mathscr{E}(t) - 2\beta\langle \mathbf{u}, \mathbf{N}\mathbf{u}\rangle \\ &\quad - \left\langle \mathbf{u}, \int_{-\infty}^{t} \mathbf{K}(t-\tau)\mathbf{u}(\tau)\, d\tau \right\rangle, \end{aligned} \tag{III.247}$$

in view of the definitions of $F(t)$, $\mathscr{E}(t)$. If we take the $\langle\cdot,\cdot\rangle$ inner-product of (III.164) with $\mathbf{u}_t$ and integrate we easily obtain

$$\begin{aligned} \mathscr{E}(t) &= \mathscr{E}(0) - \Psi(0)\int_0^t \|\mathbf{u}_\tau\|^2\, d\tau \\ &\quad - \int_0^t \left\langle \mathbf{u}_\tau, \int_{-\infty}^{\tau} \mathbf{K}(\tau-\lambda)\mathbf{u}(\lambda)\, d\lambda \right\rangle d\tau \end{aligned} \tag{III.248}$$

and so, upon substituting for $\mathscr{E}(t)$ in (III.247) we obtain

$$\begin{aligned} G_\beta(t) &\geqq -\frac{\Psi(0)}{2} F' - 2(2\beta+1)\mathscr{E}(0) - 2\beta\langle \mathbf{u}, \mathbf{N}\mathbf{u}\rangle \\ &\quad + 2(2\beta+1)\int_0^t \left\langle \mathbf{u}_\tau, \int_{-\infty}^{\tau} \mathbf{K}(\tau-\lambda)\mathbf{u}(\lambda)\, d\lambda \right\rangle d\tau \\ &\quad - \left\langle \mathbf{u}, \int_{-\infty}^{t} \mathbf{K}(t-\tau)\mathbf{u}(\tau)\, d\tau \right\rangle, \end{aligned} \tag{III.249}$$

where we have dropped a nonnegative term proportional to $\int_0^t \|\mathbf{u}_\tau\|^2\, d\tau$. If we now take the $\langle\cdot,\cdot\rangle$ inner-product of (III.164) with $\mathbf{u}(t)$ and use the definition of $F(t)$ we easily obtain the identity

$$\begin{aligned} \frac{1}{2}F'' + \frac{\Psi(0)}{2}F' &= \|\mathbf{u}_t\|^2 + \langle \mathbf{u}, \mathbf{N}\mathbf{u}\rangle \\ &\quad - \left\langle \mathbf{u}, \int_{-\infty}^{t} \mathbf{K}(t-\tau)\mathbf{u}(\tau)\, d\tau \right\rangle \end{aligned} \tag{III.250}$$

which implies that

$$(\text{III}.251)\qquad \begin{aligned}-2\beta\langle \mathbf{u}, \mathbf{N}\mathbf{u}\rangle = -\beta F'' - \beta\Psi(0)F' + 2\beta\|\mathbf{u}_t\|^2\\ -2\beta\left\langle \mathbf{u}, \int_{-\infty}^{t} \mathbf{K}(t-\tau)\mathbf{u}(\tau)\,d\tau\right\rangle.\end{aligned}$$

By substituting from (III.251) into (III.249), collecting terms, and dropping a nonnegative expression proportional to $\|\mathbf{u}_t\|^2$ we obtain the following estimate for $G_\beta(t)$:

$$(\text{III}.252)\qquad \begin{aligned}G_\beta(t) \geqq -\Psi(0)(\beta+\tfrac{1}{2})F' - \beta F'' - 2(2\beta+1)\mathscr{E}(0)\\ -(2\beta+1)\left\langle \mathbf{u}, \int_{-\infty}^{t} \mathbf{K}(t-\tau)\mathbf{u}(\tau)\,d\tau\right\rangle\\ +2(2\beta+1)\int_0^t \left\langle \mathbf{u}_\tau, \int_{-\infty}^{\tau} \mathbf{K}(\tau-\lambda)\mathbf{u}(\lambda)\,d\lambda\right\rangle d\tau.\end{aligned}$$

By combining this last estimate with (III.246) we arrive at the differential inequality

$$(\text{III}.253)\qquad \begin{aligned}FF'' - (\beta+1)F'^2 \geqq -2\Psi(0)(\beta+\tfrac{1}{2})FF' - 2\beta FF''\\ -4(2\beta+1)\mathscr{E}(0)F - 2(2\beta+1)F\left\langle \mathbf{u}, \int_{-\infty}^{t} \mathbf{K}(t-\tau)\mathbf{u}(\tau)\,d\tau\right\rangle\\ +4(2\beta+1)F\int_0^t \left\langle \mathbf{u}_\tau, \int_{-\infty}^{\tau} \mathbf{K}(\tau-\lambda)\mathbf{u}(\lambda)\,d\lambda\right\rangle d\tau,\end{aligned}$$

which is equivalent to

$$(\text{III}.254)\qquad \begin{aligned}FF'' - \left(\frac{\beta+1}{2\beta+1}\right)F'^2 \geqq -\Psi(0)FF' - 4F\mathscr{E}(0)\\ -2F\left\langle \mathbf{u}, \int_{-\infty}^{t} \mathbf{K}(t-\tau)\mathbf{u}(\tau)\,d\tau\right\rangle\\ +4F\int_0^t \left\langle \mathbf{u}_\tau, \int_{-\infty}^{\tau} \mathbf{K}(\tau-\lambda)\mathbf{u}(\lambda)\,d\lambda\right\rangle d\tau,\end{aligned}$$

this estimate being valid on $[0, \infty)$ for all $\beta > 0$. In view of our hypothesis that $\mathscr{E}(0) < 0$ we can rewrite (III.254) in the form

$$(\text{III}.255)\qquad \begin{aligned}FF'' - \left(\frac{\beta+1}{2\beta+1}\right)F'^2 \geqq -\Psi(0)FF' + 2F\Big(2|\mathscr{E}(0)| - \left\langle \mathbf{u}, \int_{-\infty}^{t} \mathbf{K}(t-\tau)\mathbf{u}(\tau)\,d\tau\right\rangle\\ +2\int_0^t \left\langle \mathbf{u}_\tau, \int_{-\infty}^{\tau} \mathbf{K}(\tau-\lambda)\mathbf{u}(\lambda)\,d\lambda\right\rangle d\tau\Big).\end{aligned}$$

We now want to bound the expressions in (III.255) involving the operator $\mathbf{K}(\cdot)$; prior to that, however, let us note that the familiar identity

$$\left\langle \mathbf{u}_\tau, \int_{-\infty}^{\tau} \mathbf{K}(\tau-\lambda)\mathbf{u}(\lambda)\,d\lambda \right\rangle = \frac{d}{d\tau}\left\langle \mathbf{u}(\tau), \int_{-\infty}^{\tau} \mathbf{K}(\tau-\lambda)\mathbf{u}(\lambda)\,d\lambda \right\rangle - \left\langle \mathbf{u}(\tau), \int_{-\infty}^{\tau} \mathbf{K}_\tau(\tau-\lambda)\mathbf{u}(\lambda)\,d\lambda \right\rangle - \langle \mathbf{u}(\tau), \mathbf{K}(0)\mathbf{u}(\tau)\rangle$$

enables us to recast (III.255) in the form

$$\begin{aligned} FF'' - \left(\frac{\beta+1}{2\beta+1}\right)F'^2 &\geqq -\Psi(0)FF' + 2F\Big(2|\mathscr{E}(0)| - 2\int_0^t \langle \mathbf{u}(\tau), \mathbf{K}(0)\mathbf{u}(\tau)\rangle\,d\tau \\ &\quad - 2\left\langle \mathbf{f}, \int_{-\infty}^{0} \mathbf{K}(-\tau)\mathbf{U}(\tau)\,d\tau \right\rangle + \left\langle \mathbf{u}, \int_{-\infty}^{t} \mathbf{K}(t-\tau)\mathbf{u}(\tau)\,d\tau \right\rangle \\ &\quad - 2\int_0^t \left\langle \mathbf{u}(\tau), \int_{-\infty}^{\tau} \mathbf{K}_\tau(\tau-\lambda)\mathbf{u}(\lambda)\,d\lambda \right\rangle d\tau\Big). \end{aligned} \tag{III.256}$$

For the integrals on the right-hand side of (III.256) we now have the following series of estimates:

$$\begin{aligned} &\left|\left\langle \mathbf{f}, \int_{-\infty}^{0} \mathbf{K}(-\tau)\mathbf{u}(\tau)\,d\tau \right\rangle\right| \\ &\qquad \leqq \gamma\|\mathbf{f}\| \sup_{[-\tau_\infty,0)} \|\mathbf{U}_{\tau_\infty}\|_+ \int_{-\infty}^{0} \|\mathbf{K}(-\tau)\|_{\mathscr{L}_s(H_+,H_-)}\,d\tau \\ &\qquad \leqq \gamma\Big(\sup_{[-\tau_\infty,0)} \|\mathbf{u}\|_+\Big)^2 \int_0^{\infty} \|\mathbf{K}(\tau)\|_{\mathscr{L}_s(H_+,H_-)}\,d\tau \\ &\qquad \leqq \gamma N^2 \|\mathscr{K}\|_{L_1[0,\infty)} \end{aligned}$$

by virtue of our hypothesis that $\mathbf{u} \in \mathscr{N}_\infty$. Thus,

$$-\left\langle \mathbf{f}, \int_{-\infty}^{0} \mathbf{K}(-\tau)\mathbf{u}(\tau)\,d\tau \right\rangle \geqq -\gamma N^2 \|\mathscr{K}\|_{L_1[0,\infty)}, \qquad t>0. \tag{III.257}$$

Also,

$$\begin{aligned} &\left|\left\langle \mathbf{u}, \int_{-\infty}^{t} \mathbf{K}(t-\tau)\mathbf{u}(\tau)\,d\tau \right\rangle\right| \\ &\qquad \leqq \Big(\sup_{[-\tau_\infty,\infty)} \|\mathbf{u}(t)\|_+\Big)^2 \int_{-\infty}^{t} \|\mathbf{K}(t-\tau)\|_{\mathscr{L}_s(H_+,H_-)}\,d\tau \\ &\qquad \leqq \gamma N^2 \|\mathscr{K}\|_{L_1[0,\infty)}, \end{aligned}$$

so that, for $t>0$

$$\left\langle \mathbf{u}, \int_{-\infty}^{t} \mathbf{K}(t-\tau)\mathbf{u}(\tau)\, d\tau \right\rangle \geqq -\gamma N^2 \|\mathscr{K}\|_{L_1[0,\infty)}. \tag{III.258}$$

Finally,

$$\begin{aligned}
&\left|\int_0^t \left\langle \mathbf{u}(\tau), \int_{-\infty}^{\tau} \mathbf{K}_\tau(\tau-\lambda)\mathbf{u}(\lambda)\, d\lambda \right\rangle d\tau\right| \\
&\qquad \leqq \int_0^\infty \left(\|\mathbf{u}(\tau)\| \int_{-\tau_\infty}^{\tau} \|\mathbf{K}_\tau(\tau-\lambda)\|_{\mathscr{L}_s(H_+,H_-)} \|\mathbf{u}(\lambda)\|_+\, d\lambda \right) d\tau \\
&\qquad \leqq \gamma N^2 \int_0^\infty \int_0^{\tau+\tau_\infty} \|\mathbf{K}_\rho(\rho)\|_{\mathscr{L}_s(H_+,H_-)}\, d\rho\, d\tau \\
&\qquad = \gamma N^2 \int_0^\infty (\hat{\mathscr{K}}(\rho)|_0^{\tau+\tau_\infty})\, d\tau \\
&\qquad = \gamma N^2 \int_{\tau_\infty}^\infty \hat{\mathscr{K}}(\lambda)\, d\lambda \leqq \gamma N^2 \|\hat{\mathscr{K}}\|_{L_1[0,\infty)},
\end{aligned}$$

by virtue of our assumption that $\hat{\mathscr{K}}(0)=0$. Thus, for $0 \leqq t < \infty$,

$$-\int_0^t \left\langle \mathbf{u}(\tau), \int_{-\infty}^{\tau} \mathbf{K}_\tau(\tau-\lambda)\mathbf{u}(\lambda)\, d\lambda \right\rangle d\tau \geqq -\gamma N^2 \|\hat{\mathscr{K}}\|_{L_1[0,\infty)}. \tag{III.259}$$

We now combine the differential inequality (III.256) with the estimates (III.257)–(III.259) and employ the hypothesis (III.268) so as to obtain

$$\begin{aligned}
FF'' - \left(\frac{\beta+1}{2\beta+1}\right)F'^2 &\geqq -\Psi(0)FF' \\
&\quad + 2F(2|\mathscr{E}(0)| - 3\gamma N^2 \{\|\mathscr{K}\|_{L_1[0,\infty)} + \|\hat{\mathscr{K}}\|_{L_1[0,\infty)}\})
\end{aligned} \tag{III.260}$$

for $0 \leqq t < \infty$, and any $\beta > 0$. The stated result, i.e., (III.244) now follows directly from (III.260) if we use the hypothesis (III.243) relative to $|\mathscr{E}(0)|$. Q.E.D.

As a direct consequence of the above theorem, we have the following result, from which the previously stated conclusions about the asymptotic behavior of $\|\mathbf{u}(t)\|$, as $t \to +\infty$, follow directly:

COROLLARY. *Under the conditions which prevail in the above theorem, for any solution* $\mathbf{u} \in \mathscr{N}_\infty$ of (III.164)–(III.166), *all* $t>0$, *and any* α, $\frac{1}{2} < \alpha < 1$

$$\|\mathbf{u}(t)\|^2 \geqq \|\mathbf{f}\|^2 \left[1 + \frac{2(1-\alpha)\langle \mathbf{f}, \mathbf{g}\rangle}{\Psi(0)\|\mathbf{f}\|^2}(1-e^{-\Psi(0)t})\right]^{1/(1-\alpha)}. \tag{III.261}$$

Proof. For any $\alpha > 0$ (compare with the analysis, (II.144)–(II.149)):

$$[F^{(1-\alpha)}]''(t) = (1-\alpha)F^{-\alpha-1}(t)[F(t)F''(t) - \alpha F'^2(t)]. \tag{III.262}$$

Therefore, by (III.244), with $\tilde{\alpha} \equiv (\beta+1)/(2\beta+1) < \frac{1}{2}$, for $0 < \beta < \infty$, it follows that

$$(1-\tilde{\alpha})F^{-\tilde{\alpha}-1}(FF'' - \tilde{\alpha}F'^2) \geqq -\Psi(0)(1-\tilde{\alpha})F^{-\tilde{\alpha}}F' \tag{III.263}$$

or

$$\begin{aligned}[F^{(1-\tilde{\alpha})}]''(t) &\geqq -\Psi(0)(1-\tilde{\alpha})F^{-\tilde{\alpha}}F' \\ &= -\Psi(0)[F^{(1-\tilde{\alpha})}]'(t).\end{aligned} \tag{III.264}$$

As β ranges over $(0, \infty)$, $\tilde{\alpha}$ ranges over $(\frac{1}{2}, 1)$. Two successive integrations of (III.264) now yield, for all $t \geqq 0$, $\alpha \in (\frac{1}{2}, 1)$,

$$\begin{aligned}F^{(1-\tilde{\alpha})}(t) &\geqq F^{(1-\tilde{\alpha})}(0) + \frac{(1-\tilde{\alpha})F^{-\tilde{\alpha}}(0)F'(0)}{\Psi(0)}(1-e^{-\Psi(0)t}) \\ &= F^{(1-\tilde{\alpha})}(0)\left[1 + \frac{(1-\tilde{\alpha})F'(0)}{\Psi(0)F(0)}(1-e^{-\Psi(0)t})\right],\end{aligned} \tag{III.265}$$

from which the stated estimate (III.261), with $\alpha = \tilde{\alpha}$, follows if we take the $(1-\tilde{\alpha})$th root on both sides of (III.265) and use the definition of $F(t)$. Q.E.D.

Remarks. From (III.261) it follows directly that

$$\lim_{t\to+\infty} \|\mathbf{u}(t)\|^2 \geqq \|\mathbf{f}\|^2\left[1 + \left(\frac{2(1-\alpha)\langle \mathbf{f}, \mathbf{g}\rangle}{\Psi(0)\cdot\|\mathbf{f}\|^2}\right)\right]^{1/(1-\alpha)} \tag{III.266}$$

if $\Psi(0) > 0$, for all $\alpha \in (\frac{1}{2}, 1)$. Therefore

$$\lim_{\Psi(0)\to+\infty} \lim_{t\to+\infty} \|\mathbf{u}(t)\|^2 = \lim_{t\to+\infty} \lim_{\Psi(0)\to+\infty} \|\mathbf{u}(t)\|^2 \geqq \|\mathbf{f}\|^2 \tag{III.267}$$

showing that $\|\mathbf{u}(t)\|$ is asymptotically bounded away from zero, even as the damping factor $\Psi(0) \to +\infty$ in (III.164). From the definition of $\tilde{\alpha}$, it is clear that $\tilde{\alpha} \to 1^-$ as $\beta \to 0^+$; allowing $\alpha \to 1^-$ in (III.261) and noting the elementary fact that $\lim_{\lambda\to 0^+}[1+\lambda x]^{1/\lambda} = e^x$, $0 < x < \infty$, we obtain from (III.261) the growth estimate

$$\|\mathbf{u}(t)\|^2 \geqq \|\mathbf{f}\|^2 \exp\left[\left(\frac{2\langle \mathbf{f}, \mathbf{g}\rangle}{\Psi(0)\|\mathbf{f}\|^2}\right)(1-e^{-\Psi(0)t})\right] \tag{III.268}$$

for $0 \leqq t < \infty$, so that as $t \to +\infty$

$$\lim_{t\to+\infty} \|\mathbf{u}(t)\|^2 \geqq \|\mathbf{f}\|^2 \exp\left(\frac{2\langle \mathbf{f}, \mathbf{g}\rangle}{\Psi(0)\|\mathbf{f}\|^2}\right). \tag{III.269}$$

For the operator $\mathbf{K}(\cdot)$, which is defined by (III.162$_2$), we have already computed a set of conditions on the kernel functions $\Phi(t)$ and $\Psi(t)$ which insure that the hypothesis (III.168) is satisfied; those conditions are given by (III.215), i.e.,

$$\ddot{\Psi}(0) \geqq 0 \text{ and } \Phi(0) < 0,$$
$$\text{with } |\Phi(0)| > \frac{a_0\gamma^2\ddot{\Psi}(0)}{b_0}.$$

By (III.216), and the definitions (III.237) of $\mathscr{K}(\cdot)$, $\hat{\mathscr{K}}(\cdot)$, it follows that $\mathscr{K}(\cdot)$, $\hat{\mathscr{K}}(\cdot) \in L_1[0, \infty)$ with $\hat{\mathscr{K}}(0) = 0$ if $\Phi(t)$, $\Psi(t)$ satisfy the first set of conditions in (III.219) and

$$\text{(III.270)} \quad \begin{aligned} &\int_0^\infty \int_0^t |\Psi^{(3)}(\tau)|\, d\tau\, dt < \infty, \qquad \int_0^\infty \int_0^t |\dot{\Phi}(\tau)|\, d\tau\, dt < \infty, \\ &\int |\Psi^{(3)}(t)|\, dt\Big|_{t=0} = 0, \qquad \int |\dot{\Phi}(t)|\, dt\Big|_{t=0} = 0. \end{aligned}$$

Finally, by (III.208)

$$\langle \mathbf{v}, \mathbf{N}\mathbf{v} \rangle = -\dot{\Psi}(0)[c_0\|\mathbf{v}\|_+^2 + \|\mathbf{v}\|^2], \qquad \mathbf{v} \in H_+,$$

where $\mathbf{N}$ is defined by (III.162$_1$), so that

$$\text{(III.271)} \quad \begin{aligned} \mathscr{E}(0) &= \tfrac{1}{2}\|\mathbf{g}\|^2 - \tfrac{1}{2}\langle \mathbf{f}, \mathbf{N}\mathbf{f} \rangle \\ &= \tfrac{1}{2}\|\mathbf{g}\|^2 + \frac{\dot{\Psi}(0)}{2}(c_0\|\mathbf{f}\|_+^2 + \|\mathbf{f}\|^2) \\ &= \tfrac{1}{2}\|\mathbf{g}\|^2 + \frac{b_0}{2a_0}\|\mathbf{f}\|_+^2 + \frac{\dot{\Psi}(0)}{2}\|\mathbf{f}\|^2. \end{aligned}$$

It follows that $\mathscr{E}(0) < 0$ if and only if

$$\text{(III.272)} \quad \dot{\Psi}(0) < -\frac{1}{\|\mathbf{f}\|^2}\left(\|\mathbf{g}\|^2 + \left(\frac{b_0}{a_0}\right)\|\mathbf{f}\|_+^2\right).$$

If (III.272) is satisfied then

$$\text{(III.273)} \quad |\mathscr{E}(0)| = \frac{|\dot{\Psi}(0)|}{2}\|\mathbf{f}\|^2 - \frac{1}{2}\left(\|\mathbf{g}\|^2 + \left(\frac{b_0}{a_0}\right)\|\mathbf{f}\|_+^2\right).$$

Therefore, the initial energy $\mathscr{E}(0)$ satisfies the hypothesis (III.243) if $\dot{\Psi}(0) < 0$ with

$$\text{(III.274)} \quad \begin{aligned} |\dot{\Psi}(0)| > &\frac{1}{\|\mathbf{f}\|^2}\left(\|\mathbf{g}\|^2 + \left(\frac{b_0}{a_0}\right)\|\mathbf{f}\|_+^2\right) \\ &+ 3\gamma N^2(\|\mathscr{K}\|_{L_1[0,\infty)} + \|\hat{\mathscr{K}}\|_{L_1[0,\infty)}). \end{aligned}$$

Finally, by the definitions of $\mathcal{K}(\cdot)$ and $\hat{\mathcal{K}}(\cdot)$ and the estimates (III.216)

$$\|\mathcal{K}\|_{L_1[0,\infty)} \leqq \gamma^2 \int_0^\infty |\ddot{\Psi}(t)|\,dt + \left(\frac{b_0}{a_0}\right)\int_0^\infty |\Phi(t)|\,dt$$

$$\|\hat{\mathcal{K}}\|_{L_1[0,\infty)} \leqq \gamma^2 \int_0^\infty \int_0^\tau |\Psi^{(3)}(\lambda)|\,d\lambda\,d\tau + \left(\frac{b_0}{a_0}\right)\int_0^\infty \int_0^\tau |\dot{\Phi}(\lambda)|\,d\lambda\,d\tau$$

from which it follows that the hypothesis (III.243), relative to the initial energy $\mathscr{E}(0)$, will be satisfied if $\dot{\Psi}(0) < 0$ with

$$\text{(III.275)} \qquad |\dot{\Psi}(0)| > \frac{1}{\|\mathbf{f}\|^2}\left(\|\mathbf{g}\|^2 + \left(\frac{b_0}{a_0}\right)\|\mathbf{f}\|_+^2\right) + 3\gamma N^2\left(\frac{b_0}{a_0}\right)\left(\int_0^\infty |\Phi(t)|\,dt + \int_0^\infty \int_0^\tau |\dot{\Phi}(\lambda)|\,d\lambda\,d\tau\right) + 3\gamma^3 N^2\left(\int_0^\infty |\ddot{\Psi}(t)|\,dt + \int_0^\infty \int_0^\tau |\Psi^{(3)}(\lambda)|\,d\lambda\,d\tau\right).$$

Chapter IV

Some Recent Directions in Research on Nonlinear Integrodifferential Equations

In the last two chapters we have derived, using logarithmic convexity and concavity arguments, various stability and growth estimates for solutions of ill-posed initial-history boundary value problems associated with some partial-integrodifferential equations of viscoelasticity and electromagnetic theory; along the way we have given an indication of some of the pertinent results on stability, global existence, and asymptotic stability of solutions to the corresponding well-posed problems for the linear theories considered, as well as for some nonlinear problems in viscoelasticity. In this final chapter we will concentrate on some recent results for both well-posed and ill-posed initial-history value problems associated with some nonlinear integrodifferential equations. In § 1 we consider a natural nonlinear generalization of the integrodifferential equation (II.23) and show that, in certain special cases, a logarithmic convexity argument can again be applied so as to deduce information on the stability and growth of solutions which are restricted to lie in a specified class of uniformly bounded functions. In the second section we return to the nonlinear integrodifferential equation (II.285) which was considered by Slemrod [139]–[141] in his studies on the stability and instability of solutions to initial-history boundary value problems for a nonlinear viscoelastic fluid; in particular, we reproduce Slemrod's Riemann invariant argument which yields finite-time breakdown of solutions to the initial-value problem associated with the equivalent damped hyperbolic system (II.291), when the initial data are sufficiently large in an appropriate sense. In § 3 we review some recent work of Nohel [122] on the global existence, uniqueness, and continuous dependence of solutions to the pure initial-value problem associated with the nonlinear viscoelastic integrodifferential equation (II.229); like the argument in the second section of this chapter, Nohel's results are based on the concept of Riemann invariant and may be contrasted with the energy arguments of Dafermos and Nohel [43] which we presented in Chapter II. Finally, as almost all of the integrodifferential equations considered in the present work have been second-order, i.e., hyperbolic equations, we make a few remarks in the last section concerning recent research efforts on more problems which lead to first-order nonlinear integrodifferential equations; specifically, we consider some problems of heat conduction in nonlinear heat conductors with memory which lead to nonlinear integrodifferential equations of the first order.

1. Convexity arguments and growth estimates for solutions to a class of nonlinear integrodifferential equations.

In Chapter II we derived a series of growth estimates for solutions of the abstract initial-value problem

$$\text{(IV.1)}\qquad \begin{aligned} &\mathbf{u}_{tt} - \mathbf{N}\mathbf{u} + \int_0^t \mathbf{K}(t-\tau)\mathbf{u}(\tau)\,d\tau = \mathbf{0}, \qquad 0 \leqq t < T,\\ &\mathbf{u}(0) = \mathbf{f}, \qquad \mathbf{u}_t(0) = \mathbf{g}, \end{aligned}$$

where $\mathbf{u} \in C^1([0, T); H_+^\lambda)$ is such that $\mathbf{u}_t \in C^1([0, T); H_+^\lambda)$ and $\mathbf{u}_{tt} \in C([0, T); H_-)$ with H_+ (with inner-product $\langle \cdot, \cdot \rangle_+$) any subspace of a real Hilbert space H (with inner-product $\langle \cdot, \cdot \rangle$) such that $H_+ \subset H$ algebraically and topologically and H_- the dual of H_+ via the inner product of H; our estimates were obtained for solutions to (IV.1) which belong to uniformly bounded subsets $H_+^\lambda \subset H_+$ of the form

$$\text{(IV.2)}\qquad H_+^\lambda = \{\mathbf{v} \in H_+ \mid \|\mathbf{v}\|_+ \leqq \lambda;\ \lambda \in \mathcal{R}^+\}.$$

In (IV.1), $\mathbf{f}, \mathbf{g} \in H_+^\lambda$, $\mathbf{N} \in \mathcal{L}_S(H_+^\lambda, H_-)$, and $\mathbf{K}(t) \in L^2([0, T); \mathcal{L}_S(H_+^\lambda, H_-))$, such that the strong operator derivative $\mathbf{K}_t$ exists in $\mathcal{L}_S(H_+^\lambda, H_-)$, where $\mathcal{L}_S(H_+^\lambda, H_-)$ denotes the space of bounded symmetric linear operators from H_+^λ into H_-.

The results of Chapter II were applied in that chapter to linear viscoelastic materials and in Chapter III our results were shown to apply to rigid nonconducting material dielectrics which are governed by the constitutive relations of Maxwell-Hopkinson as well as to dielectrics of holohedral isotropic type.

In this section we will demonstrate that it is possible to generalize the results of Chapter II to classes of uniformly bounded solutions to[1] abstract initial-

[1] Without definiteness assumptions on $\mathbf{N}$ and $\mathbf{K}$ in (IV.1) and analogous assumptions on $\mathbf{L}$ and the nonlinear map $\mathbf{G}$ in (IV.3) the initial-value problems associated with (IV.1) and (IV.3) are ill-posed and it is not possible, in general, to prove results concerning the existence, uniqueness, stability, or asymptotic stability of solutions. As was the case in Chapter II, the purpose of the present work is to show that the stability results of F. John [69] for bounded classes of solutions to ill-posed initial boundary value problems for partial differential equations, can be extended to solutions of (nonlinear) integrodifferential equations which lie in appropriate classes of bounded functions. Our growth estimates are also valid, of course, for those well-posed problems where sufficient additional hypotheses on the relevant operators are made which insure the existence of solutions in the prescribed bounded class.

value problems associated with the nonlinear integrodifferential equation

$$\text{(IV.3)} \qquad \mathbf{u}_{tt} - \mathbf{L}\mathbf{u} + \int_0^t \mathbf{G}(t-\tau; \mathbf{u}(\tau))\, d\tau = \mathbf{0}, \qquad 0 \leqq t < T,$$

where $\mathbf{L} \in \mathscr{L}_S(H_+^\lambda, H_-)$ while $\mathbf{G}: [0, T) \times H_+^\lambda \to H_-$ is continuous on H_+^λ, for each fixed $t \in [0, T)$, and of class C^1 on $[0, T)$ for each fixed $\mathbf{v} \in H_+$. To this end let $\lambda \in \mathscr{R}^+$ be arbitrary and define H_+^λ as in (IV.2). We consider solutions $\mathbf{u} \in C^2([0, T); H_+^\lambda)$ of (IV.3) subject to the initial conditions in (IV.1) and we begin by defining a real-valued function $F(t; \beta, t_0)$ via

$$\text{(IV.4)} \qquad F(t; \beta t_0) = \|\mathbf{u}(t)\|^2 + \beta(t+t_0)^2, \qquad 0 \leqq t < T,$$

where β, t are arbitrary nonnegative real numbers. We then have the following:

THEOREM (Bloom [16]). *Let* $\mathbf{u} \in C^2([0, T); H_+^\lambda)$ *be any solution of* (IV.3), (IV.1_2) *and suppose that*

(i) *there exist bounded real-valued functions* $\phi_\lambda(t), \psi_\lambda(t), 0 \leqq t < T$, *with* $\phi_\lambda(t) > 0$, $\psi_\lambda(t) > 0$ *for all* $t \in [0, T)$, *such that*

$$\text{(IV.5a)} \qquad \|\mathbf{G}(t, \mathbf{v})\|_- \leqq \phi_\lambda(t)\|\mathbf{v}\|_+, \qquad \mathbf{v} \in H_+^\lambda, \quad t \in [0, T),$$

$$\text{(IV.5b)} \qquad \|\mathbf{G}_t(t, \mathbf{v})\|_- \leqq \psi_\lambda(t)\|\mathbf{v}\|_+, \qquad \mathbf{v} \in H_+^\lambda, \quad t \in [0, T),$$

(ii) *there exists a constant* $\rho_\lambda > T \cdot \sup_{0 \leqq t \leqq T} \psi_\lambda(t)$ *such that*

$$\text{(IV.5c)} \qquad -\langle \mathbf{v}, \mathbf{G}(0, \mathbf{v})\rangle \geqq \rho_\lambda \|\mathbf{v}\|_+^2 \quad \forall \mathbf{v} \in H_+^\lambda.$$

Then for each t, $0 \leqq t < T$, $F(t; \beta, t_0)$ *satisfies*

$$\text{(IV.6)} \qquad FF'' - F'^2 \geqq -2F\left(2\mathscr{E}(0) + \beta + \lambda^2 \sup_{[0,T)} \int_0^t \phi_\lambda(t-\tau)\, d\tau\right),$$

where $\mathscr{E}(0) = \frac{1}{2}\|\mathbf{g}\|^2 - \frac{1}{2}\langle \mathbf{f}, \mathbf{L}\mathbf{f}\rangle$ *is the initial (total) energy.*[2]

Proof. From the definition of $F(t; \beta, t_0)$ we have, for $0 \leqq t < T$,

$$\text{(IV.7)} \qquad F'(t; \beta, t_0) = 2\langle \mathbf{u}, \mathbf{u}_t\rangle + 2\beta(t+t_0)$$

and

$$\text{(IV.8)} \qquad F''(t; \beta, t_0) = 2\langle \mathbf{u}_t, \mathbf{u}_t\rangle + 2\beta + 2\langle \mathbf{u}, \mathbf{L}\mathbf{u}\rangle - 2\left\langle \mathbf{u}, \int_0^t \mathbf{G}(t-\tau; \mathbf{u}(\tau))\, d\tau\right\rangle,$$

where we have substituted for $\mathbf{u}_{tt}$ in (IV.8) according to (IV.3). If we use the definitions $\mathscr{K}(t) = \frac{1}{2}\|\mathbf{u}_t\|^2$ (kinetic energy), $\mathscr{P}(t) = -\frac{1}{2}\langle \mathbf{u}, \mathbf{L}\mathbf{u}\rangle$ (potential energy)

[2] Hypotheses (IV.5a, b) simply indicate that for each t, $0 \leqq t < T$, $\mathbf{G}$ and $\mathbf{G}_t$ are bounded (nonlinear maps) from H_+^λ into H_-, while the hypothesis expressed by (IV.5c) is simply the direct generalization, to the nonlinear situation, of the basic hypothesis $-\langle \mathbf{v}, \mathbf{K}(0), \mathbf{v}\rangle \geqq \kappa\|\mathbf{v}\|_+^2$, for κ sufficiently large, that was employed in Chapter 1 to treat the linear problem (IV.1).

and $\mathscr{E}(t)=\mathscr{K}(t)+\mathscr{P}(t)$ (total energy) then we may rewrite (IV.8) in the form

$$\text{(IV.9)}\qquad \begin{aligned} F''(t;\beta,t_0)&=4\mathscr{K}(t)-4(\mathscr{E}(0)-\mathscr{K}(t))\\ &\quad-4(\mathscr{E}(t)-\mathscr{E}(0))-2\left\langle \mathbf{u}(t),\int_0^t \mathbf{G}(t-\tau;\mathbf{u}(\tau))\,d\tau\right\rangle. \end{aligned}$$

However,

$$\text{(IV.10)}\qquad \begin{aligned} \mathscr{E}(t)-\mathscr{E}(0)&=-\left\langle \mathbf{u}(t),\int_0^t \mathbf{G}(t-\tau;\mathbf{u}(\tau))\,d\tau\right\rangle\\ &\quad+\int_0^t \langle \mathbf{u}(\tau),\mathbf{G}(0;\mathbf{u}(\tau))\rangle\,d\tau\\ &\quad+\int_0^t \left\langle \mathbf{u}(\tau),\int_0^t \mathbf{G}_\tau(\tau-\sigma;\mathbf{u}(\sigma))\,d\sigma\right\rangle d\tau, \end{aligned}$$

and therefore

$$\text{(IV.11)}\qquad \begin{aligned} FF''-F'^2&=4F(2\mathscr{K}(t)+\beta)-F'^2-2F(2\mathscr{E}(0)+\beta)\\ &\quad+2F\left\langle \mathbf{u}(t),\int_0^t \mathbf{G}(t-\tau;\mathbf{u}(\tau))\,d\tau\right\rangle-4F\int_0^t\langle \mathbf{u}(\tau),\mathbf{G}(0;\mathbf{u}(\tau))\rangle\,d\tau\\ &\quad-4F\int_0^t\left\langle \mathbf{u}(\tau),\int_0^\tau \mathbf{G}_\tau(\tau-\sigma;\mathbf{u}(\sigma))\,d\sigma\right\rangle d\tau. \end{aligned}$$

From the definition of $F(t;\beta,t_0)$ (and the Schwarz inequality)

$$\text{(IV.12)}\qquad H(t;\beta,t_0)\equiv F(t;\beta,t_0)(\|\mathbf{u}_t\|^2+\beta)-\tfrac{1}{4}F'^2(t;\beta,t_0)\geqq 0,$$

so (IV.11) may be replaced by the differential inequality

$$\text{(IV.13)}\qquad \begin{aligned} FF''-F'^2&\geqq-2F(2\mathscr{E}(0)+\beta)+2F\left\langle \mathbf{u}(t),\int_0^t \mathbf{G}(t-\tau;\mathbf{u}(\tau))\,d\tau\right\rangle\\ &\quad-4F\int_0^t\langle \mathbf{u}(\tau),\mathbf{G}(0;\mathbf{u}(\tau))\rangle\,d\tau\\ &\quad-4F\int_0^t\left\langle \mathbf{u}(\tau),\int_0^\tau \mathbf{G}_\tau(\tau-\sigma;\mathbf{u}(\sigma))\,d\sigma\right\rangle d\tau. \end{aligned}$$

We now bound, from below, the sum of the last three expressions on the right-hand side of (IV.13). As $\|\mathbf{w}\|_-=\sup_{\mathbf{v}\in H_+}(|\langle \mathbf{v},\mathbf{w}\rangle|/\|\mathbf{v}\|_+)$, for $\mathbf{u}\in H_+^\lambda$

$$\text{(IV.14)}\qquad \begin{aligned} \left|\left\langle \mathbf{u}(t),\int_0^t \mathbf{G}(t-\tau;\mathbf{u}(\tau))\,d\tau\right\rangle\right|&\leqq\|\mathbf{u}(t)\|_+\int_0^t\|\mathbf{G}(t-\tau;\mathbf{u}(\tau))\|_-\,d\tau\\ &\leqq\|\mathbf{u}(t)\|_+\int_0^t\phi_\lambda(t-\tau)\|\mathbf{u}(\tau)\|_+\,d\tau\\ &\leqq\lambda^2\sup_{0\leqq t<T}\left(\int_0^t\phi_\lambda(t-\tau)\,d\tau\right). \end{aligned}$$

Thus

$$\text{(IV.15)}\qquad \left\langle \mathbf{u}(\tau), \int_0^t \mathbf{G}(t-\tau; \mathbf{u}(\tau))\,d\tau \right\rangle \geqq -\lambda^2 \cdot \sup_{0\leqq t<T} \left(\int_0^t \phi_\lambda(t-\tau)\,d\tau \right).$$

Next, it follows directly from (IV.5c) that

$$\text{(IV.16)}\qquad -\int_0^t \langle \mathbf{u}(\tau), \mathbf{G}(0; \mathbf{u}(\tau))\,d\tau \geqq \rho_\lambda \int_0^t \|\mathbf{u}(\tau)\|_+^2\,d\tau.$$

Finally,

$$\text{(IV.17)}\qquad \begin{aligned} \left| \left\langle \mathbf{u}(\tau), \int_0^\tau \mathbf{G}(\tau-\sigma; \mathbf{u}(\sigma))\,d\sigma \right\rangle \right| &\leqq \|\mathbf{u}(\tau)\|_+ \int_0^\tau \|\mathbf{G}_\tau(\tau-\sigma; \mathbf{u}(\sigma))\|_-\,d\sigma \\ &\leqq \sup_{0\leqq\eta\leqq\tau} \psi_\lambda(\eta)\|\mathbf{u}(\tau)\|_+ \int_0^\tau \|\mathbf{u}(\sigma)\|_+\,d\sigma \end{aligned}$$

and, therefore

$$\text{(IV.18)}\qquad \begin{aligned} &\left| \int_0^t \left\langle \mathbf{u}(\tau), \int_0^\tau \mathbf{G}_\tau(\tau-\sigma; \mathbf{u}(\sigma))\,d\sigma \right\rangle d\tau \right| \\ &\quad\leqq \int_0^t \left(\sup_{0\leqq\eta\leqq\tau} \psi_\lambda(\eta)\|\mathbf{u}(\tau)\|_+ \int_0^\tau \|\mathbf{u}(\sigma)\|_+\,d\sigma \right) d\tau \\ &\quad\leqq \sup_{0\leqq\tau\leqq T} \psi_\lambda(\tau) \left(\int_0^t \|\mathbf{u}(\tau)\|_+\,d\tau \right)^2 \\ &\quad\leqq t\left(\sup_{0\leqq\tau\leqq T} \psi_\lambda(\tau) \right) \int_0^t \|\mathbf{u}(\tau)\|_+^2\,d\tau. \end{aligned}$$

Thus, for each t, $0\leqq t<T$,

$$\text{(IV.19)}\qquad \int_0^t \left\langle \mathbf{u}(t), \int_0^\tau \mathbf{G}_\tau(\tau-\sigma; \mathbf{u}(\sigma))\,d\sigma \right\rangle d\tau \geqq -T\left(\sup_{0\leqq\tau<T} \psi_\lambda(\tau)\right) \int_0^t \|\mathbf{u}(\tau)\|_+^2\,d\tau.$$

Combining (IV.13) with the estimates (IV.15), (IV.16), and (IV.19) and employing the hypothesis $\rho_\lambda > T \sup_{0\leqq t<T} \psi_\lambda(t)$ we are led to (IV.6). Q.E.D.

Remark. Let $\lambda \in \mathcal{R}^+$ be arbitrary and $\mathbf{u} \in C^2([0, T); H_+^\lambda)$ any solution of (IV.3), (IV.1_2). Let $\mathbf{G}:[0, T)\times H_+^\lambda \to H_-$ satisfy hypothesis (i) and (ii) of the above theorem. If $\mathscr{E}(0)$ satisfies

$$\text{(IV.20)}\qquad \mathscr{E}(0) \leqq -\tfrac{1}{2}\lambda^2 \sup_{0\leqq t<T} \int_0^t \phi_\lambda(t-\tau)\,d\tau,$$

then on any interval $[0, T)$ such that $\|\mathbf{u}(t)\|>0$

$$\text{(IV.21)}\qquad \frac{d^2}{dt^2}(\ln \|\mathbf{u}(t)\|^2) \geqq 0, \qquad 0\leqq t<T.$$

(As in Chapter II, the result is trivial, for if (IV.20) is satisfied take $\beta = t_0 = 0$. Then $F(t; \beta, t_0)$ reduces to $\|\mathbf{u}(t)\|^2$ and the right-hand side of the estimate (IV.6), being nonnegative, may be deleted.)

Remark. From (IV.21) it follows, as in Chapter II, that

$$\|\mathbf{u}(t)\|^2 \geqq \|\mathbf{f}\|^2 \exp\left\{\frac{\langle 2\mathbf{f}, \mathbf{g}\rangle t}{\|\mathbf{f}\|^2}\right\} \tag{IV.22}$$

for all t, $0 \leqq t < T$, whenever $\|f\| \neq 0$; this latter estimate shows that $\|\mathbf{u}(t)\|^2$ is bounded from below by an exponentially increasing function of t if $\langle \mathbf{f}, \mathbf{g}\rangle > 0$ and that $\|\mathbf{u}(t)\|^2$ can not decay any faster than an exponentially decreasing function of t if $\langle \mathbf{f}, \mathbf{g}\rangle < 0$. If $\mathbf{g} = \mathbf{0}$ then $\|\mathbf{u}(t)\|^2 \geqq \|\mathbf{f}\|^2$ on $[0, T)$.

COROLLARY. *Let $\lambda \in \mathcal{R}^+$ be arbitrary and $\mathbf{u} \in C^2([0, T); H_+)$ any solution of* (IV.3), (IV.1_2). *Let $\mathbf{G}: [0, T) \times H_+^\lambda \to H_-$ satisfy hypothesis* (i) *and* (ii) *of the theorem. If $\mathscr{E}(0)$ satisfies*

$$\mathscr{E}(0) > -\tfrac{1}{2}\lambda^2 \sup_{0 \leqq t < T} \int_0^t \phi_\lambda(t-\tau)\, d\tau, \tag{IV.23}$$

then there exists a real-valued function $\pi_\beta(\cdot)$, defined on $(0, \infty)$ such that for all t, $0 \leqq t < T$,

$$F(t; \beta, t_0) F''(t; \beta, t_0) - F'^2(t; \beta, t_0) \geqq -\pi_\beta(t_0) F^2(t; \beta, t_0) \tag{IV.24}$$

for all $\beta > 0$, $t_0 > 0$.

Proof. For any $\beta > 0$, $\zeta > 0$ define

$$\pi_\beta(\zeta) = \frac{2(2\mathscr{E}(0) + \beta + \lambda^2 \sup_{[0,T)} \int_0^t \phi_\lambda(t-\tau)\, d\tau)}{\beta \zeta^2}. \tag{IV.25}$$

In view of (IV.23), $\pi_\beta(\zeta) > 0$ for any $\beta > 0$, $\zeta > 0$. Thus, from the definition of $F(t; \beta, t_0)$ it is clear that

$$\pi_\beta(t_0)\beta t_0^2 \leqq \pi_\beta(t_0) F(t; \beta, t_0). \tag{IV.26}$$

Combining (IV.25) with $\zeta = t_0$, and (IV.26) we have

$$-2\left(2\mathscr{E}(0) + \beta + \lambda^2 \sup_{[0,T)} \int_0^t \phi_\lambda(t-\tau)\, d\tau\right) \geqq -\pi_\beta(t_0) F(t; \beta, t_0) \tag{IV.27}$$

for all t, $0 \leqq t < T$, and all $\beta > 0$, $t_0 > 0$; the corollary now follows directly from (IV.6).

Remark. Stability estimates can be easily derived from (IV.24). We begin by integrating (IV.24), with $\beta > 0$, $t_0 > 0$ arbitrary, according to the "secant property" for convex functions and obtain

$$F(t; \beta, t_0) \leqq F(0; \beta, t_0) + \frac{t}{T}(F(T; \beta, t_0) - F(0; \beta, t_0)), \tag{IV.28}$$

where

$$F(t; \beta, t_0) \equiv \ln\left[\exp\left(\tfrac{1}{2}\pi_\beta(t_0) t^2\right) F(t; \beta, t_0)\right]. \tag{IV.29}$$

Thus,

$$\text{(IV.30)} \qquad F(t;\beta,t_0) \leqq e^{-\pi_\beta(t_0)t^2} F(0;\beta,t_0)^{1-t/T}(e^{\pi_\beta(t_0)T^2}F(T;\beta,t_0))^{t/T}$$

for all t, $0 \leqq t < T$. We now choose $\beta = \varepsilon/t_0^2$ ($\varepsilon > 0$ as yet arbitrary) in (IV.30); we want to extract the limit in (IV.27) as $t_0 \to +\infty$. From the definitions of $F(t;\beta,t_0)$ and $\pi_\beta(\zeta)$ we have

$$\text{(IV.31)} \qquad \lim_{t_0\to+\infty} F\left(t;\frac{\varepsilon}{t_0^2},t_0\right) = \|\mathbf{u}(t)\|^2 + \varepsilon, \qquad 0 \leqq t < T,$$

and

$$\text{(IV.32)} \qquad \lim_{t_0\to+\infty} \pi_{(\varepsilon/t_0^2)}(t_0) = \frac{2}{\varepsilon}\left(2\mathscr{E}(0) + \lambda^2 \sup_{[0,T)} \int_0^t \phi_\lambda(t-\tau)\,d\tau\right) \equiv \frac{\pi}{\varepsilon}.$$

Thus, setting $\beta = \varepsilon/t_0^2$ in (IV.30) and letting $t_0 \to +\infty$ we obtain

$$\text{(IV.33)} \qquad \|\mathbf{u}(t)\|^2 \leqq e^{-\pi t^2/\varepsilon}(\|\mathbf{f}\|^2 + \varepsilon)^{1-t/T}(e^{\pi T^2/\varepsilon}[\|\mathbf{u}(T)\|^2 + \varepsilon])^{t/T}.$$

However, as $H_+ \subset H$, topologically as well as algebraically, $\|\mathbf{u}(T)\|^2 \leqq \omega^2\|\mathbf{u}(T)\|_+^2 \leqq (\omega\lambda)^2$, where $\omega > 0$ is the embedding constant for the inclusion map $i: H_+ \to H_-$. We now choose $\varepsilon = \|\mathbf{u}_0\|^2$, $\mathbf{u}_0 \neq 0$ and select $C = C(\lambda, T)$ so large that

$$\text{(IV.34)} \qquad \max(\|\mathbf{f}\|^2, \omega^2\lambda^2) \leqq \tfrac{1}{2}C(\lambda,T)\exp\left(\frac{-\pi T^2}{\|\mathbf{f}\|^2}\right).$$

In this manner we obtain from (IV.33) an estimate of the form

$$\text{(IV.35)} \qquad \|\mathbf{u}(t)\|^2 \leqq A\|\mathbf{f}\|^{2(1-t/T)} \exp\left(\frac{-\pi t^2}{\|\mathbf{f}\|^2}\right), \qquad 0 \leqq t < T,$$

which is valid for all t, $0 \leqq t < T$, whenever $\mathbf{f} \neq 0$. The estimate (IV.35) indicates that for a nonlinear map $\mathbf{G}: [0,T) \times H_+^\lambda \to H_-$ satisfying hypothesis (i) and (ii) of the theorem, and an initial energy $\mathscr{E}(0)$ satisfying (IV.23), solutions $\mathbf{u} \in C^2([0,T); H_+^\lambda)$ of (IV.3), (IV.1_1) are bounded from above, for $0 \leqq t < T$, by an exponentially decreasing function of t whenever $\mathbf{f} \neq \mathbf{0}$. Lower bounds for $\|\mathbf{u}(t)\|^2$, in the case where $\mathscr{E}(0)$ satisfies (IV.23), may be obtained, in an analogous manner, by beginning with the estimate

$$\text{(IV.36)} \qquad \mathscr{F}(t;\beta,t_0) \geqq \mathscr{F}(0;\beta,t_0) + t\dot{\mathscr{F}}(0;\beta,t_0), \qquad 0 \leqq t < T.$$

Substituting for $\mathscr{F}(t;\beta,t_0)$ in (IV.36) from (IV.29), setting $\beta = \|\mathbf{f}\|^2/t_0^2$, and extracting the limit as $t_0 \to +\infty$ we obtain the lower bound

$$\frac{\|\mathbf{u}(t)\|^2 + \|\mathbf{f}\|^2}{2} \geqq \|\mathbf{f}\|^2 \exp\left[\frac{(-\pi t^2 + \langle \mathbf{f}, \mathbf{g}\rangle t)}{\|\mathbf{f}\|^2}\right],$$

provided $\mathbf{f} \neq \mathbf{0}$. In particular, if $\mathbf{g} = \mathbf{0}$, then

$$\text{(IV.37)} \qquad \|\mathbf{u}(t)\|^2 \geqq \|\mathbf{f}\|^2[2\exp(-\pi t^2/\|\mathbf{f}\|^2) - 1], \qquad 0 \leqq t < T_0,$$

where $T_0 = \min(T, \sqrt{(\ln 2)/\pi}\,\|\mathbf{f}\|)$.

We now wish to examine one particular example of the type of situation to which the analysis of the preceding section applies. To that end we will employ the function spaces $H = L_2(\Omega)$, $H_+ = H_0^1(\Omega)$, and $H_- = H^{-1}(\Omega)$ which were introduced in Chapter II, $\Omega = [0, 1]$ and we consider the problem

$$\text{(IV.38a)} \qquad \frac{\partial^2 u(x,t)}{\partial t^2} - \frac{\partial^2 u(x,t)}{\partial x^2} + \int_0^t \mathscr{g}(t-\tau; u(x,\tau)) \frac{\partial^2 u(x,\tau)}{\partial x^2} \, d\tau = 0,$$

$$\text{(IV.38b)} \qquad u(x,0) = f(x), \qquad u_t(x,0) = g(x),$$

for $(x, t) \in (0, 1) \times [0, T)$, where $u \in C^2([0, T); H_0^1(\Omega; \lambda))$ with

$$\text{(IV.39)} \quad H_0^1(\Omega, \lambda) = \left\{ v \in H_0^1(\Omega) \,\middle|\, \sqrt{\int_0^1 \left(\frac{\partial v}{\partial x}\right)^2 dx} < \lambda, \lambda \in \mathscr{R}^+ \right\}.$$

We assume that $f, g \in H_0^1(\Omega; \lambda)$ and that the nonlinear map $\mathscr{g}$ is defined on $[0, T) \times H_0^1(\Omega; \lambda)$ and satisfies the following hypothesis: $\mathscr{g}$ is of class C^1 in t for each $v \in H_0^1(\Omega; \lambda)$, of class C^1 in v for each fixed t, $0 \leqq t < T$, and (with $\mathscr{g}_v(t, v) \equiv \partial \mathscr{g}(t, v)/\partial v$)

(i′) there exist bounded real-valued functions $A(t) > 0, \cdots, D(t) > 0$, defined for $0 \leqq t < T$, such that $\mathscr{g}^2(t, v) \leqq A^2(t)$,

$$\mathscr{g}_v^2(t, v) \leqq B^2(t), \qquad \left[\frac{\partial \mathscr{g}}{\partial t}(t, v)\right]^2 \leqq C^2(t), \quad \text{and} \quad \left[\frac{\partial \mathscr{g}_v}{\partial t}(t, v)\right]^2 \leqq D^2(t)$$

for all $v \in H_0^1(\Omega; \lambda)$ and each t, $0 \leqq t < T$;

(ii′) $\mathscr{g}(0, v) = \hat{\rho}_\lambda \geqq T \cdot \sup_{0 \leqq t < T} (C(t) + \lambda D(t))$, $v \in H_0^1(\Omega; \lambda)$.

Given that $\mathscr{g}$ satisfies hypotheses (i′) and (ii′) above we demonstrate that the map $\mathbf{G}: [0, T) \times H_0^1(\Omega; \lambda) \to H^{-1}(\Omega)$, which is defined by $\mathbf{G}(t, \cdot) = \mathscr{g}(t, \cdot)\partial^2/\partial x^2$, $0 \leqq t < T$, satisfies hypotheses (i) and (ii) of our theorem. We begin by estimating, for arbitrary $v \in H_0^1(\Omega; \lambda)$, $t \in [0, T)$,

$$\begin{aligned}
\|\mathbf{G}(t, v)\|_- &= \sup_{w \in H_0^1(\Omega)} \frac{|\langle w, \mathbf{G}(t, v)\rangle|}{\|w\|_+} \\
&= \sup_{w \in H_0^1(\Omega)} \frac{\left|\int_0^1 w(x) \mathscr{g}(t, v(x)) \partial^2 v/\partial x^2 \, dx\right|}{\sqrt{\int_0^1 (\partial w/\partial x)^2 \, dx}} \\
&\leqq \sup_{w \in H_0^1(\Omega)} \frac{1}{\sqrt{\int_0^1 (\partial w/\partial x)^2 \, dx}} \left(\left| \int_0^1 w(x) \frac{\partial}{\partial x}\left[\mathscr{g}(t, v(x)) \frac{\partial v}{\partial x}\right] dx \right| \right. \\
&\qquad \left. + \left| \int_0^1 w(x) \mathscr{g}_v(t, v) \left(\frac{\partial v}{\partial x}\right)^2 dx \right| \right) \\
&\leqq \sup_{w \in H_0^1(\Omega)} \frac{1}{\sqrt{\int_0^1 (\partial w/\partial x)^2 \, dx}} \left(\left| \int_0^1 \mathscr{g}(t, v(x)) \frac{\partial w}{\partial x} \frac{\partial v}{\partial x} \, dx \right| \right. \\
&\qquad \left. + \left| \int_0^1 w(x) \mathscr{g}_v(t, v(x)) \left(\frac{\partial v}{\partial w}\right)^2 dx \right| \right),
\end{aligned} \tag{IV.40}$$

where, by virtue of the standard trace theorem, $w \in H_0^1(\Omega)$ vanishes at $x=0$ and $x=1$. Applying the Schwarz inequality to (IV.40) yields

(IV.41)
$$\begin{aligned}\|\mathbf{G}(t,v)\|_- &\leqq \sup_{w\in H^0(\Omega)} \frac{1}{\sqrt{\int_0^1 (\partial w/\partial x)^2\,dx}}\left(\sqrt{\int_0^1 g^2\left(\frac{\partial w}{\partial x}\right)^2 dx}\sqrt{\int_0^1 \left(\frac{\partial v}{\partial x}\right)^2 dx}\right.\\ &\qquad\left.+\sqrt{\int_0^1 w^2(x)\left(\frac{\partial v}{\partial x}\right)^2 dx}\sqrt{\int_0^1 g_v^2\left(\frac{\partial v}{\partial x}\right)^2 dx}\right)\\ &\leqq \sqrt{\int_0^1\left(\frac{\partial v}{\partial x}\right)^2 dx}\left(A(t)+B(t)\sup_{w\in H_0^1(\Omega)}\frac{\sqrt{\int_0^1 w^2(x)(\partial v/\partial x)^2\,dx}}{\sqrt{\int_0^1(\partial w/\partial x)^2\,dx}}\right),\end{aligned}$$

by virtue of the hypotheses on g and g_v; however,

(IV.42)
$$\begin{aligned}\sqrt{\int_0^1 w^2(x)\left(\frac{\partial v}{\partial x}\right)^2 dx} &\leqq \sqrt{\sup_{0\leqq x\leqq 1} w^2(x)\int_0^1\left(\frac{\partial v}{\partial x}\right)^2 dx}\\ &\leqq \sqrt{\int_0^1\left(\frac{\partial w}{\partial v}\right)^2 dx\cdot\int_0^1\left(\frac{\partial v}{\partial x}\right)^2 dx}\\ &\leqq \lambda\sqrt{\int_0^1\left(\frac{\partial w}{\partial x}\right)^2 dx}\end{aligned}$$

as $v\in H_0^1(\Omega;\lambda)$. Employing (IV.42) in (IV.41_2) we obtain,

(IV.43)
$$\|\mathbf{G}(t,v)\|_- \leqq (A(t)+\lambda B(t))\sqrt{\int_0^1\left(\frac{\partial v}{\partial x}\right)^2 dx} \equiv \hat{\phi}_\lambda(t)\|v\|_+$$

for any $\mathbf{v}\in H_0^1(\Omega;\lambda)$ and all t, $0\leqq t<T$. In a similar fashion, it follows from our assumptions on $\partial g/\partial t$ and $\partial g_v/\partial t$ that

(IV.44)
$$\|\mathbf{G}_t(t,v)\|_- \leqq \psi_\lambda(t)\|v\|_+,\qquad 0\leqq t<T,\quad v\in H_0^1(\Omega;\lambda),$$

where $\psi_\lambda(t)\equiv C(t)+\lambda D(t)$, $0\leqq t<T$. Finally, for any $v\in H_0^1(\Omega;\lambda)$,

(IV.45)
$$-\langle v,\mathbf{G}(0,v)\rangle = -\hat{\rho}_\lambda\int_0^1 v\frac{\partial^2 v}{\partial x^2}\,dx = \hat{\rho}_\lambda\int_0^1\left(\frac{\partial v}{\partial x}\right)^2 dx = \hat{\rho}_\lambda\|\ \|_+^2,$$

where $\hat{\rho}_\lambda \geqq T\sup_{0\leqq t<T}(C(t)+\lambda D(t)) = T\sup_{0\leqq t<T}\hat{\psi}_\lambda(t)$. This completes the demonstration that $\mathbf{G}(t,\cdot)=g(t,\cdot)\partial^2/\partial x^2$, with g satisfying hypotheses (i′) and (ii′) above, satisfies hypotheses (i) and (ii) of our theorem. The various growth estimates obtained above may, therefore, be applied to (weak) solutions of the initial-value problem (IV.38); we remark in passing that, for the problem at hand,

$$\mathscr{E}(0)=\frac{1}{2}\int_0^1 [f^2(x)+g^2(x)]\,dx>0$$

so that condition (IV.19), with $\hat{\phi}_\lambda$ in place of ϕ_λ, is never satisfied while condition (IV.23) is always satisfied.

Example. As a particular example of (IV.38a) we consider the nonlinear integrodifferential equation ($\hat{\rho}_\lambda > 0$ arbitrary, $\lambda > 0$)

$$\text{(IV.46)} \qquad \frac{\partial^2 u(x,t)}{\partial t^2} - \frac{\partial^2 u(x,t)}{\partial x^2} + \hat{\rho}_\lambda \int_0^t e^{+\hat{g}(t-\tau)u^2(x,\tau)} \left(\frac{\partial^2 u(x,\tau)}{\partial x^2} \right) d\tau = 0$$

on $(0,1)\times[0,T)$, where $\hat{g} \in C^1([0,T))$ and $\hat{g}'$ are bounded and $\hat{g}$ satisfies $\hat{g}(0)=0$, $\hat{g}'(t)>0$, $0 \leqq t < T$. Thus

$$\text{(IV.47)} \qquad g(t,v) = \hat{\rho}_\lambda \exp(-\hat{g}(t)v^2), \qquad \mathbf{v} \in H_0^1(\Omega;\lambda).$$

It is a simple exercise to verify that $g(t,v)$, as defined by (IV.47), satisfies hypothesis (i′) above with $A(t)=\hat{\rho}_\lambda$, $B(t)=2\lambda\hat{\rho}_\lambda\hat{g}(t)$, $C(t)=\lambda^2\hat{\rho}_\lambda\hat{g}'(t)$, and $D(t)=\lambda\hat{\rho}_\lambda\hat{g}'(t)\sqrt{2(1+\lambda^4\hat{g}^2(t))}$. Hypothesis (ii′) above will then be satisfied if

$$\text{(IV.48)} \qquad \frac{1}{\lambda^2 T} \geqq \sup_{[0,T)} [\hat{g}'(t)(1+\sqrt{2(1+\lambda^4\hat{g}^2(t))})],$$

e.g., for $g(t)=t$, $0 \leqq t < T$, is equivalent to

$$\text{(IV.49)} \qquad \frac{1}{2T} \geqq 1+\sqrt{2(1+\lambda^4 T^2)}.$$

2. Instability of solutions to initial-history boundary value problems for nonlinear viscoelastic fluids.

In Chapter I we considered the nonlinear heat conduction problem (I.52) and showed, by means of a concavity argument that, under appropriate hypotheses on the initial temperature distribution $u_0(x)$ and the nonlinearity f, classical smooth solutions $u(\cdot,t)$ could not exist for $t \in [0,\infty)$. In this section we wish to consider the problem of shearing motions of a nonlinear viscoelastic fluid in which the shearing stress is related to the shear rate $v_x(v(x,t)=\dot{u}(x,t)$, the velocity at the point x at time t) by (II.284); we will show that initial-history boundary value problems associated with the governing nonlinear integrodifferential equation of motion for this problem are ill-posed, in the sense that smooth solutions must break down in finite-time, if the initial velocity history is sufficiently large (in a sense to be made precise). This problem has been considered by Slemrod in [139]–[141]; the argument is based on the concept of the Riemann invariant and is closely related to the earlier work of Lax [86], and the more recent work of Nishida [121], on undamped and damped systems of quasi-linear hyperbolic equations, respectively. A similar argument based on the use of Riemann invariants has been used by Nohel [122] to discuss the problem of global existence and asymptotic stability of solutions of the nonlinear viscoelasticity equations (II.229) which result from a constitutive assumption that is closely related to (II.284); the relationship between the two constitutive hypotheses has been discussed in Chapter II, § 5, with particular emphasis on the special case where the relaxation function in (II.229) is of the form $g(t) = e^{-\alpha t}$. In the next section we will present Nohel's Riemann invariant argument for the problem with governing nonlinear evolution equation (II.229) and will contrast these results with the analogous results obtained by Dafermos and Nohel [43] for initial-history boundary value problems associated with (II.229) by means of energy arguments alone; this latter analysis has already been presented in § 5 of Chapter II.

For the nonlinear viscoelastic fluid considered by Slemrod [139]–[141], in which the shearing stress is given by (II.284), we have already indicated that the corresponding evolution equation is given by (II.285) or, in view of the fact that the nonlinearity σ in (II.285) is assumed to be an odd (real valued, analytic) function on the real line, by

$$v_t(x, t) = \sigma\left(\int_0^\infty e^{-\alpha\tau} v_x(x, t-\tau)\, d\tau\right)_x, \tag{IV.50}$$

where, without loss of generality, we have taken the constant mass density $\rho \equiv 1$. We also assume, as in [139], that the fluid is confined between two parallel walls of infinite extent at $x = 0$ and $x = h$ and that the top wall moves with constant velocity V; the boundary conditions are taken to be of the no-slip variety so that associated with (IV.50) we have

$$v(0, t) = 0, \quad v(h, t) = V, \qquad t \geqq 0. \tag{IV.51}$$

The system (IV.50), (IV.51) then clearly admits the steady rectilinear flow $v(x) = Vx/h$ and, thus, to study the stability of this steady flow against shearing perturbations we define the velocity field

$$\hat{v}(x, t) = v(x, t) - \frac{Vx}{h}. \tag{IV.52}$$

It is a straightforward matter to verify that $\hat{v}$ satisfies the system

$$\hat{v}_t(x, t) = \sigma\left(\int_0^\infty e^{-\alpha\tau} \hat{v}_x(x, t-s)\, ds + \frac{V}{\alpha h}\right)_x, \tag{IV.53}$$

$$\hat{v}(0, t) = 0, \quad \hat{v}(h, t) = 0, \tag{IV.54}$$

to which we append the prescription of a smooth velocity history

$$\hat{v}(x, \tau) = \hat{v}_0(x, \tau), \qquad -\infty < \tau \leqq 0. \tag{IV.55}$$

Following Slemrod [139] we now set

$$\kappa = \frac{V}{\alpha h},$$

$$\hat{\sigma}(\zeta) = \sigma(\zeta + \kappa) - \sigma(\kappa), \tag{IV.56}$$

$$r(x, t) = \int_0^\infty e^{-\alpha\tau} \hat{v}_t(x, t-\tau)\, d\tau,$$

$$w(x, t) = \int_0^\infty e^{-\alpha\tau} \hat{v}_x(x, t-\tau)\, d\tau,$$

and, as per the analysis in Chapter II ((II.287)–(II.291)), we find that the system (IV.53)–(IV.55) for the perturbed flow $\hat{v}(x, t)$ is equivalent to the system

$$\begin{aligned} w_t(x, t) &= r_x(x, t), \\ r_t(x, t) &= \hat{\sigma}(w(x, t))_x - \alpha r(x, t), \end{aligned} \tag{IV.57}$$

$$r(0, t) = 0, \quad r(h, t) = 0, \qquad t \geqq 0, \tag{IV.58}$$

$$\left.\begin{aligned} r(x, 0) &= r_0(x) \\ w(x, 0) &= w_0(x) \end{aligned}\right\}, \qquad 0 \leqq x \leqq h, \tag{IV.59}$$

where $r_0(x)$, $w_0(x)$ are obtained directly from the definitions of $r(x, t)$, $w(x, t)$ in (IV.56) and the prescribed velocity history (IV.55). We assume that the constitutive function σ is truly nonlinear in the sense that $\sigma''(\zeta^*) \neq 0$ for some real number ζ^* and choose the velocity V of the top wall at $x = h$ to be $V = \alpha h \zeta^*$, in which case (IV.56) yields $\hat{\sigma}(0) = 0$, $\hat{\sigma}''(0) \neq 0$. If we rewrite the quasi-linear system (IV.57) in the form

$$\begin{aligned} r_t &= 0 \cdot r_x + \hat{\sigma}'(w) w_x - \alpha r, \\ w_t &= 1 \cdot r_x + 0 \cdot w_x \end{aligned} \tag{IV.57$'$}$$

then it is well-known that our nonlinear system is strictly hyperbolic if the associated matrix

$$\begin{bmatrix} 0 & \hat{\sigma}' \\ 1 & 0 \end{bmatrix} \tag{IV.58$'$}$$

has real distinct eigenvalues; this, in turn, is easily seen to be equivalent to the condition that $\hat{\sigma}' > 0$ (which will be assumed in all that follows).

Following Slemrod [139] we now reformulate (IV.57)–(IV.59) as a pure initial value problem on the real line by extending r_0 and w_0 as odd and even periodic functions, respectively, with respect to $x = 0$, with periods $2h$. We then define $w(x, t) = w(-x, t)$, $r(x, t) = -r(-x, t)$, for $-h \leqq x \leqq 0$, and $w(x + 2kh, t) = w(x, t)$, $r(x + 2kh, t) = r(x, t)$, for $-h \leqq x \leqq h$, where k is an arbitrary integer. It is then easily verified that the pair of functions (r, w) will be a solution of

(IV.57)–(IV.59) if and only if the defined extended pair (r, w) is a solution of (IV.57), for $-\infty < x < \infty$, which satisfies the extended initial data.

In order to introduce the concept of Riemann invariant we now proceed as follows: we first compute the eigenvalues and associated eigenvectors of (IV.58$'$) in the standard elementary fashion, i.e., we consider the system of algebraic equations generated by

$$\begin{bmatrix} 0 & \hat{\sigma}' \\ 1 & 0 \end{bmatrix} \begin{bmatrix} e_1 \\ e_2 \end{bmatrix} = \lambda \begin{bmatrix} e_2 \\ e_1 \end{bmatrix},$$

namely,

$$\text{(IV.60)} \qquad \begin{aligned} \hat{\sigma}' e_2 &= \lambda e_1, \\ e_1 &= \lambda e_2, \end{aligned}$$

or

$$\hat{\sigma}' e_2 = \lambda^2 e_2,$$

so that the real eigenvalues are given by

$$\text{(IV.61)} \qquad \lambda_1 = \gamma = \sqrt{\hat{\sigma}'(w)}, \quad \lambda_2 = \mu = -\sqrt{\hat{\sigma}'(w)}.$$

These eigenvalues determine the characteristic curves associated with the quasi-linear system (IV.57), namely, the curves which are given by the system of ordinary differential equations

$$\text{(IV.62)} \qquad \frac{dx_i}{dt} = \lambda_i = \mp\sqrt{\hat{\sigma}'(w(x_i, t))}.$$

The eigenvectors associated with λ_i, $i = 1, 2$ are then given by

$$\text{(IV.63)} \qquad \mathbf{e} = \begin{bmatrix} \mp\sqrt{\hat{\sigma}'(w)} \\ 1 \end{bmatrix}.$$

If we rewrite (IV.57) in the form

$$\text{(IV.64)} \qquad \begin{aligned} w_t - r_x &= 0, \\ r_t - \hat{\sigma}'(w) w_x &= -\alpha r, \end{aligned}$$

and multiply the first equation in this system by $\sqrt{\hat{\sigma}'(w)}$ and the second by one and then add, we easily obtain

$$\text{(IV.65)} \qquad (r_t - \sqrt{\hat{\sigma}'(w) r_x}) + \sqrt{\hat{\sigma}'(w)}\,(w_t - \sqrt{\hat{\sigma}'(w)}\, w_x) = -\alpha r.$$

We now let $' = \partial/\partial t - \sqrt{\hat{\sigma}'(w)}\, \partial/\partial x$ denote differentiation along the left-hand characteristic curve x_1 determined by

$$\frac{dx_1}{dt} = \lambda_1 = -\sqrt{\hat{\sigma}'(w)};$$

i.e., for any sufficiently differentiable function $Z(x, t)$ evaluated along x_1,

$$\frac{d}{dt} Z(x_1, t) = \frac{\partial Z}{\partial t} + \frac{\partial Z}{\partial x}\frac{dx_1}{dt} = Z_t + \lambda_1 Z_x = Z^{\prime}.$$

Then, (IV.65) becomes

$$r^{\prime} + \sqrt{\hat{\sigma}'(w)}\, w^{\prime} = -\alpha r, \tag{IV.66}$$

and in a completely analogous fashion we find that

$$r^{\backprime} - \sqrt{\hat{\sigma}'(w)}\, w^{\backprime} = -\alpha r, \tag{IV.67}$$

where ${}^{\backprime} = \partial/\partial t + \sqrt{\hat{\sigma}'(w)}\,\partial/\partial x$ denotes differentiation along the right-hand characteristic x_2 which is characterized by

$$\frac{dx_2}{dt} = \lambda_2 = +\sqrt{\hat{\sigma}'(w)}.$$

If we now choose $\mathscr{r} = \mathscr{r}(r, w)$ so that

$$\mathscr{r}_r = 1, \qquad \mathscr{r}_w = \sqrt{\hat{\sigma}'(w)},$$

i.e.,

$$\mathscr{r}(r, w) = r + \int_0^w \sqrt{\hat{\sigma}'(z)}\, dz, \tag{IV.68}$$

then (IV.66) assumes the form

$$\mathscr{r}^{\prime} = -\alpha r. \tag{IV.69}$$

In an analogous fashion, choosing $\mathscr{s} = \mathscr{s}(r, w)$ so that

$$\mathscr{s}_r = 1, \quad \mathscr{s}_w = -\sqrt{\hat{\sigma}'(w)},$$

i.e.,

$$\mathscr{s}(r, w) = r - \int_0^w \sqrt{\hat{\sigma}'(z)}\, dz, \tag{IV.70}$$

we obtain from (IV.67)

$$\mathscr{s}^{\backprime} = -\alpha r. \tag{IV.71}$$

From (IV.68), (IV.70) we further note that $2r = \mathscr{r} + \mathscr{s}$ so that (IV.69), (IV.71) may be rewritten in the form of the diagonal system

$$\begin{aligned} \mathscr{r}^{\prime} &= -\frac{\alpha}{2}(\mathscr{r} + \mathscr{s}), \\ \mathscr{s}^{\backprime} &= -\frac{\alpha}{2}(\mathscr{r} + \mathscr{s}). \end{aligned} \tag{IV.72}$$

The functions $\mathscr{r}(r, w)$, $\mathscr{s}(r, w)$ which are defined, respectively, by (IV.68), (IV.70), and which, as a consequence of the original quasi-linear hyperbolic system (IV.57) satisfy (IV.72) along the left and right-hand characteristics, are called the Riemann invariants associated with the system (IV.57). In the undamped case $\alpha = 0$, (IV.72) reduces to the statement $\mathscr{r}' = \mathscr{s}^{\backprime} = 0$, i.e., the Riemann invariants $\mathscr{r}, \mathscr{s}$ are, in fact, invariant along the left and right-hand characteristics of the corresponding undamped quasi-linear hyperbolic system, i.e., (IV.57) with $\alpha = 0$. As a consequence of (IV.61) and (IV.68), (IV.70) we note that

$$\text{(IV.73)} \qquad \gamma = \gamma(\mathscr{r} - \mathscr{s}), \qquad \mu = \mu(\mathscr{r} - \mathscr{s}).$$

Our aim at this point in the analysis is to study the behavior of the solutions to the system (IV.72) of equations satisfied by the Riemann Invariants $\mathscr{r}, \mathscr{s}$ with associated initial data

$$\text{(IV.74)} \qquad \mathscr{r}(x, 0) = \mathscr{r}_0(x), \qquad \mathscr{s}(x, 0) = \mathscr{s}_0(x),$$

which are assumed to be smooth functions and which may be obtained in terms of the initial datum $r_0(x)$, $w_0(x)$ by means of (IV.68), (IV.70), i.e.,

$$\begin{aligned} \mathscr{r}(x, 0) &= \mathscr{r}(r(x, 0), w(x, 0)) \\ &= \mathscr{r}(r_0(x), w_0(x)) \\ &= r_0(x) + \int_0^{w_0(x)} \sqrt{\hat{\sigma}'(z)}\, dz, \end{aligned}$$

with a similar expression for $\mathscr{s}(x, 0)$ in terms of $r_0(x)$, $w_0(x)$. Because the transformation effected by (IV.68), (IV.70) from $(r, w) \in \mathscr{R}^1 \times \mathscr{R}^1$ to $(\mathscr{r}, \mathscr{s}) \in \mathscr{R}^1 \times \mathscr{R}^1$ is one-to-one, if $(\mathscr{r}, \mathscr{s})$ breaks down in finite time in the sense that $(\mathscr{r}(\cdot, t), \mathscr{s}(\cdot, t)) \notin C^1(\mathscr{R}^1) \times C^1(\mathscr{R}^1)$, $\forall t \in [0, \infty)$ then the same will be true for $(r(\cdot, t), w(\cdot, t))$ and the desired finite time breakdown of classical smooth solutions to the system (IV.53)–(IV.55) will then follow as an immediate consequence of the equations (IV.56) defining (r, w) in terms of the perturbed velocity field $\hat{v}(\cdot, t)$.

Preliminary to the statement of the finite time breakdown result for the system (IV.72), (IV.74) we need the result contained in the following lemma (which is due essentially to Nishida [121]):

LEMMA. *Let* $|\mathscr{r}_0| = \sup_{x \in \mathscr{R}^1} |\mathscr{r}_0(x)|$, $|\mathscr{s}_0| = \sup_{x \in \mathscr{R}^1} |\mathscr{s}_0(x)|$. *As long as smooth solutions of* (IV.72), (IV.74) *exist, the a priori estimate*

$$\text{(IV.75)} \qquad |\mathscr{r}(x, t)| + |\mathscr{s}(x, t)| \leqq |\mathscr{r}_0| + |\mathscr{s}_0|$$

is valid.

Proof. We integrate the characteristic differential equations (IV.62) so as to introduce the characteristic curves

$$\text{(IV.76)}\qquad \begin{aligned} x_1 &= x_1(\beta, t) = \beta + \int_0^t \gamma\, d\tau, \qquad \beta \in \mathscr{R}^1,\\ x_2 &= x_2(\delta, t) = \delta + \int_0^t \mu\, d\tau, \qquad \delta \in \mathscr{R}^1, \end{aligned}$$

where γ, μ are given by (IV.61). We then rewrite the system (IV.72) in the equivalent form

$$\text{(IV.77)[3]}\qquad \begin{aligned} \frac{d}{dt}\mathscr{r}(x_1(\beta, t), t)) &= -\frac{\alpha}{2}(\mathscr{r}(x_1(\beta, t), t) + \mathscr{s}(x_1(\beta, t), t)),\\ \frac{d}{dt}\mathscr{s}(x_2(\delta, t), t)) &= -\frac{\alpha}{2}(\mathscr{r}(x_2(\delta, t), t) + \mathscr{s}(x_2(\delta, t), t)), \end{aligned}$$

and integrate this system by introducing the integrating factor $\exp[(\alpha/2)t]$; we easily obtain

$$\text{(IV.88)}\qquad \begin{aligned} e^{\alpha t/2}\mathscr{r}(x_1(\beta, t), t) &= \mathscr{r}_0(\beta) - \frac{\alpha}{2}\int_0^t e^{\alpha\tau/2}\mathscr{s}(x_1(\beta, \tau), \tau)\, d\tau,\\ e^{\alpha t/2}\mathscr{s}(x_2(\delta, t), t) &= \mathscr{s}_0(\delta) - \frac{\alpha}{2}\int_0^t e^{\alpha\tau/2}\mathscr{r}(x_2(\delta, \tau), \tau)\, d\tau. \end{aligned}$$

If we now set

$$\text{(IV.89)}\qquad S(t) = \sup_{x\in\mathscr{R}^1} e^{\alpha t/2}(|\mathscr{s}(x, t)|), \qquad R(t) = \sup_{x\in\mathscr{R}^1} e^{\alpha t/2}(|\mathscr{r}(x, t)|),$$

then (IV.88) yields

$$\text{(IV.90)}\qquad \begin{aligned} e^{\alpha t/2}(|\mathscr{r}(x_1(\beta, t), t)|) &\leqq |\mathscr{r}_0| + \frac{\alpha}{2}\int_0^t S(\tau)\, d\tau,\\ e^{\alpha t/2}(|\mathscr{s}(x_2(\delta, t), t)|) &\leqq |\mathscr{s}_0| + \frac{\alpha}{2}\int_0^t R(\tau)\, d\tau. \end{aligned}$$

In view of the fact that $r(x, t)$, $w(x, t)$ are periodic in x for each $t > 0$, and the definitions of the Riemann invariants $\mathscr{r}(x, t)$, $\mathscr{s}(x, t)$ are in terms of r and w it follows that $\mathscr{r}$ and $\mathscr{s}$ are periodic in x, for each fixed $t > 0$. Hence for each fixed $t > 0 \exists \bar{x}_1, \bar{x}_2$ such that

$$\text{(IV.91)}\qquad e^{\alpha t/2}(|\mathscr{r}(\bar{x}_1, t)|) = R(t), \quad e^{\alpha t/2}(|\mathscr{s}(\bar{x}_2, t)|) = S(t).$$

Now, as the system (IV.57) has been assumed to be strictly hyperbolic, i.e., $\hat{\sigma}'(w) > 0$, it follows from (IV.61) that $\mu < 0 < \gamma$. Therefore if we trace backwards along the characteristic curves (IV.76) we may find $\bar{\beta} = \bar{\beta}(t)$, $\bar{\delta} = \bar{\delta}(t)$

[3]Equation numbers (IV.78) to (IV.87) have been intentionally omitted.

such that

$$\bar{x}_1 = x_1(\bar{\beta}(t), t), \qquad \bar{x}_2 = x_2(\bar{\delta}(t), t). \tag{IV.92}$$

Choosing, for each $t > 0$, $\beta = \bar{\beta}$, $\delta = \bar{\delta}$ in (IV.90), and employing (IV.91), we obtain from (IV.90)

$$\begin{aligned} R(t) &\leqq |r_0| + \frac{\alpha}{2}\int_0^t S(\tau)\,d\tau, \\ S(t) &\leqq |s_0| + \frac{\alpha}{2}\int_0^t R(\tau)\,d\tau, \end{aligned} \tag{IV.93}$$

and, addition of these inequalities then yields the estimate

$$W(t) \leqq |r_0| + |s_0| + \frac{\alpha}{2}\int_0^t W(\tau)\,d\tau, \tag{IV.94}$$

where $W(t) = R(t) + S(t)$. Application of the Gronwall inequality to (IV.94) then yields the estimate

$$W(t) \leqq (|r_0| + |s_0|)e^{\alpha t/2} \tag{IV.95}$$

and the required result, i.e. (IV.75), now follows directly from (IV.95) and the definitions of $W(t)$, $R(t)$, and $S(t)$. Q.E.D.

We are now in a position to state and prove the basic result concerning the finite time breakdown of smooth solutions to the system (IV.57)–(IV.59):

THEOREM (Slemrod [139], [141]). *Suppose that $|r_0|$, $|s_0|$ are sufficiently small and $\hat{\sigma}'(0) > 0$, $\hat{\sigma}''(0) > 0$. If either $r_{0,x}$ or $s_{0,x}$ is positive and sufficiently large at some $x \in \mathscr{R}^1$ then* (IV.57)–(IV.59) *has a solution (w, r) in $C^1(\mathscr{R}^1) \times C^1(\mathscr{R}^1)$ for only a finite time. A similar result holds when $\hat{\sigma}''(0) < 0$ and either $r_{0,x}$ or $s_{0,x}$ is sufficiently negative at some point $x \in \mathscr{R}^1$.*

Proof. (We only sketch the proof in outline form; complete details may be found in either [139] or [141]). As $\hat{\sigma}'(0) > 0$, it follows from the above lemma that w will stay uniformly near zero if we choose $|r_0|$, $|s_0|$ sufficiently small and hence $\hat{\sigma}'(w) > 0$; thus, under our hypotheses, the system (IV.57) remains strictly hyperbolic. If we differentiate the first equation in (IV.72) with respect to x we obtain

$$r_{tx} + \mu_x r_x + \mu r_{xx} = -\frac{\alpha}{2}(r_x + s_x). \tag{IV.96}$$

However, $\mu_x = \mu_r r_x + \mu_s s_x = -\gamma_r(r_x - s_x)$ and therefore (IV.96) may be put in the form

$$r_{tx} - \gamma_r r_x^2 + \gamma_r r_x s_x + \mu r_{xx} = -\frac{\alpha}{2}(r_x + s_x). \tag{IV.97}$$

Now, directly from the definitions of r', s' we note that $(r' - s')/2\mu = s_x$, and thus substitution in (IV.97) produces the equation

$$\mathscr{r}_{tx} - \gamma_{\mathscr{r}}\mathscr{r}_x^2 + \gamma_{\mathscr{r}}\mathscr{r}_x\left(\frac{\mathscr{r}' - \mathscr{s}'}{2\gamma}\right) + \mu\,\mathscr{r}_{xx} = -\frac{\alpha}{2}\mathscr{r}_x - \frac{\alpha}{4\mu}(\mathscr{r}' - \mathscr{s}'), \tag{IV.98}$$

which, in view of the fact that $(\log \gamma)' = \gamma'/\gamma$, can be written as

$$(\mathscr{r}_x)' = \gamma_{\mathscr{r}}\mathscr{r}_x^2 - \left(\frac{\log\gamma}{2}\right)'\mathscr{r}_x - \frac{\alpha}{2}\mathscr{r}_x - \frac{\alpha}{4\mu}(\mathscr{r}' - \mathscr{s}'). \tag{IV.99}$$

In obtaining (IV.99) from (IV.98) we have also used the fact that $\gamma = \gamma(\mathscr{r} - \mathscr{s})$. By proceeding in an analogous fashion we obtain for $(\mathscr{s}_x)$ the expression

$$(\mathscr{s}_x)^{`} = \gamma_{\mathscr{r}}\mathscr{s}_x^2 - \left(\frac{\log\gamma}{2}\right)^{`}\mathscr{s}_x - \frac{\alpha}{2}\mathscr{s}_x - \frac{\alpha}{4\gamma_{\mathscr{r}}}(\mathscr{r}^{`} - \mathscr{s}^{`}). \tag{IV.100}$$

We now multiply (IV.99) and (IV.100) through by the integrating factor $\sqrt{\gamma}$ and obtain

$$\begin{aligned} \theta' &= \gamma^{-1/2}\gamma_{\mathscr{r}}\theta^2 - \frac{\alpha}{2}\theta - \frac{\alpha}{4}\gamma^{-1/2}(\mathscr{r}' - \mathscr{s}'),\\ \psi^{`} &= \gamma^{-1/2}\gamma_{\mathscr{r}}\psi^2 - \frac{\alpha}{2}\psi - \frac{\alpha}{4}\gamma^{-1/2}(\mathscr{r}^{`} - \mathscr{s}^{`}), \end{aligned} \tag{IV.101}$$

where $\theta = \mathscr{r}_x\gamma^{1/2}$, $\psi = \mathscr{s}_x\gamma^{1/2}$. The system (IV.101) can then be rewritten in the form

$$\begin{aligned} \theta' &= \gamma^{-1/2}\gamma_{\mathscr{r}}\theta^2 - \frac{\alpha}{2}\theta + \alpha\mathscr{f}',\\ \psi^{`} &= \gamma^{-1/2}\gamma_{\mathscr{r}}\psi^2 - \frac{\alpha}{2}\psi + \alpha\mathscr{f}^{`}, \end{aligned} \tag{IV.102}$$

if we introduce $\mathscr{f} = \int_0^{\mathscr{r}-\mathscr{s}} (\gamma^{-1/2}(\zeta)/4)\,d\zeta$. The final form for our system is now obtained by setting

$$\hat\theta = \theta - \alpha\mathscr{f},$$
$$\hat\psi = \psi - \alpha\mathscr{f},$$

namely,

$$\begin{aligned} \hat\theta' &= \gamma^{-1/2}\gamma_{\mathscr{r}}\hat\theta^2 + 2\alpha\gamma^{-1/2}\gamma_{\mathscr{r}}\hat\theta\mathscr{f} + \gamma^{-1/2}\gamma_{\mathscr{r}}\alpha^2\mathscr{f}^2 - \frac{\alpha}{2}\hat\theta - \frac{\alpha^2}{2}\mathscr{f},\\ \hat\psi^{`} &= \gamma^{-1/2}\gamma_{\mathscr{r}}\hat\psi^2 + 2\alpha\gamma^{-1/2}\gamma_{\mathscr{r}}\hat\theta\mathscr{f} + \gamma^{-1/2}\gamma_{\mathscr{r}}\alpha^2\mathscr{f}^2 - \frac{\alpha}{2}\hat\psi - \frac{\alpha^2}{2}\mathscr{f}. \end{aligned} \tag{IV.103}$$

At this point in the proof we will need to make use of the following lemma, a proof of which may be found in [82]:

LEMMA. *Let $\mathscr{F}:(t_0, \infty)\times \mathscr{R}^1 \to \mathscr{R}^1$ be continuous and suppose that $x(t)$ is the unique solution to the scalar differential equation*

$$\frac{dx}{dt} = \mathscr{F}(x, t)$$

with $x(t_0) = x_0$ on $t_0 \leqq t < t_0 + c$. Suppose, also, that there exists $y(t) \in C^1(t_0, t_0+c)$ such that

$$\frac{dy}{dt} \geqq \mathscr{F}(y, t)$$

with $y(t_0) = x_0$. Then $y(t) \geqq x(t)$ on $t_0 \leqq t < t_0 + c$.

In order to apply the idea of the above lemma to the system (IV.103), which is equivalent to the system of equations (IV.99) and (IV.100) for $(t_x)'$ and $(s_x)^{`}$, respectively, we set

$$\text{(IV.104)} \qquad \begin{aligned} &\varepsilon = \inf \frac{\hat{\sigma}''(w)}{4(\hat{\sigma}'(w))^{5/4}} > 0, \qquad F = \inf\left(-\frac{\alpha^2}{2}\ell\right), \\ &\Gamma = \sup\left|\frac{\alpha}{2}(4\gamma^{-1/2}\gamma_{,}\ell - 1)\right|. \end{aligned}$$

It can then be shown that

$$\text{(IV.105)} \qquad \begin{aligned} &\gamma^{-1/2}\gamma_{,} \geqq \varepsilon, \\ &\frac{\alpha}{2}[4\gamma^{-1/2}\gamma_{,}\ell - 1]\hat{\theta} \geqq -\frac{\varepsilon}{2}\hat{\theta}^2 - \frac{\Gamma^2}{2\varepsilon}, \\ &-\frac{\alpha^2}{2}\ell \geqq F. \end{aligned}$$

If we call the right-hand side of (IV.103_1) $\mathscr{F}(\hat{\theta}, t)$, it follows from (IV.105) that

$$\text{(IV.106)} \qquad \mathscr{F}(\hat{\theta}, t) \geqq \mathscr{G}(\hat{\theta}, t) \equiv \frac{\varepsilon}{2}\hat{\theta}^2 - \frac{\Gamma^2}{2\varepsilon} + F.$$

By the above lemma it therefore sufficies to compare the solutions of (IV.103_1), i.e.,

$$\text{(IV.107)} \qquad \hat{\theta}' = \mathscr{F}(\hat{\theta}, t),$$

with those of

$$\text{(IV.108)} \qquad \tilde{\theta}' = \mathscr{G}(\tilde{\theta}, t),$$

where $\tilde{\theta}(x, 0) = \hat{\theta}(x, 0)$. The explicit solution to (IV.108) is given by

$$\text{(IV.109)} \qquad \begin{aligned} &\frac{1}{\tilde{\theta}(x_1(\beta, t), t) + G} = \frac{e^{G\varepsilon t}}{\tilde{\theta}(\beta, 0) + G} + \frac{1}{2G}(1 - e^{G\varepsilon t}), \\ &G \equiv \left[\left(\frac{\Gamma}{\varepsilon}\right)^2 - \frac{2F}{\varepsilon}\right]^{1/2}, \end{aligned}$$

from which it can be shown that $\tilde{\theta}(x_1(\beta, t), t)$ (and hence, by the above lemma $\hat{\theta}(x_1(\beta, t), t)$) goes to infinity in finite time if for some $x \in \mathcal{R}^1$

$$\text{(IV.110)} \qquad 0 < 2G[\gamma^{1/2}(x, 0)\imath_{0'x}(x) - \alpha \not{f}(\imath_0(x) - \jmath_0(x)) + G]^{-1}.$$

An analysis of the bracketed expression in (IV.110) shows that if $\imath_{0,x}(x)$ is sufficiently large (IV.110) will be satisfied which, in turn, implies that $\hat{\theta}(x_1(\beta, t), t)$ goes to infinity in finite time; this fact, coupled with the definitions of $\hat{\theta}$ in terms of θ, and θ in terms of $\imath_x$ shows that $\imath_x(x, t)$ also goes to infinity in finite time and a similar analysis yields the analogous result for $\jmath_x(x, t)$. Thus $(\imath(\cdot, t), \jmath(\cdot, t)) \notin C^1(\mathcal{R}^1) \times C^1(\mathcal{R}^1)$ for all time and in view of the one-to-one correspondence between $(\imath, \jmath)$ and solutions (r, w) of (IV.57), $(r(\cdot, t), w(\cdot, t)) \notin C^1(\mathcal{R}^1) \times C^1(\mathcal{R}^1)$ for all time when $|\imath_0|$, $|\jmath_0|$ are sufficiently small and $\imath_{0,x}$ or $\jmath_{0,x}$ is positive and sufficiently large at some point $x \in \mathcal{R}^1$. From the relations which exist between $\imath_0(x)$, $\jmath_0(x)$ and $r_0(x)$, $w_0(x)$ it follows that this finite time breakdown result for smooth solutions (r, w) of (IV.57) holds under analogous assumptions on the initial data r_0, w_0. Q.E.D.

Remarks. From (IV.53)–(IV.56) it follows that the perturbed velocity field in the nonlinear viscoelastic fluid satisfies

$$\text{(IV.111)} \qquad \begin{cases} \hat{v}_t = \hat{\sigma}'(w)w_x, \\ \hat{v}_x = r_x + \alpha w, \end{cases}$$

where the last equation is obtained by differentiation of (IV.56$_3$) with respect to x followed by an integration by parts with respect to τ. Under the hypotheses of the above theorem it follows directly from (IV.111) that $\hat{v}$ cannot be of class C^1 in (x, t) for all $t \in [0, \infty)$ even if the initial velocity history is smooth.

3. Global existence of solutions for nonlinear viscoelastic initial-history value problems: Riemann invariant arguments.

In the remarks at the end of Chapter II, we noted that the pure initial-history value problem associated with the nonlinear viscoelastic integrodifferential equation (II.229), with $\mathcal{g}(t)$ normalized so that $\mathcal{g}(0) = 1$, may be brought into the equivalent form (II.311) where $k(t)$ is the resolvent kernel associated with $\mathcal{g}'(t)$ and the nonhomogeneous contribution $\Gamma(x, t)$ is given by (II.312). By carrying out the indicated partial differentiation on the left-hand side of (II.311), and then effecting an integration by parts, we may further bring the initial-value problem associated with (II.229) into the form of an initial-value problem for a damped, nonlinear, nonhomogeneous wave equation, namely,

$$\text{(IV.112)} \qquad \begin{aligned} &u_{tt} + \alpha u_t - \sigma(u_x)_x = G, \qquad -\infty < x < \infty, \quad t \geqq 0, \\ &u(x, 0) = f(x), \qquad u_t(x, 0) = g(x), \qquad -\infty < x < \infty, \end{aligned}$$

where $\alpha = k(0) > 0$ and $G = G(x, t)$ is given by (II.300a). If we further set $r = u_t$, $w = u_x$ then (IV.112) assumes the form of an initial-value problem for the damped, nonhomogeneous system (II.302) which, for the sake of convenience, we repeat below, i.e, for $t \geqq 0$, $-\infty < x < \infty$,

$$\text{(IV.113)} \qquad \begin{aligned} &w_t(x, t) - r_x(x, t) = 0, \\ &r_t(x, t) - \sigma'(w) w_x(x, t) + \alpha r(x, t) = G(x, t), \end{aligned}$$

with associated initial data

$$\text{(IV.114)} \qquad r(x, 0) = g(x), \quad w(x, 0) = f'(x), \qquad -\infty < x < \infty.$$

In Chapter II we considered the pure initial-value problem for (II.229), in the form (II.311), by means of an energy argument due to Dafermos and Nohel [43] and estaiblished global existence of smooth solutions when the data f, g, G are sufficiently small (in a sense which is made precise by the analysis presented there); in essence, those results, under appropriate further restrictions on $G(x, t)$ also establish global existence of solutions to the initial-value problem for the damped, nonhomogeneous system (IV.113) when the data is sufficiently small. In the last section of this chapter, on the other hand, we proved, using a Riemann invariant argument of Slemrod [139], [141] that classical smooth solutions of hyperbolic damped systems of the form (IV.113), with $G \equiv 0$, must break down in finite time if the initial data are sufficiently large (again, in a sense which is made precise by the analysis in § 3). As we have previously indicated, the problem of whether or not the (smooth) solutions of the initial-value problem associated with the damped, nonhomogeneous, hyperbolic system (IV.113) break down in finite time, for sufficiently large initial data has been posed by MacCamy [104] and, is to the best of our knowledge, still open. In this section we wish to reconsider the problem of global existence of solutions for the damped, nonhomogeneous problem (IV.113), (IV.114) by using a Riemann Invariant argument due to Nohel [122] which is patterned on the Riemann Invariant argument of Slemrod [139], [141] for the associated damped homogenous problem considered in the previous section; Nohel's arguments are very closely related to those originally employed by MacCamy in [104] in his treatment of the initial-boundary value problem associated with (II.229) but, as we shall see, Nohel's approach [122] also permits us to study the continuous dependence of solutions of (IV.113), and hence of the equivalent nonlinear integrodifferential equation (II.229) [also the equivalent damped, nonhomogeneous wave equation (IV.112)] on the data f, g, and G.

In what follows we assume that $\sigma : \mathcal{R}^1 \to \mathcal{R}^1$, $G : \mathcal{R}^1 \times [0, \infty) \to \mathcal{R}^1$ and $f, g : \mathcal{R}^1 \to \mathcal{R}^1$ are given smooth functions with $\sigma \in C^2(\mathcal{R}^1)$, $\sigma(0) = 0$, $\sigma''(\zeta) \neq 0$ for some $\zeta \in \mathcal{R}^1$, and $\sigma'(\zeta) \geqq \varepsilon > 0$, $\forall \zeta \in \mathcal{R}^1$, and some $\varepsilon > 0$. We also assume that G and the initial data f, g in (IV.114) satisfy

$$\text{(IV.115)}\qquad \begin{aligned} &G, G_x \in C([0,\infty)\times\mathcal{R}^1),\\ &\mathcal{G}_1 = \sup_{x\in\mathcal{R}^1} |G(x,t)| \in L^\infty(0,\infty)\cap L^1(0,\infty),\\ &\mathcal{G}_2 = \sup_{x\in\mathcal{R}^1} |G_x(x,t)| \in L^\infty(0,\infty),\\ &f\in\mathcal{B}^1(\mathcal{R}^1), \quad g\in\mathcal{B}^1(\mathcal{R}^1), \end{aligned}$$

where $\mathcal{B}^m$ denotes the set of all real-valued functions with bounded, continuous derivatives up to and including the order m. As per the analysis of § 3, the system (IV.113) will be strictly hyperbolic in the region $\Omega = \{(w, r) | w \in \mathcal{R}^1, r \in \mathcal{R}^1\}$ if the eigenvalues $\mu = -\sqrt{\sigma'(w)}$, $\gamma = \sqrt{\sigma'(w)}$ of the associated matrix $\begin{pmatrix} 0 & \sigma' \\ 1 & 0 \end{pmatrix}$ are real and distinct; this will certainly be true under our assumption that $\sigma'(\zeta) \geqq \varepsilon > 0$ $\forall \zeta \in \mathcal{R}^1$ and some $\varepsilon > 0$. As in § 3 we now diagonalize the damped, nonhomogeneous system (IV.113) by introducing the Riemann invariants $\imath$, $\jmath$ which are again defined by (IV.68) and (IV.70), respectively, but with $\hat{\sigma}$ in those expressions now replaced by σ; a direct computation, analogous to the one which led to the diagonal system (IV.72) for the case where $G \equiv 0$, now shows that (IV.113), (IV.114) is equivalent to the system

$$\text{(IV.116)}\qquad \left.\begin{aligned} &\imath_t + \mu \imath_x + \frac{\alpha}{2}(\imath + \jmath) = G,\\ &\jmath_t + \gamma \jmath_x + \frac{\alpha}{2}(\imath + \jmath) = G, \end{aligned}\right\} \quad 0 \leqq t < \infty, \quad x \in \mathcal{R}^1,$$

$$\imath(x,0) = \imath_0(x), \quad \jmath(x,0) = \jmath_0(x), \qquad x \in \mathcal{R}^1$$

where, as before, $\mu = \mu(\imath - \jmath)$, $\gamma = \gamma(\imath - \jmath)$ and it is easily shown that

$$\text{(IV.117)}\qquad \begin{aligned} \imath_0(x) &= g(x) + \int_0^{f'(x)} \sqrt{\sigma'(\zeta)}\, d\zeta,\\ \jmath_0(x) &= g(x) - \int_0^{f'(x)} \sqrt{\sigma'(\zeta)}\, d\zeta. \end{aligned}$$

In view of our assumptions on σ, and the hypotheses relative to f, g it follows directly from (IV.117) that $\imath_0$, $\jmath_0 \in \mathcal{B}^1(\mathcal{R}^1)$. Also, if $(\imath, \jmath)$ is a smooth $\mathcal{B}^1$ solution of the system (IV.116) on $\mathcal{R}^1 \times [0, \infty)$, then $u(x, t)$, as determined by $u_t = r(\imath, \jmath)$, $u_x = w(\imath, \jmath)$, where (r, w) will be the corresponding smooth $\mathcal{B}^2$ solution of (IV.113), (IV.114), will be a smooth $\mathcal{B}^2$ solution of the Cauchy problem associated with the nonhomogeneous, nonlinear, damped wave equation (IV.112). Our aim, therefore, will be to prove that for sufficiently small initial-data, the Cauchy problem for the strictly hyperbolic diagonal system in (IV.116) possesses a unique global smooth solution which depends continuously on the data r_0, s_0, and G. As per the energy argument analysis of Dafermos and Nohel [43] which was presented in Chapter II, the argument here consists of combining a local existence theorem for solutions of the system (IV.116), with a priori estimates on r, s, r_x, and s_x, and a standard continuation theorem, to obtain the existence of a unique smooth global solution to (IV.116).

The required local existence theorem for systems of the form (IV.116) is well known, and thus, as in [122], we merely state the result below as the following:

LEMMA. *Suppose that $r_0, s_0 \in \mathscr{B}^1(\mathscr{R}^1)$ and that $G, G_x \in \mathscr{B}^0(\mathscr{R}^1)$ for $t \in [0, T]$ where $T > 0$. Then there exists T_1, $0 < T_1 \leqq T$ such that the initial-value problem* (IV.116) *has a unique smooth solution $(r(\cdot, t), s(\cdot, t)) \in \mathscr{B}^1(\mathscr{R}^1) \times \mathscr{B}^1(\mathscr{R}^1)$ for $t \in [0, T_1]$.*

Following Nohel [122] we now state and prove the first of the required set of a priori estimates; the proofs of both sets of estimates involve analysis which is very closely related to that effected in § 3 and, thus, the proof of both sets of estimates will merely be sketched in outline form; the reader may consult [122] for the details or may easily fill in the details himself by simply following the related analysis of § 3:

LEMMA. *Let the assumptions of the above lemma hold and assume, in addition, that $\mathscr{G}_1(t) = \sup_{x \in \mathscr{R}^1} |G(x, t)| \in L^1(0, \infty)$. Then for as long as the $\mathscr{B}^1 \times \mathscr{B}^1$ solution $(r(\cdot, t), s(\cdot, t))$ of* (IV.116) *exists we have*

$$\text{(IV.118)} \qquad \sup_{x \in \mathscr{R}^1} |r(x, t)| \leqq M_0, \qquad \sup_{x \in \mathscr{R}^1} |s(x, t)| \leqq M_0,$$

where

$$\text{(IV.119)} \qquad \begin{cases} M_0 = |r_0| + |s_0| + 2 \displaystyle\int_0^\infty \mathscr{G}_1(\xi)\, d\xi, \\ |r_0| = \displaystyle\sup_{x \in \mathscr{R}^1} |r_0(x)|, \ |s_0| = \displaystyle\sup_{x \in \mathscr{R}^1} |r_0(x)| \end{cases}$$

as in § 3.

Proof. As in § 3, we introduce the right and left-hand characteristics which are defined by (IV.76) and rewrite the system of equations (IV.116) in the equivalent form

$$\text{(IV.120)} \qquad \begin{aligned} \frac{d}{dt} r(x_1(\beta, t), t) &= -\frac{\alpha}{2}\left(r(x_1(\beta, t), t) + s(x_1(\beta, t), t)\right) \\ &\quad + G(x_1(\beta, t), t), \\ \frac{d}{dt} s(x_2(\delta, t), t) &= -\frac{\alpha}{2}\left(r(x_2(\delta, t), t) + s(x_2(\delta, t), t)\right) \\ &\quad + G(x_2(\delta, t), t), \end{aligned}$$

which, of course, is just the system (IV.77) when $G\equiv 0$. If we again define $S(t)$, $R(t)$ by (IV.89) and set $W(t)=S(t)+R(t)$, as in § 3, then integration of the equations in (IV.120) (using the integrating factor $e^{(\alpha/2)t}$ again, and the initial conditions $r(x_1(\beta,0),0)=r_0(\beta)$, $s(x_2(\delta,0),0)=s_0(\delta)$) followed by an analysis which is completely analogous to the one which took us in § 3 from the integrated set of equations (IV.88), for the case where $G\equiv 0$, to the inequality (IV.94), yields, in this case, the estimate

$$W(t)\leqq |r_0|+|s_0|+\frac{\alpha}{2}\int_0^t W(\tau)\,d\tau+2\int_0^t e^{(\alpha/2)\tau}\mathscr{G}_1(\tau)\,d\tau. \tag{IV.121}$$

An application of Gronwall's inequality to (IV.121) then produces the estimate

$$W(t)\leqq (|r_0|+|s_0|)\,e^{(\alpha/2)t}+2\,e^{(\alpha/2)t}\int_0^t \mathscr{G}_1(\tau)\,d\tau \tag{IV.122}$$

and the desired result, i.e., (IV.118), then follows directly from (IV.122), (IV.119), and the definitions of $W(t)$, $R(t)$, and $S(t)$. Q.E.D.

Our second set of a priori estimates is contained in the statement of the following

LEMMA. *Let the assumptions of the above local existence theorem be satisfied and assume that $G(x,t)$ satisfies the conditions in* (IV.115). *Set*

$$\begin{aligned} N_1 = {} & |r_0|+|s_0|+\sup_{x\in\mathscr{R}^1}|r_0'(x)|+\sup_{x\in\mathscr{R}^1}|s_0'(x)| \\ & +\|\mathscr{G}_1\|_{L^1(0,\infty)}+\|\mathscr{G}_1\|_{L^\infty(0,\infty)}+\|\mathscr{G}_2\|_{L^\infty(0,\infty)}. \end{aligned} \tag{IV.123}$$

Then for as long as the $\mathscr{B}^1(\mathscr{R}^1)$ solution $(r(\cdot,t), s(\cdot,t))$ of (IV.116) *exists, if N_1 is sufficiently small, there exists $M_1=M_1(N_1)>0$ with $M_1(N_1)\to 0$ as $N_1\to 0$ such that*

$$\sup_{x\in\mathscr{R}^1}|r_x(x,t)|\leqq M_1,\qquad \sup_{x\in\mathscr{R}^1}|s_x(x,t)|\leqq M_1. \tag{IV.124}$$

Proof. We will simply sketch a few steps of the proof and refer the reader to [122] for the remaining details. If we differentiate the first equation in (IV.116) we easily obtain the direct generalization of (IV.97), namely,

$$r_{tx}+\mu r_{xx}=-\mu_r r_x^2-\mu_s r_x s_x-\frac{\alpha}{2}(r_x+s_x)+G_x. \tag{IV.125}$$

In comparing (IV.125), with $G\equiv 0$, with (IV.97) we should note that the relations $\mu=\mu(r-s)$, $\gamma=\gamma(r-s)$, $\mu=-\gamma$ have been employed in (IV.125), i.e., $\gamma_r=-\mu_r=\mu_s$. From the second equation in (IV.116) we obtain

$$s_x=\frac{s'}{2\mu}+\frac{\alpha}{4\mu}(r+s)-\frac{G}{2\mu}, \tag{IV.126}$$

where, as before $'=\partial/\partial t+\mu\,\partial/\partial x$. If we set $h=\frac{1}{2}\ln[-\mu(r-s)]$, differentiate h

along the μ characteristic, using $\mu_{\jmath}=-\mu_{\imath}$ and then substitute from the resulting expression, and (IV.126), into (IV.125) we obtain, after some simplification the differential equation

$$(e^h \imath_x)' + \left(\frac{\alpha}{2}+\mu_{\imath}\imath_x\right)e^h \imath_x = z' + e^h G_x, \tag{IV.127}$$

where

$$z = z(\imath-\jmath) = \int_0^{\imath-\jmath} \frac{\alpha}{4\mu(\zeta)} e^{h(\zeta)}\, d\zeta. \tag{IV.128}$$

Setting $\nu(t)=\imath_x(x_1(\beta,t),t)\exp[h(x_1(\beta,t),t)]$ and integrating (IV.127) along the μ characteristic then yields

$$\begin{aligned}\nu(t) = \nu(0)\exp\left[-\int_0^t \mathcal{K}(\tau)\,d\tau\right]\\ +\int_0^t \mathcal{P}(\zeta)\exp\left[-\int_\zeta^t \mathcal{K}(\tau)\,d\tau\right]d\zeta,\end{aligned} \tag{IV.129}$$

where

$$\begin{aligned}\mathcal{K}(t) &= \frac{\alpha}{2}+[\mu_{\imath}(\imath(x_1(\beta,t),t)-\jmath(x_1(\beta,t),t))]\cdot[\imath_x(x_1(\beta,t),t)],\\ \mathcal{P}(t) &= z'(x_1(\beta,t),t)+[G_x(x_1(\beta,t),t)]\cdot\exp[h(x_1(\beta,t),t)].\end{aligned} \tag{IV.130}$$

The desired bound on $|\imath_x(x,t)|$ now follows from (IV.129) and the definition of $\nu(t)$ once it has been established that $\mathcal{K}(t)\geqq \alpha/4$; this last result follows, in turn, from the fact that for N_1 sufficiently small $|\mu_{\imath}\imath_x|\leqq \alpha/4$ uniformly in (x,t) for any solution of (IV.116). In fact, for any solution of (IV.116) it can be shown that

$$|\mu_{\imath}\imath_x| \leqq K_1(N_1)K_2(N_2) \tag{IV.131}$$

uniformly in (x,t) where $K_1, K_2\to 0$ as $N_1\to 0$. An argument similar to the one briefly sketched above establishes the required a priori bound on $|\jmath_x(x,t)|$. Q.E.D.

Once we have established the local existence theorem for the system (IV.116), and the above a priori estimates for $\imath$, $\jmath$, $\imath_x$ and $\jmath_x$, it then follows from (IV.116), and the assumption that $\|\mathcal{G}_1\|_{L^\infty(0,\infty)}<\infty$ that, provided N_1 is sufficiently small, $\imath_t$, $\jmath_t$ satisfy uniform a priori estimates for as long as the unique smooth solution exists. The desired global existence theorem for the system (IV.116) is then a direct consequence of the local existence theorem, the established a priori estimates, and a standard continuation argument analogous to the one employed in Chapter II in the course of our discussion of the results obtained by Dafermos and Nohel [43]; the unique $\mathcal{B}^1(\mathcal{R}^1)\times\mathcal{B}^1(\mathcal{R}^1)$ solution $(\imath(\cdot,t),\jmath(\cdot,t))$ of (IV.116) on $[0,\infty)$, which exists when N_1 is chosen sufficiently small, will then satisfy the a priori estimates delineated in (IV.118), (IV.124) for $t \in [0, \infty)$.

In view of the fact that the global existence result for the system (IV.116) is already implict in the earlier work of MacCamy [104], the more important result in [122] would seem to be the following continuous data dependence theorem:

THEOREM (Nohel [122]). *Let the initial data* $\mathfrak{r}_0(\cdot)$, $\mathfrak{s}_0(\cdot) \in \mathscr{B}^1(\mathscr{R}^1)$ *and assume that the constant* N_1, *which is defined by* (IV.123), *is so small that, as per the above global existence theorem, the Cauchy problem* (IV.116) *has a unique solution* $(\mathfrak{r}(\cdot, t), \mathfrak{s}(\cdot, t))$ *in* $\mathscr{B}^1(\mathscr{R}^1) \times \mathscr{B}^1(\mathscr{R}^1)$ *for* $0 \leqq t < \infty$. *We assume, of course, that* $G(\cdot, t)$ *satisfies* (IV.115). *Let* $\bar{\mathfrak{r}}_0(\cdot)$, $\bar{\mathfrak{s}}_0(\cdot)$, *and* $\bar{G}(\cdot, t)$ *be a second set of data functions and* $(\bar{\mathfrak{r}}(\cdot, t), \bar{\mathfrak{s}}(\cdot, t))$ *the corresponding* $\mathscr{B}^1(\mathscr{R}^1) \times \mathscr{B}^1(\mathscr{R}^1)$ *solution of* (IV.116) on $[0, \infty)$. *Then, if we define*

$$\chi(t) = \sup_{x \in \mathscr{R}^1} |\mathfrak{r}(x, t) - \bar{\mathfrak{r}}(x, t)| + \sup_{x \in \mathscr{R}^1} |\mathfrak{s}(x, t) - \bar{\mathfrak{s}}(x, t)|, \tag{IV.132}$$

$\exists M_2 = M_2(\sigma, M_0) > 0$, *such that for* $0 \leqq t < \infty$

$$\chi(t) \leqq e^{2M_2M_1t}\left(\chi(0) + \int_0^t e^{-2M_2M_1\tau} \sup_{x \in \mathscr{R}^1} |G(x, \tau) - \bar{G}(x, \tau)|\, d\tau\right), \tag{IV.133}$$

where M_0, M_1 *are the bounds in the a priori estimates* (IV.118), (IV.124).

Proof. If $(\mathfrak{r}(\cdot, t), \mathfrak{s}(\cdot, t))$, $(\bar{\mathfrak{r}}(\cdot, t), \bar{\mathfrak{s}}(\cdot, t))$ are $\mathscr{B}^1(\mathscr{R}^1) \times \mathscr{B}^1(\mathscr{R}^1)$ solutions of (IV.116) then

$$\begin{aligned} (\mathfrak{r} - \bar{\mathfrak{r}})_t + \mu \mathfrak{r}_x - \bar{\mu} \bar{\mathfrak{r}}_x &= -\frac{\alpha}{2}[(\mathfrak{r} - \bar{\mathfrak{r}}) + (\mathfrak{s} - \bar{\mathfrak{s}})] + G - \bar{G}, \\ (\mathfrak{s} - \bar{\mathfrak{s}})_t + \gamma \mathfrak{s}_x - \bar{\gamma} \bar{\mathfrak{s}}_x &= -\frac{\alpha}{2}[(\mathfrak{r} - \bar{\mathfrak{r}}) + (\mathfrak{s} - \bar{\mathfrak{s}})] + G - \bar{G}, \end{aligned} \tag{IV.134a}$$

holds on $\mathscr{R}^1 \times [0, \infty)$, subject to the initial data

$$\left.\begin{aligned} \mathfrak{r}(x, 0) - \bar{\mathfrak{r}}(x, 0) &= \mathfrak{r}_0(x) - \bar{\mathfrak{r}}_0(x) \\ \mathfrak{s}(x, 0) - \bar{\mathfrak{s}}(x, 0) &= \mathfrak{s}_0(x) - \bar{\mathfrak{s}}_0(x) \end{aligned}\right\}, \qquad x \in \mathscr{R}^1. \tag{IV.134b}$$

In (IV.134a) $\bar{\mu} = \mu(\bar{\mathfrak{r}} - \bar{\mathfrak{s}})$, $\bar{\gamma} = \gamma(\bar{\mathfrak{r}} - \bar{\mathfrak{s}})$. By setting

$$\begin{aligned} \mu \mathfrak{r}_x - \bar{\mu} \bar{\mathfrak{r}}_x &= \mu(\mathfrak{r} - \bar{\mathfrak{r}})_x + (\mu - \bar{\mu})\bar{\mathfrak{r}}_x, \\ \gamma \mathfrak{s}_x - \bar{\gamma} \bar{\mathfrak{s}}_x &= \gamma(\mathfrak{s} - \bar{\mathfrak{s}})_x + (\gamma - \bar{\gamma})\bar{\mathfrak{s}}_x, \end{aligned} \tag{IV.135}$$

we may rewrite (IV.134a) in the equivalent form

$$\begin{aligned} (\mathfrak{r} - \bar{\mathfrak{r}})_t + \mu(\mathfrak{r} - \bar{\mathfrak{r}})_x &= -\frac{\alpha}{2}[(\mathfrak{r} - \bar{\mathfrak{r}}) + (\mathfrak{s} - \bar{\mathfrak{s}})] + G - \bar{G} - (\mu - \bar{\mu})\bar{\mathfrak{r}}_x, \\ (\mathfrak{s} - \bar{\mathfrak{s}})_t + \gamma(\mathfrak{s} - \bar{\mathfrak{s}})_x &= -\frac{\alpha}{2}[(\mathfrak{r} - \bar{\mathfrak{r}}) + (\mathfrak{s} - \bar{\mathfrak{s}})] + G - \bar{G} - (\gamma - \bar{\gamma})\bar{\mathfrak{s}}_x. \end{aligned} \tag{IV.136}$$

However, by the man value theorem $\exists\theta_1, \theta_2, 0<\theta_1, \theta_2<1$ such that

$$
\text{(IV.137)}\quad
\begin{aligned}
\mu(r-s)-\mu(\bar{r}-\bar{s}) &= \mu-\bar{\mu} \\
&= \frac{d\mu}{d\zeta}(\bar{r}-\bar{s}+\theta_1[r-s-(\bar{r}-\bar{s})])(r-s-(\bar{r}-\bar{s})),\\
\gamma(r-s)-\gamma(\bar{r}-\bar{s}) &= \gamma-\bar{\gamma} \\
&= \frac{d\gamma}{d\zeta}(\bar{r}-\bar{s}+\theta_2[r-s-(\bar{r}-\bar{s})])(r-s-(\bar{r}-\bar{s})),
\end{aligned}
$$

so that (IV.136) can be rewritten in the form

$$
\text{(IV.138)}\quad
\begin{aligned}
(r-\bar{r})'+\frac{\alpha}{2}(r-\bar{r}) &= -\frac{\alpha}{2}(s-\bar{s})-\frac{d\mu}{d\zeta}(\cdot)(r-\bar{r}-(s-\bar{s}))\bar{r}_x+G-\bar{G},\\
(s-\bar{s})'+\frac{\alpha}{2}(s-\bar{s}) &= -\frac{\alpha}{2}(r-\bar{r})-\frac{d\gamma}{d\zeta}(\cdot)(r-\bar{r}-(s-\bar{s}))\bar{s}_x+G-\bar{G}.
\end{aligned}
$$

In view of our assumptions on σ,

$$
\frac{d\mu}{d\zeta}=-\tfrac{1}{2}(\sigma'(\zeta))^{-1/2}\sigma''(\zeta)
$$

and, thus, in view of the a priori estimates represented by (IV.118), (IV.124), $\exists M_2=M_2(\sigma, M_0)$ such that

$$
\text{(IV.139)}\quad \left|\frac{d\mu}{d\zeta}(\cdot)\bar{r}_x\right|\leqq M_2M_1,\qquad \left|\frac{d\gamma}{d\zeta}(\cdot)\bar{s}_x\right|\leqq M_2M_1,
$$

for all $(x, t)\in\mathscr{R}^1\times[0,\infty)$. We now integrate (IV.138$_1$) along a μ characteristic and (IV.138$_2$) along a γ-characteristic, employing the uniform bounds in (IV.139), so as to obtain the estimates

$$
\text{(IV.140)}\quad
\begin{aligned}
e^{\alpha t/2}\sup_{x\in\mathscr{R}^1}|r(x,t)-\bar{r}(x,t)| &\leqq \sup_{x\in\mathscr{R}^1}|r_0(x)-\bar{r}_0(x)|\\
&\quad+\int_0^t e^{\alpha t/2}\sup_{x\in\mathscr{R}^1}|G(x,\tau)-\bar{G}(x,\tau)|\,d\tau\\
&\quad+\int_0^t e^{\alpha\tau/2}\left(\frac{\alpha}{2}+M_2M_1\right)\sup_{x\in\mathscr{R}^1}|s(x,\tau)-\bar{s}(x,\tau)|\,d\tau\\
&\quad+M_1M_2\int_0^t e^{\alpha\tau/2}\sup_{x\in\mathscr{R}^1}|r(x,\tau)-\bar{r}(x,\tau)|\,d\tau
\end{aligned}
$$

and

$$e^{\alpha t/2} \sup_{x\in\mathcal{R}^1} |\mathfrak{z}(x,t)-\bar{\mathfrak{z}}(x,t)| \leqq \sup_{x\in\mathcal{R}^1} |\mathfrak{z}_0(x)-\bar{\mathfrak{z}}_0(x)|$$

(IV.141)
$$+\int_0^t e^{\alpha\tau/2} \sup_{x\in\mathcal{R}^1} |G(x,\tau)-\bar{G}(x,\tau)|\,d\tau$$
$$+\int_0^t e^{\alpha\tau/2}\left(\frac{\alpha}{2}+M_2M_1\right) \sup_{x\in\mathcal{R}^1} |\mathfrak{r}(x,\tau)-\bar{\mathfrak{r}}(x,\tau)|\,d\tau$$
$$+M_1M_2\int_0^t e^{\alpha\tau/2} \sup_{x\in\mathcal{R}^1} |\mathfrak{z}(x,\tau)-\bar{\mathfrak{z}}(x,\tau)|\,d\tau.$$

If we add (IV.140) and (IV.141) together and use the definition (IV.132) of $\chi(t)$ we easily obtain the estimate

(IV.142)
$$\chi(t)\,e^{\alpha t/2} \leqq \chi(0)+2\int_0^t e^{\alpha\tau/e} \sup_{x\in\mathcal{R}^1} |G(x,\tau)-\bar{G}(x,\tau)|\,d\tau$$
$$+\int_0^t \left(\frac{\alpha}{2}+2M_1M_2\right) e^{\alpha\tau/2}\chi(\tau)\,d\tau$$

for $0\leqq t<\infty$. The desired continuous dependence result for the $\mathcal{B}^1(\mathcal{R}^2)\times\mathcal{B}^1(\mathcal{R}^1)$ solution $(\mathfrak{r}(\cdot,t),\mathfrak{z}(\cdot,t))$ of the Cauchy problem (IV.116), i.e., (IV.133), now follows directly from (IV.142) via an application of Gronwall's inequality. Q.E.D.

In view of the equivalence of the Cauchy problems (IV.116), (IV.113)–(IV.114), and (IV.112) we may conclude our discussion of Nohel's results in [122] by stating the following theorem concerning the global existence, uniqueness, and continuous dependence of solutions to the initial value problem for the damped, nonlinear, nonhomogeneous wave equation in (IV.112):

THEOREM (Nohel [122]). *Let the assumptions on σ and the assumptions in* (IV.115) *relative to f, g, and G be satisfied and set*

(IV.143)
$$M=\sup_{x\in\mathcal{R}^1}|f'(x)|+\sup_{x\in\mathcal{R}^2}|g(x)|+\sup_{x\in\mathcal{R}^1}|f''(x)|$$
$$+\|\mathcal{G}_1\|_{L^1(0,\infty)}+\|\mathcal{G}_1\|_{L^\infty(0,\infty)}+\|\mathcal{G}_2\|_{L^\infty(0,\infty)}.$$

If M is sufficiently small, then the Cauchy problem (IV.112) *has a unique solution $u(\cdot,t)\in\mathcal{B}^2(\mathcal{R}^1)$ on $[0,\infty)$; furthermore $u(x,t)$ satisfies the a priori estimates*

(IV.144)
$$\left.\begin{array}{l}\sup_{x\in\mathcal{R}^1}|u_x(x,t)|\\ \sup_{x\in\mathcal{R}^1}|u_t(x,t)|\end{array}\right\}\leqq \sup_{x\in\mathcal{R}^1}|f'(x)|+\sup_{x\in\mathcal{R}}|g(x)|+2\int_0^\infty \mathcal{G}_1(\zeta)\,d\zeta\equiv M_0$$

and, $\exists M_1 = M_1(M) > 0$ *with* $M_1 \to 0$ *as* $M \to 0$ *such that*

$$\text{(IV.145)} \qquad \{\sup_{x\in\mathcal{R}^1} |u_{xx}(x,t)|, \sup_{x\in\mathcal{R}^1} |u_{xt}(x,t)|, \sup_{x\in\mathcal{R}^1} |u_{tt}(x,t)|\} \leqq M_1,$$

for $0 \leqq t < \infty$. *If* (IV.115) *is satisfied by a second set of data functions* $\bar{f}$, $\bar{g}$, *and* $\bar{G}$, *and* $\bar{u}(\cdot, t) \in \mathcal{B}^2(\mathcal{R}^1)$ *on* $[0, \infty)$ *denotes the corresponding global solution of* (IV.112), *then* $\exists M_2 = M_2(\sigma, M_0) > 0$ *such that*

$$\text{(IV.146)} \qquad \Lambda(t) \equiv \sup_{x\in\mathcal{R}^1} |u_x(x,t) - \bar{u}_x(x,t)| + \sup_{x\in\mathcal{R}^1} |u_t(x,t) - \bar{u}_t(x,t)|$$

satisfies the continuous dependence estimate

$$\text{(IV.147)} \quad \Lambda(t) \leqq e^{2M_1M_2t}\left(\Lambda(0) + \int_0^t e^{-2M_1M_2\tau} \sup_{x\in\mathcal{R}^1} |G(x,\tau) - \bar{G}(x,\tau)|\, d\tau\right)$$

for $0 \leqq t < \infty$.

Remark. If we combine (IV.147) with the elementary fact that $u(x,t) = f(x) + \int_0^t u_\tau(x,\tau)\, d\tau$ and use the definition of $\Lambda(t)$, i.e. (IV.146), we easily obtain the continuous dependence estimate on $[0, \infty)$:

$$\text{(IV.148)} \quad \begin{aligned} \sup_{x\in\mathcal{R}^1} |u(x,t) - \bar{u}(x,t)| &\leqq \sup_{x\in\mathcal{R}^1} |f(x) - \bar{f}(x)| \\ &\quad + \frac{e^{2M_1M_2t}}{M_1M_2}\left(\sup_{x\in\mathcal{R}^1} |f'(x) - \bar{f}'(x)| + \sup_{x\in\mathcal{R}^1} |g(x) - \bar{g}(x)| \right. \\ &\quad \left. + \int_0^t e^{-2M_1M_2\tau} \sup_{x\in\mathcal{R}^1} |G(x,\tau) - \bar{G}(x,\tau)|\, d\tau\right). \end{aligned}$$

4. Remarks on recent research on first-order nonlinear integrodifferential equations.

In this concluding section, we wish to discuss, briefly, some recent directions in research on nonlinear first-order integrodifferential equations, particularly those which come up in problems of heat conduction in nonlinear materials with memory. It has been impossible, within the context of this rather limited treatise, to present an adequate discussion of many topics which should be included in any comprehensive discussion of the qualitative behavior of solutions to linear and nonlinear integrodifferential equations; therefore, we have limited our study to a consideration of some recent results on ill-posed problems for, essentially, second-order linear and nonlinear equations which arise in viscoelasticity and the theory of nonconducting rigid dielectrics with memory. In what follows we offer a brief survey of recent trends in research on integrodifferential equations which are, for the most part, nonlinear and of the first-order; more specifically, the work to be referenced below deals with:

(A) Global existence and asymptotic stability of solutions to scalar nonlinear first-order Volterra integrodifferential equations and to abstract nonlinear first-order integrodifferential equations in Hilbert and Banach spaces,

(B) Problems connected with nonlinear first-order integrodifferential equations arising in theories of nonlinear heat conduction in materials with memory.

For further information concerning the operator theoretic concepts (e.g., maximal monotone operator) referenced below, the reader may consult the excellent survey articles of Goldstein [56], Evans [50], Crandall [35], and Brezís [26] or the recent treatise of Barbu [6]; for a comprehensive treatment of some of the qualitative work on solutions of nonlinear Volterra integral and integrodifferential equations (up through 1971) the reader may wish to consult the monograph by R. K. Miller [114].

(A) *Abstract and scalar first-order nonlinear equations.* We begin with some simple concepts which are operator-theoretic in nature and which have found recent widespread application in the theory of nonlinear semigroups and applications to nonlinear partial differential equations ([6]), [27], [35], [50], [56], and the recent treatise by A. Pazy [128]); some of the ideas have already been introduced in Chapter II but we shall now endeavor to be somewhat more precise.

Let X be a real Banach space with norm $\|(\cdot)\|$. An operator in X (possibly multivalued) will be a map $\mathbf{A}: X \to 2^X$ (the set of all subsets of X). By the domain of $\mathbf{A}$ we understand $\mathcal{D}(\mathbf{A}) = \{\mathbf{x} \in X | \{\mathbf{Ax}\} \neq \Phi_e\}$ where Φ_e, of course, denotes the empty-set. $\mathbf{A}$ is single-valued if $\{\mathbf{Ax}\}$ is a singleton for all $\mathbf{x} \in \mathcal{D}(\mathbf{A})$. More generally, $\mathbf{A}$ may be multivalued, in which case corresponding to $\mathbf{x} \in X$ there will be a set $\{\mathbf{Ax}\} \subseteq X$ and for fixed $\mathbf{x} \in X$ we will write $\mathbf{f} \in \mathbf{Ax}$, meaning $\mathbf{f}$ belongs to the set of values $\{\mathbf{Ax}\}$. In this case $\mathbf{A}^{-1}\mathbf{f} = \{\mathbf{x} \in X | \mathbf{f} \in \mathbf{Ax}\}$. We also have the obvious definitions $(\lambda\mathbf{A})\mathbf{x} = \{\lambda\mathbf{f} | \mathbf{f} \in \mathbf{Ax}\}$ and $(\mathbf{A}_1 + \mathbf{A}_2)\mathbf{x} = \{\mathbf{f}_1 + \mathbf{f}_2 | \mathbf{f}_1 \in \mathbf{A}_1\mathbf{x}, \mathbf{f}_2 \in \mathbf{A}_2\mathbf{x}\}$. A mapping $\mathbf{B}: X \to X$ is said to be a *contraction* if for $\mathbf{y}_i \in \mathbf{Bx}_i$, $i = 1, 2$, $\|\mathbf{y}_1 - \mathbf{y}_2\| \leqq \|\mathbf{x}_1 - \mathbf{x}_2\|$; as $\mathbf{x}_1 = \mathbf{x}_2$ clearly implies that $\mathbf{y}_1 = \mathbf{y}_2$, a contraction is always single-valued. We then have:

DEFINITION. Let $\mathbf{A}$ be an operator in X. Then $\mathbf{A}$ is said to be *accretive* if $(\mathbf{I} + \lambda\mathbf{A})^{-1}$ is a contraction $\forall\lambda \geqq 0$.

By using the definition of contraction we may easily reformulate the definition of an accretive operator in X, i.e., suppose that $(\mathbf{I} + \lambda\mathbf{A})^{-1}$ is a contraction $\forall\lambda \geqq 0$. Let, for $i = 1, 2$, $\mathbf{z}_i \in (\mathbf{I} + \lambda\mathbf{A})^{-1}\mathbf{x}_i$ then $\mathbf{x}_i \in (\mathbf{I} + \lambda\mathbf{A})\mathbf{z}_i$ or, equivalently, $\mathbf{x}_i = \mathbf{z}_i + \lambda\mathbf{y}_i$, for some $\mathbf{y}_i \in \mathbf{Az}_i$, $i = 1, 2$. As $(\mathbf{I} + \lambda\mathbf{A})^{-1}$ is a contraction we must then have

$$\begin{aligned} \|\mathbf{z}_1 - \mathbf{z}_2\| &\leqq \|\mathbf{x}_1 - \mathbf{x}_2\| \\ &= \|(\mathbf{z}_1 + \lambda\mathbf{y}_1) - (\mathbf{z}_2 + \lambda\mathbf{y}_2)\|, \end{aligned} \tag{IV.149}$$

$\forall\lambda \geqq 0$ and $\mathbf{y}_i \in \mathbf{Az}_i$, $i = 1, 2$; the converse is also easily seen to be true, and thus we have the equivalent:

DEFINITION. Let $\mathbf{A}$ be an operator in X. Then $\mathbf{A}$ is said to be *accretive* if $\forall \mathbf{z}_i \in \mathscr{D}(\mathbf{A})$, $i=1, 2$, (IV.149) holds $\forall \lambda \geqq 0$, $\mathbf{y}_i \in \mathbf{A}\mathbf{z}_i$, $i=1, 2$.

We also need the following:

DEFINITION. Let $\mathbf{A}$ be an accretive operator in X. Then $\mathbf{A}$ is said to be *m-accretive* if for some $\lambda > 0$ (and, hence, as it turns out, $\forall \lambda > 0$) $R(\mathbf{I}+\lambda \mathbf{A}) = X$.

Remark. If $\mathbf{A}: X \to X$ is an accretive operator in X then $-\mathbf{A}$ is said to be dissipative; if $\mathbf{A}$ is dissipative in X and $R(\mathbf{I}-\lambda \mathbf{A}) = X$ for some $\lambda > 0$ (and, hence, $\forall \lambda > 0$) then $\mathbf{A}$ is said to be m-dissipative.

In the special case where $X = H$, a real Hilbert space with inner-product $\langle \cdot, \cdot \rangle$ and associated natural norm $\|(\cdot)\|$, accretive and m-accretive maps in H are usually called monotone and maximal monotone operators, respectively. In this case, if $\mathbf{A}: H \to H$ is accretive (monotone) so that (IV.149) holds for $\mathbf{z}_i \in \mathscr{D}(\mathbf{A})$, $\mathbf{y}_i \in \mathbf{A}\mathbf{z}_i$, $i=1, 2$, and $\lambda \geqq 0$ then it is easily seen that an equivalent condition is that

$$\text{(IV.150)} \qquad \langle \mathbf{y}_1 - \mathbf{y}_2, \mathbf{z}_1 - \mathbf{z}_2 \rangle \geqq 0,$$

$\forall \mathbf{z}_1, \mathbf{z}_2 \in \mathscr{D}(\mathbf{A})$, $\forall \mathbf{y}_1 \in \mathbf{A}\mathbf{z}_1$, $\forall \mathbf{y}_2 \in \mathbf{A}\mathbf{z}_2$. We, therefore, have the following:

DEFINITION. Let $\mathbf{A}: H \to H$ be an operator (possibly multivalued) in H. Then $\mathbf{A}$ is said to be *monotone* if (IV.150) holds $\forall \mathbf{z}_i \in \mathscr{D}(A)$, $\forall \mathbf{y}_i \in \mathbf{A}\mathbf{z}_i$, $i=1, 2$. $\mathbf{A}$, a monotone operator in H, is said to be *maximal monotone* if $R(\mathbf{I}+\lambda \mathbf{A}) = H$, for some $\lambda > 0$ (and, hence, $\forall \lambda > 0$); an equivalent condition is that $\mathbf{A}$ does not admit a proper monotone extension in H.

The final definition we will need, in order to delineate the list of results we have in mind concerning abstract, first-order, integrodifferential equations, is that of the subdifferential of a map ϕ on a real Hilbert space H. Thus, let $\phi: H \to (-\infty, \infty]$ be a proper ($\phi \not\equiv \infty$) convex, lower semicontinuous function and set

$$\mathscr{D}(\phi) = \{\mathbf{x} \in H \,|\, \phi(\mathbf{x}) < +\infty\}.$$

For any $\mathbf{x} \in \mathscr{D}(\phi)$ the *subdifferential* of ϕ at $\mathbf{x}$ is defined to be the (multivalued) map

$$\text{(IV.151)} \qquad \partial\phi(\mathbf{x}) = \{\mathbf{f} \in H \,|\, \phi(\mathbf{y}) - \phi(\mathbf{x}) \geqq \langle \mathbf{f}, \mathbf{y}-\mathbf{x} \rangle \forall \mathbf{y} \in \mathscr{D}(\phi)\}.$$

Remarks. It may be shown that the set of values $\{\partial\phi(\mathbf{x})\}$, for a given $\mathbf{x} \in H$, is closed and convex (and may, of course, be empty) and that the map $\mathbf{x} \to \partial\phi(\mathbf{x})$ is maximal monotone.

We now offer, below, a brief description of the contents of what we believe to be the more pertinent recent major contributions to the area of nonlinear first-order abstract (and scalar) integrodifferential equations.

(i) In a rather influential paper, published in 1972, MacCamy and Wong [110] consider abstract functional-differential equations of the form

$$\mathbf{u}_t(t)+\int_0^t \mathbf{A}(t-\tau)\mathbf{g}(\mathbf{u}(\tau))\,d\tau=\mathbf{f}(t,\mathbf{u}(t)), \tag{IV.152}$$

where $\mathbf{u}(t)$ is an element of a Hilbert space H, $\mathbf{A}(t)$ a family of bounded symmetric operators on H, and $\mathbf{g}$ an operator with domain contained in H; the nonlinear operator $\mathbf{g}$ may be unbounded. It is further assumed that $\mathbf{A}(t)$ is strongly positive, namely, that there exists a semigroup $e^{\mathbf{B}t}$, where $\mathbf{B}$ is symmetric and satisfies $\langle\boldsymbol{\zeta},\mathbf{B}\boldsymbol{\zeta}\rangle\leqq-\lambda\|\boldsymbol{\zeta}\|^2$, for some $\lambda>0$, such that $\mathbf{A}'(t)=\mathbf{A}(t)-e^{\mathbf{B}t}$ is positive, i.e., such that for all smooth $\mathbf{v}(t)$

$$\int_0^T\left\langle\mathbf{v}(t),\int_0^t\mathbf{A}'(t-\tau)\mathbf{v}(\tau)\right\rangle d\tau\geqq0\qquad\forall T>0. \tag{IV.153}$$

Under the latter assumption of strong positivity relative to $\mathbf{A}$, the authors [110] show that, if $\mathbf{g}$, $\mathbf{f}$ are suitably restricted, then any solution which is weakly bounded and uniformly continuous must tend to zero weakly. The abstract integrodifferential equation (IV.152) has as its scalar prototype the equation

$$\dot u(t)=-\int_0^t a(t-\tau)g(u(\tau))\,d\tau+f(t,u(t)) \tag{IV.154}$$

with the most common assumptions relative to the function $a(t)$ being either that $a(t)$ is completely monotone, namely, that

$$a'(t)\not\equiv0,\qquad(-1)^k a^{(k)}(t)\geqq0,\qquad k=0,1,2,\cdots, \tag{IV.155}$$

or that (IV.155) holds for $k=0,1,\cdots,n$, $n<\infty$; such conditions with $n=2$ (Hannsgen [66]) or $n=3$ (Levin [87]) have been used to study the asymptotic behavior of solutions to the associated nonperturbed integrodifferential equation, i.e., (IV.154) with $f\equiv0$; the complete monotonicity condition for this case seems to have been first employed by Levin and Nohel [88] where the additional assumption that $ug(u)>0$, $u\neq0$, is also used. The essential element, in the analysis of the abstract equation (IV.152) which is to be found in [110], is a certain weak stability principle which may be formulated as follows: Let $C([0,\infty);H)$ be the space of all continuous functions on $[0,\infty)$ with values in H. Let $\tilde{\mathbf{A}}=\{\mathbf{A}(t)|t\in[0,\infty)\}$ be a strongly continuous one-parameter family of bounded linear operators on H and define, for fixed $T>0$, the functional $Q_{\mathbf{A}}[\mathbf{v};t]$ on $C([0,\infty);H)$ by

$$Q_{\mathbf{A}}[\mathbf{v};T]=\int_0^T\left\langle\mathbf{v}(t),\int_0^t\mathbf{A}(t-\tau)\mathbf{v}(\tau)\,d\tau\right\rangle dt. \tag{IV.156}$$

The one-parametric family $\tilde{A}$ defines a positive kernel, in the terminology of [110], if $Q_{\tilde{\mathbf{A}}}[\mathbf{v};T]\geqq0$, $\forall\mathbf{v}\in C([0,\infty);H)$ and $T\geqq0$. If we introduce the uniformly continuous one-parameter semigroup $\tilde{\mathbf{S}}$ generated by the symmetric linear operator $\mathbf{S}$ (assumed to have domain dense in H and to satisfy $\langle \zeta, \mathbf{S}\zeta \rangle \geqq \sigma\|\zeta\|^2$, $\forall \zeta \in \mathscr{D}(\mathbf{S})$ and some $\sigma > 0$), namely, $\tilde{\mathbf{S}} = \{e^{-\mathbf{S}t} | t \in [0, \infty)\}$, and let $Q_{\tilde{\mathbf{S}}}[\mathbf{v}; T]$ be the functional associated with $\tilde{\mathbf{S}}$, i.e.,

$$Q_{\tilde{\mathbf{S}}}[\mathbf{v}; T] = \int_0^t \left\langle \mathbf{v}(t), \int_0^t e^{-\mathbf{S}(t-\tau)}\mathbf{v}(\tau)\, d\tau \right\rangle dt, \tag{IV.157}$$

then the idea of strong positivity of $\mathbf{A}(t)$ intorduced earlier is easily seen to be equivalent to the requirement that for all $T > 0$

$$Q_{\tilde{\mathbf{A}}}[\mathbf{v}; T] \geqq Q_{\tilde{\mathbf{S}}}[\mathbf{v}; T] \quad \forall \mathbf{v} \in C([0, \infty); H). \tag{IV.158}$$

An element $\mathbf{v}(t) \in C([0, \infty); H)$ is termed weakly stable (in [110]) if, for each $\mathbf{w} \in H$, the function $\langle \mathbf{v}(t), \mathbf{w} \rangle$ is bounded and uniformly continuous in t, i.e., $\mathbf{v}(t) \in C([0, \infty); H)$ is weakly stable if $\mathbf{v}(t)$ is weakly bounded and weakly uniformly continuous. The weak stability principle, which forms the basis for the analysis of the asymptotic behavior of solutions to (IV.152) in [110], assumes the following form: Suppose that $\mathbf{A}(t)$ is strongly positive and $\mathbf{v}(t)$ is a weakly stable element in $C([0, \infty); H)$. If $Q_{\mathbf{A}}[\mathbf{v}; T]$ is bounded for all T, then $\mathbf{v}(t)$ converges weakly to zero. For further details and some applications to some partial differential functional equations we refer the interested reader to [110].

(ii) In [123], Nohel and Shea employ frequency domain methods, of the type introduced by Popov ([129], [130]) to study the stability of nonlinear feedback systems, in order to study the global existence, boundedness, and asymptotic behavior of solutions of the scalar Volterra integrodifferential initial value problem

$$\begin{aligned} &\dot{u}(t) + \int_0^t a(t-\xi) g(u(\xi))\, d\xi = f(t), \qquad 0 \leqq t < \infty, \\ &u(0) = u_0, \end{aligned} \tag{IV.159}$$

where it is assumed that $a(t)\, e^{-\sigma t} \in L^1(0, \infty)\ \forall \sigma > 0$, and $f(t) \in L^1(0, \infty)$, $g'(\zeta) \in C(-\infty, \infty)$. In order to delineate the specific results proven in [123], and to make note of the connections with the concepts of positivity and strong positivity introduced above, we set $\Lambda = \{\lambda \in \mathbb{C} | \operatorname{Re} \lambda > 0\}$ and define

$$\begin{aligned} G(x) &= \int_0^x g(\zeta)\, d\zeta, \\ \hat{a}(s) &= \int_0^\infty e^{-st} a(t)\, dt, \qquad s \in \Lambda, \\ U(i\tau) &= \liminf_{\substack{s \to i\tau \\ s \in \Lambda}} \operatorname{Re} \hat{a}(s), \qquad -\infty \leqq \tau \leqq \infty, \end{aligned} \tag{IV.160}$$

where by $U(i\tau)$ at $\tau=+\infty$ is meant the $\lim_{\sigma\to 0^+,\eta\to+\infty}\inf \operatorname{Re}\hat{a}(\sigma+i\eta)$ with an analogous definition of $U(i\tau)$ at $\tau=-\infty$; we also set

$$U_0(i\tau)=\lim_{\sigma\to 0^+}\inf \operatorname{Re}\hat{a}(\sigma+i\tau). \tag{IV.161}$$

Then it is proven in [123] that:

(α) If $U(i\tau)\geqq 0$, $-\infty\leqq\tau\leqq\infty$, $|g(\zeta)|\leqq M[1+|G(\zeta)|]$, for some $M>0$, and $\liminf_{\zeta\to\pm\infty} G(\zeta)>-\infty$, then (IV.159) has at least one solution $u(t)$ on $0\leqq t<\infty$, and for any solution $u(t)$ the function $g(u(t))$ is bounded. If, in addition, $\limsup_{\zeta\to\pm\infty} G(\zeta)=+\infty$ then all solutions of (IV.159) are bounded.

(β) Let $u(t)$ be a bounded solution of (IV.159) and assume that $U(i\tau)\geqq 0$, $-\infty\leqq\tau\leqq\infty$, $U_0(i\tau)\geqq\eta/(1+\tau^2)$, for some $\eta>0$, a.e. on $-\infty<\tau<\infty$, and that $a(t)$ is a function of bounded variation on $[1,\infty)$; then $\lim_{t\to\infty} g(u(t))=0$ and $\lim_{t\to\infty}(\dot{u}(t)-f(t))=0$. Furthermore, if either $g(\zeta)$ does not vanish on any interval or $ta(t)\in L^1(0,\infty)$, then $\lim_{t\to\infty} u(t)\equiv\hat{u}$ exists.

While the results in (β) amount to an analysis of bounded solutions of (IV.159), the hypotheses which are relevant in this case do not preclude the existence of unbounded solutions of (IV.159).

In the present situation (scalar equation) it is easy to see that the conditions $Q_{\hat{A}}[\mathbf{v};T]\geqq 0$, $\forall\mathbf{v}\in C([0,\infty);H)$, $\forall T>0$, where $Q_{\hat{A}}[\mathbf{v};T]$ is given by (IV.156), reduce to the statement that for every $v\in C[0,\infty)$ and for every $T>0$,

$$\int_0^T v(t)\int_0^t a(t-\zeta)v(\zeta)\,d\zeta\,dt\geqq 0, \tag{IV.162}$$

in which case $a(t)\in L^1_{\mathrm{Loc}}(0,\infty)$ is said to be of positive type. In terms of the functions U, $\hat{a}$, introduced in (IV.160) it may be verified that the conditions

$$\begin{aligned} &a(t) \text{ is of positive type,}\\ &\operatorname{Re}\hat{a}(s)\geqq 0 \quad \forall s\in\Lambda,\\ &U(i\tau)\geqq 0, \qquad -\infty\leqq\tau\leqq\infty, \end{aligned} \tag{IV.163}$$

are all equivalent. The concept of strong positivity which was previously introduced, i.e., (IV.158), reduces in the present scalar situation to the following idea: $a(t)$ is *strongly positive* if there exists a constant δ such that $b(t)=a(t)-\delta e^{-t}$ is of positive type; the condition vis à vis $U_0(i\tau)$ in (β) above may be motivated by noting that it can be shown that the kernel $a(t)$ is strongly positive if and only if $a(t)$ is of positive type and $\exists\eta>0$ such that $U_0(i\tau)\geqq\eta/(1+\tau^2)$ a.e. on $-\infty<\tau<\infty$. In particular, any twice-differentiable function $a(t)$ satisfying $(-1)^k a^{(k)}(t)\geqq 0$, $0<t<\infty$, $k=0,1,2$, with $a'(t)\not\equiv 0$, is strongly positive; for proofs of the results stated in (α) and (β) above, as well as various generalizations, we refer the reader to [123].

(iii) Many authors in recent years, and most notably London [103], Barbu [7], [8], MacCamy [107]–[109], and Crandall and Nohel [36], have considered nonlinear Volterra equations of the form

$$\mathbf{u}(t)+\int_0^t a(t-\tau)\mathbf{g}(\mathbf{u}(\tau))\,d\tau \ni \mathbf{f}(t), \qquad t \geqq 0, \tag{IV.164}$$

where a, $\mathbf{g}$, and $\mathbf{f}$ are given and $\mathbf{u}$, the unknown function, takes values in a real Hilbert space H with inner-product $\langle \cdot, \cdot \rangle$ and natural norm $\|(\cdot)\|$. The kernel function $a(t)$ is real-valued and defined on $\mathscr{R}^1$ while $\mathbf{f}: \mathscr{R}^1 \to H$. The nonlinear function $\mathbf{g}$ is, in general, multivalued with $\mathscr{D}(\mathbf{g}) \subset H$, $\mathscr{R}(\mathbf{g}) \subset H$. A solution of (IV.164) on $[0, T)$ is then a function $\mathbf{u}:[0, T) \to H$ and satisfying $\mathbf{u} \in L^2([0, T); H)$, $\mathbf{u}(t) \in \mathscr{D}(\mathbf{g})$ a.e. on $[0, T)$, and

$$\mathbf{u}(t)+\int_0^t a(t-\tau)\mathbf{w}(\tau)\,d\tau = \mathbf{f}(t), \qquad 0 \leqq t \leqq T, \tag{IV.165}$$

for some $\mathbf{w} \in L^2([0, T); H)$ such that $\mathbf{w}(t) \in \mathbf{g}(\mathbf{u}(t))$ a.e. on $[0, T)$. It has been proven by London [103] that there exists a unique solution of (IV.164) on $[0, T)$ if the following conditions are satisfied:

$$\begin{aligned} &a(0) > 0, \qquad \mathbf{g} = \partial\phi \text{ (subdifferential of } \phi) \\ &\qquad\qquad\qquad \text{for some lower semicontinuous,} \\ &\qquad\qquad\qquad \text{proper, convex } \phi: H \to (-\infty, \infty], \\ &\mathbf{f}(0) \in \mathscr{D}(\phi), \end{aligned} \tag{IV.166}$$

and

$$\begin{aligned} &a(t), \mathbf{f}(t) \text{ are absolutely continuous functions on } [0, T) \\ &\qquad \text{such that } a' \in BV[0, T), \mathbf{f}' \in L^2([0, T); H). \end{aligned} \tag{IV.167}$$

If $a(t)$, $\mathbf{f}(t)$ are absolutely continuous on $[0, \infty)$ and $\mathbf{f}' \in L^2_{\mathrm{loc}}([0, \infty); H)$ it is possible to show that there exists a unique solution on $[0, \infty)$ of the nonlinear Volterra equation (IV.164); under the above sets of conditions, it also follows that $\mathbf{u}(t)$ and $\mathbf{g}(\mathbf{u}(t))$ are locally absolutely continuous on the interval of existence and thus it follows from (IV.165) that, a.e. on the interval of existence, $\mathbf{u}(t)$ is a strong solution of the Volterra integrodifferential equation

$$\mathbf{u}'(t)+a(0)\mathbf{w}(t)+\int_0^t a'(t-\tau)\mathbf{w}(\tau)\,d\tau = \mathbf{f}'(t), \tag{IV.168}$$

where $\mathbf{w}(t) \in \mathbf{g}(\mathbf{u}(t))$ a.e. on that interval; the results in [103], as well as in the related papers referenced above, make use of the properties of maximal monotone operators in a Hilbert space (a standard reference is the monograph by Brezís [27]) and various techniques that have been developed for scalar Volterra integral equations. In comparing the existence and uniqueness theorem of London [103] with the earlier work of Barbu [7], one notes that the latter author requires that the kernel function $a(t)$ be continuous on $[0, \infty)$, locally absolutely continuous on $(0, \infty)$, that $(-1)^k a^k(t) \geqq 0$, $k = 0, 1$, a.e. for

$t>0$, and that $\operatorname{Re} \hat{a}(\lambda)>0$ when $\operatorname{Re} \lambda>0$; both authors make the same assumptions concerning $\mathbf{f}$ and $\mathbf{g}$ in (IV.164), as far as the existence of solutions goes, but $\mathbf{g}$ is required in [7] to be a strictly monotone operator, i.e., to satisfy (IV.150) $\forall \mathbf{z}_1, \mathbf{z}_2 \in \mathcal{D}(\mathbf{g})$, $\forall \mathbf{y}_1 \in \mathbf{g}(\mathbf{z}_1)$, $\mathbf{y}_2 \in \mathbf{g}(z_2)$, with the strict inequality sign, in order for the uniqueness statement in the theorem to be valid. We note, also, that Barbu [7] obtains some existence theorems, for the nonlinear Volterra equation (IV.164), which do not require that $\mathbf{g}$ be the subdifferential of a convex function ϕ; in these cases, however, additional assumptions are needed on $\mathbf{g}$ which are of the form

$$\|\mathbf{g}(\mathbf{v})\|_{V'} \leqq C_1\|\mathbf{v}\|_V, \qquad \mathbf{v} \in V, \quad C_1>0,$$

$$(\mathbf{g}(\mathbf{v}), \mathbf{v})_{V,V'} \geqq C_2\|\mathbf{v}\|_V, \qquad \mathbf{v} \in V, \quad C_2>0,$$

where V is a Hilbert space, $V \subset H$, $\mathbf{g}: V \rightarrow V'$, with V' the dual of V via the inner-product in H, and $(\cdot\,,\cdot)_{V',V}$ denotes the duality pairing on $V \times V'$. Results concerning the boundedness and asymptotic behavior of solutions to (IV.164) are also presented in [102] and [7], as well as in the various related papers referenced above; we will content ourselves with simply delineating the form which some of these results assume in the paper of Barbu [7]: as we have already indicated, Barbu [7] assumes that $a(t)$ is continuous on $[0, \infty)$, locally absolutely continuous on $[0, \infty)$, and satisfies $(-1)^k a^{(k)}(t) \geqq 0$ for $k=0, 1$, a.e. for $t>0$, and $\operatorname{Re} \hat{a}(\lambda)>0$ for $\operatorname{Re} \lambda>0$; as in London [103] it is also assumed that $\mathbf{g}=\partial \phi$ (a subdifferential), and that $\mathbf{f}(0) \in \mathcal{D}(\phi)$ with $\mathbf{f}$ absolutely continuous on $[0, \infty)$ and such that $\mathbf{f}' \in L^2((0, \infty); H)$. Under the additional hypotheses that $a(\infty)>0$ and that

$$\phi(\mathbf{v}) \rightarrow +\infty \quad \text{as } \|\mathbf{v}\| \rightarrow +\infty, \tag{IV.169}$$

Barbu [7] is able to prove that any solution $\mathbf{u}(t)$ of (IV.164) satisfies

$$\begin{aligned} &\|\mathbf{u}(t)\| \text{ is bounded on } [0, \infty), \\ &\mathbf{u}', \mathbf{g}(\mathbf{u}) \in L^2((0, \infty); H), \\ &\lim_{t \rightarrow +\infty} \phi(\mathbf{u}(t))=\phi_{\infty}=\min \{\phi(\mathbf{v}) \mid \mathbf{v} \in H\}. \end{aligned} \tag{IV.170}$$

Under the additional assumptions that

$$\begin{aligned} &\sqrt{t}\, a' \in L^1(0, \infty), \\ &\phi(\mathbf{f}) \in L^1(0, \infty), \\ &\sqrt{t}\, \mathbf{f}' \in L^2((0, \infty); H), \end{aligned} \tag{IV.171}$$

it follows that $\sqrt{t}\, \mathbf{u}'$ and $\sqrt{t}\, \mathbf{g}(\mathbf{u})$ belong to $L^2((0, \infty); H)$ and that $t \phi(\mathbf{u}(t))$ is bounded on $[0, \infty)$; for proofs of these results we refer the reader to § 3–5 of Barbu [7]. Applications of the existence, uniqueness, and stability theorems, for solutions of (IV.164), which are presented in London [103] and Barbu [7], may be found in § 6 of each of these respective papers. In a later paper [8], Barbu has generalized the results in [7] to allow for situations in which the kernel $a(t)$ in (IV.164) is of the form $a(t)=b(t)\mathbf{S}(t)$ where $b(t)$ is a real-valued positive-definite function and $\mathbf{S}(t)$ is a continuous semigroup of bounded linear operators on H; more specifically it is assumed in [8] that $a\in C^1(0,\infty)\cap L^1(0,1)$, that $(-1)^k a^{(k)}(t)\geqq 0$ *for* $k=0,1$ and all $t>0$, that $a'(t)$ is nondecreasing on $(0,\infty)$ and that the infinitesimal generator $\mathbf{A}$ of the semigroup $\mathbf{S}(t)$ is self-adjoint and dissipative, i.e., that $\langle \mathbf{Av},\mathbf{v}\rangle\leqq 0$, $\forall \mathbf{v}\in H$ so that $\mathbf{S}(t)$ is, in fact, a contraction semigroup on H. It is further assumed that $\mathbf{g}=\partial\phi$ (a subdifferential) with $\inf\{\phi(\mathbf{v})|\mathbf{v}\in H\}>-\infty$ and that $\phi(\mathbf{S}(t)\mathbf{v})\leqq\phi(\mathbf{v})$, $\forall\mathbf{v}\in H$ and all $t>0$; for a precise statement of the existence, uniqueness, and boundedness results which follow under these conditions, and for an application of those results to a nonlinear initial-boundary value problem arising in the theory of generalized heat transfer between solids and gases, namely

$$\text{(IV.172)}\qquad \begin{aligned} &\frac{\partial u}{\partial t}(\mathbf{x},t)=\Delta u(\mathbf{x},t), && \mathbf{x}\in\Omega\subseteq R^n,\quad t>0,\\ &\frac{\partial u}{\partial n}(\mathbf{x},t)\in -g(u(\mathbf{x},t)), && \mathbf{x}\in\partial\Omega,\quad t>0,\\ &u(\mathbf{x},0)=u_0(\mathbf{x}), && \mathbf{x}\in\Omega, \end{aligned}$$

where g is a maximal monotone operator from $R^1\to R^1$, and $\partial/\partial n$ is the outward normal derivative to $\partial\Omega$, we refer the reader to [8, § 4]. A paper whose results serve to extend and generalize those of Barbu, London, et al., referenced above, is that of Crandall and Nohel [36]; these authors study the initial-value problem

$$\text{(IV.173)}\qquad \begin{aligned} &\frac{d\mathbf{u}}{dt}+\mathbf{Au}\ni\mathbf{G}(\mathbf{u}), \qquad 0\leqq t<T,\\ &\mathbf{u}(0)=\mathbf{u}_0, \end{aligned}$$

where $\mathbf{A}$ is a given m-accretive (and possibly multi-valued) operator in a real Banach space X with norm $\|(\cdot)\|$ and $\mathbf{G}$ is a prescribed mapping, $\mathbf{G}:C([0,T];\mathscr{D}(A))\to L^1([0,T);X)$. The results obtained by Crandall and Nohel [36] for the initial-value problem (IV.173) lead to simple proofs of generalizations of results for initial-value problems of the form

$$\text{(IV.174)}\qquad \begin{aligned} &\frac{d\mathbf{u}}{dt}+m\mathbf{Au}(t)+a*\mathbf{Au}(t)=\mathscr{F}(t), \qquad 0<t\leqq T,\\ &\mathbf{u}(0)=\mathbf{u}_0, \end{aligned}$$

(where $m>0$ is a constant, $\mathbf{A}$ is a maximal monotone operator in a real Hilbert space H, and the kernel function $a(t)$ is real-valued) as well as for abstract integral equations of the form

$$\text{(IV.175)} \qquad \mathbf{u}(t)+b * \mathbf{Au}(t) \ni \mathscr{F}(t), \qquad 0 \leqq t \leqq T,$$

where $\mathbf{A}$ is a maximal monotone operator; the initial-value problem (IV.174) was considered by MacCamy in [105] while the integral equation (IV.175) is precisely of the form considered earlier by Barbu [7], [8], London [103], et al., if we assume that $\mathbf{Au}=\partial\phi(\mathbf{u})$, with $\partial\phi$ the subdifferential of a function $\phi: H \to (-\infty, \infty]$ which is convex, lower semicontinuous, and proper. For $\mathscr{F}$ in an appropriate class of mappings, the techniques developed in [36] can also be used to study the nonconvolution Volterra equation

$$\text{(IV.176)} \qquad \mathbf{u}(t)+\int_0^t a(t,s)\mathbf{Au}(s)\,ds \ni \mathscr{F}(t), \qquad 0 \leqq t \leqq T,$$

where $\mathbf{A}$ is m-accretive on a Banach space X; for a more precise delineation of the relationships which exist between the recent work of Crandall and Nohel, on the initial-value problem (IV.173), and the earlier work of Barbu, London, MacCamy, et al., on problems of the form (IV.174), (IV.175), we refer the reader to § 4 of [36].

(B) *Heat flow in nonlinear heat conductors with memory.* Consider a rigid heat conductor (one-dimensional heat flow) with $u(x,t)$, $\varepsilon(x,t)$, $q(x,t)$ and $r(x,t)$ denoting the temperature, internal energy, heat flux, and heat supply, respectively. According to the principle of balance of energy we must have

$$\text{(IV.177)} \qquad \frac{\partial \varepsilon}{\partial t}(x,t) = -\frac{\partial}{\partial x} q(r,t)+r(x,t).$$

The usual one-dimensional heat equation for the evolution of the temperature $u(x,t)$ is obtained from the balance equation (IV.177) by employing the constitutive hypotheses

$$\text{(IV.178)} \qquad \varepsilon(x,t)=c^2u(x,t), \qquad q(x,t)=-ku_x(x,t),$$

with $k>0$ (thermal conductivity) and $c \neq 0$. If we retain the first constitutive hypothesis, but replace the second (essentially Fourier's law of heat conduction) by $q(x,t)=-k\sigma(u_x(x,t))$, where σ is nonlinear, we obtain the nonlinear evolution equation

$$\text{(IV.179)} \qquad u_t(x,t)=\left(\frac{K}{c^2}\right)\sigma(u_x(x,t))_x+r(x,t).$$

Models of heat flow in nonlinear heat conductors, which are governed by evolution equations of the form (IV.179), are not very attractive from a physical standpoint because they predict infinite speeds of propagation for thermal disturbances. In 1968, Gurtin and Pipkin [64] proposed a theory of heat conduction in nonlinear materials with memory which generalizes the classical theory represented by the constitutive hypotheses (IV.178) and predicts that thermal disturbances propagate with finite wave speeds. The linear one-dimensional version of the theory constructed in [64] is governed by constitutive assumptions of the form

$$\text{(IV.180)}\qquad \begin{aligned} \varepsilon(x,t) &= bu(x,t) + \int_0^\infty B(\tau)u(x,t-\tau)\,d\tau, \\ q(x,t) &= -\int_0^\infty k(\tau)u_x(x,t-\tau)\,d\tau, \end{aligned}$$

i.e., the internal energy and the heat flux are linear functionals of the histories of the temperature and temperature gradient fields, respectively; the problem of studying the stability of temperature fields in linear heat conductors with memory has been considered by several authors and, in particular, one may refer to the paper [119] by Nachlinger and Nunziato (and the references cited therein) for a discussion of the stability of temperature fields in heat conductors which are governed by the three-dimensional form of the linearized constitutive hypotheses (IV.180). Our interest in this subsection will be with a particular nonlinear version of the constitutive theory (IV.180) which has been treated by MacCamy [105] and Dafermos and Nohel [43]; the nonlinear constitutive theory in question results by retaining the linear relation (IV.180_1) and replacing (IV.180_2) by the constitutive assumption

$$\text{(IV.181)}\qquad q(x,t) = -\int_0^\infty k(\tau)\sigma(u_x(x,t-\tau))\,d\tau.$$

If we assume that $u(x,\tau)=\varepsilon(x,\tau)=0$, $\tau<0$ then (IV.177), (IV.180_1), and (IV.181) yield the evolution equation

$$\text{(IV.182)}\qquad b\frac{\partial u}{\partial t} + \int_0^t B(t-\tau)u_\tau(x,\tau)\,d\tau = \int_0^t k(t-\tau)\frac{\partial}{\partial x}\sigma(u_x(x,\tau))\,d\tau + r(x,t).$$

Prescribed nonzero past histories of the temperature and internal energy fields may simply be absorbed into the heat supply expression in (IV.182). We have already indicated (Chapter II, (II.314)–(II.315)) that the Volterra integral equation

$$\text{(IV.183)}\qquad \varepsilon(x,t) = bu(x,t) + \int_0^t B(t-\tau)u(x,\tau)\,d\tau$$

has the solution

$$\text{(IV.184)}\qquad u(x,t) = \frac{1}{b}\varepsilon(x,t) + \int_0^t \rho(t-\tau)\varepsilon(x,\tau)\,d\tau,$$

where $\rho(t)$ is the resolvent kernel associated with $B(t)$. Using this fact, it is not difficult to see that (IV.182) may be put into the equivalent form

$$u_t(x, t) = \int_0^t a(t-\tau)\sigma(u_x(x, \tau))_x \, d\tau + \mathcal{F}(x, t), \tag{IV.185}$$

where

$$\begin{aligned} a(t) &= \frac{1}{b} k(t) + \int_0^t \rho(t-\tau)k(\tau)\, d\tau, \\ \mathcal{F}(x, t) &= \frac{1}{b} r(x, t) + \int_0^t \rho(t-\tau) r(x, \tau)\, d\tau. \end{aligned} \tag{IV.186}$$

In what follows, we consider the problem of establishing global existence, uniqueness, and asymptotic stability results for solutions of (IV.185) on $[0, 1]\times[0, \infty)$ subject to initial and boundary data of the form

$$\begin{aligned} u(x, 0) &= f(x), \qquad x \in [0, 1], \\ u(0, t) &= u(1, t) = 0, \qquad t > 0. \end{aligned} \tag{IV.187}$$

As was indicated at the end of Chapter II, the problem of establishing global existence, uniqueness, and asymptotic stability of solutions to the initial-boundary value problem associated with the nonlinear first-order integrodifferential equation (IV.185), may be approached via the energy estimate arguments of Dafermos and Nohel [43], in a manner analogous to the approach described for the related nonlinear viscoelastic integrodifferential evolution equation (II.229); we prefer, however, to present, in outline form, the earlier treatment of MacCamy [105], which is very similar to the analysis of the initial-boundary value problem associated with (II.229) that was presented by MacCamy in [104] as well as to the work of Nohel [122] that has been described in § 3 of this chapter. Indeed, the analysis presented in [105] is so closely related to Nohel's treatment [122] of the initial-boundary value problem associated with (II.229), that we will be able to simply reference, at a certain point in the discussion, the bulk of the analysis which is related to the derivation of the pointwise estimates (IV.118), (IV.124) in § 3.

The basis of the treatment of (IV.185) in [105] involves, essentially, the following: (i) rewrite (IV.185) in the form (compare with (II.299), (II.300a))

$$\frac{1}{a(0)} u_{tt} + k(0)u_t - \frac{\partial}{\partial x}\sigma(u_x) - R(x, t), \tag{IV.188}$$

where $k(t)$ is the resolvent kernel associated with $a(t)$, and $R(x, t)$ depends on $\mathcal{F}(x, t)$, $f(x)$, and $u(x, t)$, in a manner to be described; (ii) derive integral energy estimates for the solutions of (IV.185), (IV.187) which imply that $\sup_{[0,1]} |R(x, \cdot)| \in L^1(0, \infty)$ (compare with the hypothesis (IV.115) relative to the nonhomogeneous term $G(x, t)$ which appears in the nonlinear hyperbolic system (IV.112) considered by Nohel [122]); (iii) employ the Nishida Riemann invariant argument (in a manner completely analogous to that described in § 3 of this chapter) to show that certain time and spatial derivatives of $u(x, t)$ are bounded for as long as solutions exist; and (iv) combine the latter a priori pointwise bounds with a standard local existence theorem to obtain the global existence, uniqueness, and asymptotic stability results.

We begin by delineating the basic hypotheses in [105] relative to the kernel function $a(t)$, the data $f(x)$ and $\mathcal{F}(x, t)$, and the nonlinearity σ. For $a(t)$ we assume that

$$\text{(IV.189)} \qquad \begin{aligned} &a(0)>0, \qquad \dot{a}(0)<0, \\ &t^j a^{(k)}(t) \in L_1(0, \infty), \qquad k=0, 1, 2, \quad j \leqq 3+N \quad \text{for some } N \geqq 0 \\ &\operatorname{Re} \hat{a}(i\eta)>0 \quad \forall \eta, \end{aligned}$$

while for $\mathcal{F}(x, t)$ it is assumed that

$$\text{(IV.190)} \qquad \mathcal{F} \in C^3([0, 1]\times[0, \infty)).$$

We also set $\bar{\mathcal{F}}(t) = \sup_{x\in[0,1]} (|\mathcal{F}(x, t)|, |\mathcal{F}_t(x, t)|)$ and assume that

$$\text{(IV.191)} \qquad \bar{\mathcal{F}} \in L^1(0, \infty)\cap L^2(0, \infty)\cap L^\infty(0, \infty).$$

Finally, the nonlinearity σ is assumed to satisfy

$$\text{(IV.192)} \qquad \sigma \in C^2(-\infty, \infty), \quad \sigma(0)=0, \quad \sigma'(\zeta) \geqq \varepsilon,$$

for all ζ and some $\varepsilon>0$. The first of the two basic theorems in [105] then assumes the following form:

THEOREM (MacCamy [105]). *If $u(x, t)$ is a solution of* (IV.185), (IV.187) *then* (IV.189)–(IV.192) *imply that*

$$\text{(IV.193)} \qquad \lim_{t\to+\infty} \int_0^1 u^2(x, t)\, dx = 0.$$

In addition, if $|\sigma(\zeta)| \leqq M_1(|\zeta|+|\zeta|^\gamma)$ for some $M>0$ and γ, $1<\gamma<2$, then for any $\eta \in C^1[0, 1]$ such that $\eta(0)=\eta(1)=0$,

$$\text{(IV.194)} \qquad \lim_{t\to\infty} \int_0^1 u_t(x, t)\eta(x)\, dx = 0.$$

Remarks. Under the strengthened assumptions that (IV.189_2) hold for some $N \geqq 4$, that

$$\text{(IV.195)} \qquad t^j \bar{\mathcal{F}} \in L^1(0, \infty)\cap L^2(0, \infty)\cap L^\infty(0, \infty), \qquad j \leqq N,$$

and that

$$\text{(IV.196)} \qquad |\sigma(\zeta)| \leqq \bar{\sigma}|\zeta| \quad \forall \zeta,$$

estimates of the form

$$\text{(IV.197)} \qquad \Big(\sup_{x\in[0,1]} |u(x,t)|\Big)^2, \quad \int_0^1 u_t^2(x,t)\,dx = \mathcal{O}(t^{-N})$$

can be derived for solutions $u(x,t)$ of (IV.185), (IV.187). The further hypotheses

$$\text{(IV.198)} \qquad \begin{aligned} &\mathcal{F}(0,t)=\mathcal{F}(1,t)=\mathcal{F}_{xx}(0,t)=\mathcal{F}_{xx}(1,t)=0, \qquad t>0,\\ &f(x)\in C^3[0,1], \qquad f(0)=f(1)=f''(0)=f''(1)=0 \end{aligned}$$

are needed later on in order to convert (IV.185), (IV.187) into a pure initial-value problem to which a variant of Nishida's [121] Riemann invariant argument can be applied. The second of the two basic theorems in [105] is the global existence theorem; in order to state this result we first define

$$\text{(IV.199)} \qquad \begin{aligned} \mathcal{D} = &\sup_{x\in[0,1]} \sum_{i=0}^{2} \left|\frac{d^i f}{dx^i}(x)\right| \\ &+ \sum_{j=0}^{N} \big(\|t^j\bar{\mathcal{F}}(t)\|_{L^1(0,\infty)} + \|t^j\bar{\mathcal{F}}(t)\|_{L^2(0,\infty)} + \|t^j\bar{\mathcal{F}}(t)\|_{L^\infty(0,\infty)}\big). \end{aligned}$$

We then have the following:

THEOREM (MacCamy [105]). *Let* (IV.189$_2$) *for* $N\geqq 4$, (IV.195), *and* (IV.198) *hold. If* $\mathcal{D}$, *as defined by* (IV.199) *is sufficiently small, there exists a unique global solution of* (IV.185), (IV.187).

We will begin our discussion of the two basic theorems stated above by looking at the analysis which leads to (IV.193); this analysis depends on the derivation of certain energy estimates associated with the system (IV.185), (IV.187). We start by defining, for $\phi\in C^1[0,1]$, the norms

$$\text{(IV.200)} \qquad \begin{aligned} \|\phi\| &= \left(\int_0^1 \phi^2(x)\,dx\right)^{1/2} = \|\phi\|_{L^2(0,1)},\\ \|\phi\|_1 &= \left(\int_0^1 \phi'^2(x)\,dx\right)^{1/2} = \|\phi\|_{H^1(0,1)}, \end{aligned}$$

and recalling that $\|\phi\|\leqq\|\phi\|_1$. In view of our hypothesis relative to σ, i.e., (IV.192$_3$),

$$\text{(IV.201)} \qquad -\int_0^1 \frac{\partial}{\partial x}\sigma(u_x)u\,dx = \int_0^1 \sigma(u_x)u_x\,dx \geqq \varepsilon\int_0^1 u_x^2\,dx = \varepsilon\|u\|_1^2$$

so that the operator $h: H_0^1(0,1)\to L^2(0,1)$, defined by $h(u)(x) = -(\partial/\partial x)\sigma(u_x)$, is coercive in the $\|(\cdot)\|_1$ norm, i.e., we have $\langle h(u), u\rangle_{L^2}\geqq\varepsilon\|u\|_1^2$. Note also that the functional $\mathcal{H}(u)$ which is defined on $\mathcal{D}(h)$ via

$$\text{(IV.202)} \qquad \mathcal{H}(u) = \int_0^1\int_0^{u_x} \sigma(\zeta)\,d\zeta\,dx$$

satisfies

$$\text{(IV.203)}\qquad \begin{aligned}\langle \dot{u}, h(u)\rangle_{L^2} &= -\int_0^1 u_t \frac{\partial}{\partial x}\sigma(u_x)\,dx\\ &= \int_0^1 u_{xt}\sigma(u_x)\,dx = \frac{d}{dt}\mathcal{H}(u(\cdot,t))\end{aligned}$$

(if $u \in C^2(0,1)$ and $u(0,t)=u(1,t)=0$, $\forall t>0$) and

$$\text{(IV.204)}\qquad \mathcal{H}(u) \geqq \frac{\varepsilon}{2}\int_0^1 u_x^2\,dx = \frac{\varepsilon}{2}\|u\|_1^2.$$

These results are used in the derivation of the required energy estimates. For any $u \in C^1([0,1]\times[0,T])$, $T>0$, we define

$$\text{(IV.205)}\qquad \begin{aligned}&\|u\|(t)=\|u(\cdot,t)\|, \qquad \|u\|_1(t)=\|u(\cdot,t)\|_1,\\ &|||u|||^T = \left(\int_0^T \|u\|^2(t)\,dt\right)^{1/2},\\ &|||u|||_1^T = \left(\int_0^T \|u\|_1^2(t)\,dt\right)^{1/2},\end{aligned}$$

and demonstrate that $\exists M>0$ such that for any $T>0$ and any solution $u(x,t)$ of (IV.185), (IV.187)

$$\text{(IV.206)}\qquad \begin{aligned}&\|u_t\|(t),\quad \|u\|_1(t),\quad |||u_t|||^T,\quad \|u\|(t) \leqq M,\\ &|||u|||_1^T,\quad |||u|||^T \leqq M.\end{aligned}$$

For the purpose of illustrating the basic technique we will, in fact, only derive the first four estimates in (IV.206) and refer the interested reader to [105, § 4] for the derivation of the last two estimates in this group. If we differentiate (IV.185) through with respect to t and solve for $\sigma(u_x)_x$ in the resulting equation (in the manner of (II.313)–(II.314)) we obtain

$$\text{(IV.207)}\qquad \frac{1}{a(0)}u_{tt} + \frac{d}{dt}\int_0^t k(t-\tau)u_\tau(x,\tau)\,d\tau - \sigma(u_x)_x = \Phi,$$

where

$$\text{(IV.208)}\qquad \begin{aligned}\Phi(x,t) \equiv{}& \frac{1}{a(0)}\mathcal{F}_t(x,t) + \int_0^t k(t-\tau)\mathcal{F}_\tau(x,\tau)\,d\tau\\ &+ k(t)u_t(x,0)\\ ={}& \frac{1}{a(0)}\mathcal{F}_t(x,t) + k(0)\mathcal{F}(x,t)\\ &+ \int_0^t \dot{k}(t-\tau)f(x,\tau)\,d\tau,\end{aligned}$$

and $k(t)$ is the resolvent kernel associated with $a(t)$. As $\dot{k} \in L^1(0, \infty)$, and $\mathcal{F}(x, t)$ satisfies (IV.191), it follows that the $|||(\cdot)|||^T$ norms of the three expressions on the right-hand side of (IV.208) are bounded, independently of T; hence $|||\Phi|||^T \leqq M_2$, where M_2 does not depend on T. If we multiply (IV.207) by $u_t(x, t)$, integrate over $[0, 1] \times [0, T]$, and employ (IV.202), (IV.203), (IV.204), and the fact that $k(t)$ satisfies (compare with (III.317(iii))

$$\text{(IV.209)} \qquad \int_0^T \phi(t) \frac{d}{dt}(k * \phi)(t)\, dt \geqq \alpha \int_0^T \phi^2(t)\, dt$$

for some $\alpha > 0$, any $T > 0$, and all $\phi \in C[0, T)$, we obtain

$$\text{(IV.210)} \qquad \frac{1}{2a(0)} \|u_t\|^2(T) + \frac{\alpha}{2}(|||u_t|||^T)^2 + \frac{\varepsilon}{2}\|u\|_1^2(T) \leqq \mathscr{C}(\alpha, M_2),$$

where $\mathscr{C}$ is independent of T; the first three estimates in (IV.206) now follow directly from (IV.210) while the last one follows from the embedding of H_0^1 in L^2, i.e., from the fact that $\|u\| \leqq \|u\|_1$. The estimates in (IV.206) immediately imply that (IV.193) is satisfied, for the last two estimates in this group imply that $\|u\|(\cdot) \in L^2(0, \infty)$ while the first set of estimates imply that $\|u\|(t)$ is uniformly continuous.

Concerning the result (IV.194), it follows, again by the first set of estimates in (IV.206), that $\|u_t\|(\cdot) \in L^2(0, \infty)$ and, thus, $\chi(\cdot) = \int_0^1 u_t(x, \cdot)\eta(x)\, dx \in L^2(0, \infty)$. By virtue of the hypothesis that $|\sigma(\zeta)| \leqq M_1(|\zeta| + |\zeta|^\gamma)$ for some $M > 0$, and $\gamma \in (1, 2)$, it follows that with $\eta \in C^1[0, 1]$, $\eta(0) = \eta(1) = 0$,

$$\text{(IV.211)} \qquad \begin{aligned} \left| \int_0^1 \frac{\partial}{\partial x}\sigma(u_x)\eta\, dx \right| &= \left| \int_0^1 \sigma(u_x)\eta_x\, dx \right| \\ &\leqq M_2 \int_0^1 (|u_x| + |u_x|^\gamma)\eta_x\, dx \\ &\leqq M\left\{ \|u\|_1\|\eta\|_1 + \|u\|_1^{\gamma/2}\left(\int_0^1 |\eta_x|^{2/(2-\gamma)}\, dx \right)^{2-\gamma/2} \right\} \\ &\leqq C_1. \end{aligned}$$

If we multiply (IV.185) by η and integrate over $[0, 1]$ we obtain

$$\text{(IV.212)} \qquad \begin{aligned} \chi(t) = \int_0^t a(t-\tau) \int_0^1 \frac{\partial}{\partial x}\sigma(u_x(x, \tau))\eta(x)\, dx\, d\tau \\ + \int_0^1 \mathcal{F}(x, t)\eta(x)\, dx. \end{aligned}$$

In view of (IV.191), the second term on the right-hand side of (IV.212) is uniformly continuous (in t) and the uniform continuity of the first term then follows directly from (IV.211); thus $\chi(\cdot)$ is uniformly continuous and in $L^2(0, \infty)$ and (IV.194) follows.

Remarks. The energy estimates, which yield the conclusions of the first basic theorem above, depend, for their derivation, on the hypotheses (IV.189) relative to $a(t)$, (IV.191), and (IV.192). If in (IV.189$_2$) we require that $N>0$ and that (IV.195), (IV.196) also hold, then, as has already been indicated, energy estimates which yield the conclusions (IV.197) may be derived for solutions of the system (IV.185), (IV.187). In fact, as shown in § 5 of [105], much stronger results, which include (IV.197) as a special subcase, can be obtained; these are of the form

$$\text{(IV.213)} \qquad T^m\|u_t\|^2(T), \qquad T^m\|u\|_1^2(T), \qquad T^m\|u\|^2(T) \leqq M_m,$$

$$\left(\int_0^T\int_0^1 t^m u_t^2(x,t)\,dx\,dt\right)^{1/2} \leqq M_m,$$

$$\text{(IV.214)} \qquad \left(\int_0^T\int_0^1 t^m u_x^2(x,t)\,dx\,dt\right)^{1/2} \leqq M_m,$$

$$\left(\int_0^T\int_0^1 t^m u^2(x,t)\,dx\,dt\right)^{1/2} \leqq M_m,$$

for any $m \leqq N$ and some $M_m>0$.

Having obtained a set of integral energy estimates for solutions of (IV.185), (IV.187), and the asymptotic results (IV.193), (IV.194) as a consequence of these estimates, the analysis in [105] proceeds via the derivation of some L^∞ bounds for $u(x,t)$ and certain of its derivatives; as we have already indicated, the combination of these pointwise estimates with a standard local existence theorem then yields the global existence theorem previously stated, for $\mathscr{D}$ sufficiently small, with $\mathscr{D}$ given by (IV.199). In what follows we will assume the hypothesis of the second basic theorem of [105] (we refer to the statement of that theorem immediately following (IV.199). We begin by noting that if we set $|u|(t)=\sup_{x\in[0,1]}|u(x,t)|$ then, in view of our boundary conditions, $|u|(t)\leqq \|u\|_1(t)$. However, in view of (IV.197) we certainly have $|u|(T)=\mathcal{O}(T^{-2})$ and, thus, $|u|(\cdot)\in L^1(0,\infty)$. We now note that we may rewrite (IV.207) in the form (IV.188) with

$$\begin{aligned} R(x,t) &= \Phi(x,t) - \int_0^t \dot{k}(t-\tau)u_\tau(x,\tau)\,d\tau \\ \text{(IV.215)} \qquad &= \Phi(x,t) - \dot{k}(0)u(x,t) + \dot{k}(t)u(x,0) \\ &\quad + \int_0^t \ddot{k}(t-\tau)u(x,\tau)\,d\tau. \end{aligned}$$

However, (IV.208$_2$), (IV.191), the basic smoothness properties of the resolvent kernel $k(t)$, and the estimates above, collectively imply that $\sup_{[0,1]}|R(x,\cdot)|\in L^1(0,\infty)$. We now consider (IV.188) subject to the initial data

$$\text{(IV.216)} \qquad u(x,0)=f(x), \qquad u_t(x,0)=\mathscr{F}(x,0), \qquad 0<x<1,$$

and the boundary conditions (IV.187$_2$). In a manner entirely analogous to the reformulation of (IV.57)–(IV.59) as a pure initial-value problem, we extend the functions $f(\cdot)$, $\mathscr{F}(\cdot, t)$, and $\Phi(\cdot, t)$ to all of R^1 as odd functions which are periodic with period two in x. It then follows that $\sigma(u_x(0, t)) = \sigma(u_x(1, t)) = 0$, $t > 0$ so that $u_{xx}(0, t) = u_{xx}(1, t) = 0$, $t > 0$, and hence $u(\cdot, t)$ may be extended as an odd, C^2 function on R^1 which is periodic with period two. It therefore suffices to consider the equation (IV.188), for $-\infty < x < \infty$, with associated initial data (IV.216) on $(-\infty, \infty)$, where we have retained the same notation for the original and the extended functions; the solution of the extended initial-value problem, when restricted to $0 \leqq x \leqq 1$, will be a solution of the original initial-boundary value problem (IV.188), (IV.216), (IV.187$_2$).

Once the original initial-boundary value problem associated with (IV.188) has been extended as an initial-value problem on R^1, the proof of the associated global existence theorem proceeds precisely as in § 3 for the system (IV.112) on R^1: we set, without loss of generality, $a(0) = 1$, and define $r = u_t$, $w = u_x$ so that (IV.188) assumes the form of the system (IV.113) with $\alpha = k(0)$ and $G(x, t)$ replaced by $R(x, t)$; $R(x, t)$ is given, of course, by (IV.215), (IV.208$_2$) and, as already indicated, our energy estimates, the smoothness hypotheses relative to the data, and the properties of the resolvent kernel $k(t)$, imply that $\sup_{[0,1]} |R(x, \cdot)| \in L^1(0, \infty)$. Once the extended initial-value problem on R^1, (IV.188), (IV.216), has been transformed into the initial-value problem

$$
\text{(IV.217)} \qquad
\begin{aligned}
&w_t(x, t) - r_x(x, t) = 0, \\
&r_t(x, t) - \sigma'(w) w_x(x, t) + k(0) r(x, t) = R(x, t), \\
&r(x, 0) = \mathscr{F}(x, 0), \qquad w(x, 0) = f'(x),
\end{aligned}
$$

on $-\infty < x < \infty$, we again (following the analysis of § 3) introduce the Riemann Invariants $\mathscr{r}$, $\mathscr{s}$ which are defined by (IV.68) and (IV.70), respectively; the system (IV.217) is, thus, transformed into the equivalent system

$$
\text{(IV.218)} \qquad
\begin{aligned}
&\left.\begin{aligned}
&\mathscr{r}_t + \mu \mathscr{r}_x + \frac{k(0)}{2}(\mathscr{r} + \mathscr{s}) = R \\
&\mathscr{s}_t + \gamma \mathscr{s}_x + \frac{k(0)}{2}(\mathscr{r} + \mathscr{s}) = R
\end{aligned}\right\} \quad 0 \leqq t < \infty, \quad x \in \mathscr{R}^1, \\
&\mathscr{r}(x, 0) = \mathscr{r}_0(x), \qquad \mathscr{s}(x, 0) = \mathscr{s}_0(x),
\end{aligned}
$$

where $\mu = -\sqrt{\sigma'(w)}$, $\gamma = \sqrt{\sigma'(w)}$ and $\mathscr{r}_0(x)$, $\mathscr{s}_0(x)$ are given by

$$
\text{(IV.219)} \qquad
\begin{aligned}
&\mathscr{r}_0(x) = \mathscr{F}(x, 0) + \int_0^{f'(x)} \sqrt{\sigma'(\zeta)}\, d\zeta, \\
&\mathscr{s}_0(x) = \mathscr{F}(x, 0) - \int_0^{f'(x)} \sqrt{\sigma'(\zeta)}\, d\zeta.
\end{aligned}
$$

MacCamy [105] now proceeds in a manner completely analogous to that followed in § 3, i.e., the direct equivalents of the estimates (IV.118), (IV.124) for $\sup_{x\in\mathscr{R}^1}|\imath(x,t)|$, $\sup_{x\in\mathscr{R}^1}|\jmath(x,t)|$ and $\sup_{x\in\mathscr{R}^1}|\imath_x(x,t)|$, $\sup_{x\in\mathscr{R}^1}|\jmath_x(x,t)|$, respectively, are derived in [105] via an analysis which, step for step, coincides with that delineated in (IV.120)–(IV.123) and (IV.125)–(IV.131); the quantity $\mathscr{D}$ in [105], which is defined by (IV.199), corresponds to the quantity N_1, in the analysis of § 3, which is given by (IV.123). Once the equivalents of the estimates (IV.118), (IV.124) have been established in [105] for the solutions of the system (IV.218)—the estimates being valid on $R^1\times[0, T]$, where T is the maximal length of the time interval of existence of a solution $(\imath, \jmath)$—the definitions of $(\imath, \jmath)$, in terms of (r, w), and the fact that $r = u_t$, $w = u_x$, then yield a priori pointwise upper bounds for the derivatives u_x, u_t, u_{xx}, u_{xt}, and u_{tt} of the form

$$\text{(IV.220)}\qquad \begin{aligned}&|u_x|^T, |u_t|^T \leqq A,\\ &|u_{xx}|^T, |u_{xt}|^T, |u_{tt}|^T \leqq B,\end{aligned}$$

where $|v|^T = \sup_{x\in[0,1], t\in[0,T]}|v(x,t)|$, and the positive constants A, B are independent of T.

In order to obtain the required local existence theorem (on $R^1\times[0, T)$, for some $T>0$) for the solutions of the extended initial value problem associated with (IV.188), MacCamy [105] argues that (IV.188), with $R(x,t)$ expressed in terms of the solution $u(x,t)$ via (IV.215), (IV.208$_2$), is a linear perturbation of a nonlinear hyperbolic equation, and that the standard local existence results for the nonlinear equation, i.e. for (IV.188) with $R\equiv 0$, carry over with trivial modifications; the a priori bounds (IV.220) then apply to the derivatives of this unique, locally defined solution on $R^1\times[0, T)$ and a standard continuation argument (of the type employed in Chapter II) then yields the desired global existence theorem for the extended initial-value problem associated with (IV.188) and, hence, also for the original initial-boundary value problem (IV.188), (IV.216), (IV.187$_2$).

Remarks. As has been indicated by Nohel in [122], the global existence, uniqueness, and continuous data dependence results which we obtained in § 3 of this chapter for the initial-value problem (IV.122), can be used to obtain a local existence and uniqueness result for smooth solutions of the associated system

$$\text{(IV.221)}\qquad \begin{aligned}&y_{tt}+\alpha y_t-\sigma(y_x)_x = G(y), \qquad -\infty<x<\infty, \quad 0\leqq t<T,\\ &y(x,0)=f(x), \qquad y_t(x,0)=g(x).\end{aligned}$$

In fact, it can be shown that if $\mathscr{S}(G)$ denotes the solution of (IV.112), on $\mathscr{R}^1\times[0, T]$, then a solution of (IV.221) is a fixed point of the composition map $\mathscr{K}$ which is defined by $\mathscr{K}(y)=\mathscr{S}(G(y))$; such a fixed point can be found by using the continuous data dependence result of § 3 for smooth solutions of (IV.112) with sufficiently small data. This basic idea is, of course, already present in the work of Dafermos and Nohel [43] that we have discussed in Chapter II (where such a fixed point argument has been used to obtain a unique local solution of the nonlinear viscoelastic initial-value problem (II.311)) and it forms the basis of the argument used by Crandall and Nohel in [36] to treat the abstract functional initial-value problem (IV.173). Furthermore, within the context of the problem considered by MacCamy in [105], i.e., the initial-boundary value problem (IV.188), (IV.216), (IV.187$_2$), the global existence and continuous data dependence results of Nohel [122] could also be used to establish the existence of a unique local solution to the extended initial-value problem associated with (IV.188); indeed, (IV.188) is precisely of the form (IV.221$_1$), with $G(u)$ given by

$$\text{(IV.222)}\quad \begin{aligned} G(u)=&\frac{1}{a(0)}\mathscr{F}_t(x,t)+k(0)\mathscr{F}(x,t)+\int_0^t \dot{k}(t-\tau)f(x,\tau)\,d\tau\\ &-\dot{k}(0)u(x,t)+\dot{k}(t)u(x,0)+\int_0^t k(t-\tau)u(x,\tau)\,d\tau,\end{aligned}$$

and thus the local existence problem for the extended initial-value problem associated with (IV.188) can be handled by the fixed point technique discussed by Nohel [122] and employed by Crandall and Nohel [36].

Concluding Remarks. The work of MacCamy [105], discussed above, deals with the partially nonlinear one-dimensional set of constitutive assumptions (IV.180$_1$), (IV.181). To the best of our knowledge, evolution equations associated with either the fully nonlinear set of constitutive assumptions presented by Gurtin and Pipkin [64], or some partially nonlinear specialization, have not been studied in a three-dimensional setting. The problem of heat flow in an unbounded two-dimensional body of material with memory, which exhibits nonlinear response, has been treated by Dafermos and Nohel, in § 7 of their paper [43], by means of an energy estimate argument of the type employed in Chapter II to discuss the initial-value problem (II.311). More specifically, Dafermos and Nohel establish global existence, uniqueness, and asymptotic stability for solutions of the initial-value problem

$$\text{(IV.223)}\quad \begin{aligned} &u_t(x_1,x_2,t)+\int_0^t a(t-\tau)Au(x_1,x_2,\tau)\,d\tau=\mathscr{F}(x_1,x_2,t),\\ &\qquad 0\leqq t<\infty,\qquad \mathbf{x}=(x_1,x_2)\in R^2,\\ &u(x_1,x_2,0)=f(x_1,x_2),\end{aligned}$$

where

$$\text{(IV.224)}\quad Au=\left[\frac{\partial}{\partial x_1}(\sigma(\sqrt{u_{x_1}^2+u_{x_2}^2}))u_{x_1}+\frac{\partial}{\partial x_2}(\sigma(\sqrt{u_{x_1}^2+u_{x_2}^2}))u_{x_2}\right]$$

with $\sigma(0)>0$, $\sigma \in C^4(\mathscr{R}^1)$; the initial-value problem (IV.223) represents a natural two-dimensional generalization of the pure-initial value problem associated with the nonlinear, one-dimensional, integrodifferential evolution equation (IV.185) which results from the partially nonlinear set of constitutive assumptions (IV.180$_1$), (IV.181). It would appear that the energy method devised by Dafermos and Nohel in [43] could also be extended to handle three-dimensional problems of heat flow in nonlinear conductors with memory.

The one-dimensional linearized version of the constitutive theory presented in [105] is of the form (IV.180); as we have already indicated, Nachlinger and Nunziato [119] have studied the asymptotic stability of temperature fields in heat conductors with memory which are governed by the three-dimensional version of (IV.180), namely, by constitutive assumptions of the form

$$\text{(IV.225)}\qquad \begin{aligned} \varepsilon(\mathbf{x}, t) &= c\theta(\mathbf{x}, t) + \int_{-\infty}^{t} \alpha(t-s)\theta(\mathbf{x}, s)\, ds, \\ q(\mathbf{x}, t) &= -\int_{-\infty}^{t} \beta(t-s)\nabla_{\mathbf{x}}\theta(\mathbf{x}, s)\, ds. \end{aligned}$$

In (VI.225), c is the specific heat, $\alpha(t)$ is the energy relaxation function, $\beta(t)$ is the heat flux relaxation function, and $\theta(\mathbf{x}, t)$ denotes the deviation at time t from the equilibrium temperature field $\theta_0(\mathbf{x})$; while Nachlinger and Nunziato [119] establish certain stability results, for solutions of initial-history boundary value problems associated with the linear integrodifferential evolution equation that corresponds to (IV.225), and other authors (Hermann and Nachlinger [67]) have studied the related problem of uniqueness and wave propagation in such materials, there seems to have been very little work done until quite recently with regard to establishing existence theorems for the temperature field $\theta(\mathbf{x}, t)$; in [120], however, J. Naumann has considered the initial-history boundary value problem which is generated by the constitutive relations (IV.225), and the three-dimensional version of the balance equation (IV.177), namely, the problem

$$\text{(IV.226)}\qquad \begin{aligned} &\theta_t(\mathbf{x}, t) + a_0\theta(\mathbf{x}, t) + \int_{-\infty}^{t} a(t-s)\theta(\mathbf{x}, s)\, ds \\ &\qquad -\int_{-\infty}^{t} \beta(t-s)\nabla_{\mathbf{x}}^2\theta(\mathbf{x}, s)\, d\mathbf{x} = 0 \end{aligned}$$

in $\Omega\times[0, T)$, $\Omega\subseteq R^3$ with smooth $\partial\Omega$, $a_0=\alpha(0)$, $a(t)=\alpha'(t)$,

$$\text{(IV.227)}\qquad \begin{aligned} \theta(\mathbf{x}, t) &= 0 \quad \text{on } \Gamma_1\times[0, T), \\ q_i(\mathbf{x}, t)n_i(\mathbf{x}) &= \chi(\mathbf{x}, t) \quad \text{on } \Gamma_2\times[0, T) \end{aligned}$$

(where $\mathbf{n}(\mathbf{x})$ is the outward unit normal to $\partial\Omega$ at $\mathbf{x}\in\partial\Omega$, $\partial\Omega=\Gamma_1\cup\Gamma_2$ with

$\Gamma_1 \cap \Gamma_2 = \phi$, and χ is prescribed on $\Gamma_2 \times [0, T)$); also

$$\theta(\mathbf{x}, \tau) = \zeta(\mathbf{x}, \tau), \quad \text{on } \bar{\Omega} \times (-\infty, 0], \tag{IV.228}$$

where the initial-history ζ is prescribed on $\bar{\Omega} \times (-\infty, 0]$. Naumann [120] shows that the initial-history boundary value problem (IV.226)–(IV.228) can be modeled by an abstract first-order integrodifferential initial-value problem of the form

$$\begin{aligned} &\mathbf{u}_t + a_0 \mathbf{u} + \int_0^t a(t-s)\mathbf{u}(s)\,ds \\ &\quad + \int_0^t \beta(t-s)\mathbf{B}\mathbf{u}(s)\,ds + \partial\phi(\mathbf{u}(t)) \ni \mathbf{f}(t), \\ &\mathbf{u}(0) = \mathbf{u}_0, \end{aligned} \tag{IV.229}$$

where $a_0 \in \mathcal{R}^1$, $a(t)$, $\beta(t)$ are given real-valued functions satisfying $a(\cdot) \in C^1([0, T])$, $\beta(0) > 0$, $\beta(\cdot) \in C^2([0, T])$, $\mathbf{B}$ is a bounded linear map, from a Hilbert space $H_+ (\subset H$ a second Hilbert space) into its dual space H_-, which satisfies

$$\langle \mathbf{B}\mathbf{u}, \mathbf{v} \rangle + \lambda \|\mathbf{v}\|^2 \geqq b_0 \|\mathbf{v}\|_+^2 \quad \forall \mathbf{v} \in H_+, \quad \lambda \geqq 0, \quad b_0 > 0,$$

$$\langle \mathbf{B}\mathbf{v}, \mathbf{w} \rangle = \langle \mathbf{B}\mathbf{w}, \mathbf{v} \rangle \quad \forall \mathbf{v}, \mathbf{w} \in H_+$$

(it is assumed, in the usual way, that H_+ is densely and compactly embedded in H and $\partial\phi$ denotes the subdifferential of a proper, convex, lower semicontinuous functional $\phi : H_+ \to (-\infty, \infty]$). It is also assumed that

$$\int_0^t \left\langle \int_0^s \beta(s-\tau)\mathbf{B}\mathbf{u}(\tau)\,d\tau, \mathbf{u}(s) \right\rangle ds \geqq 0$$

for all $t \in [0, T]$ and any $\mathbf{u} \in L^2([0, T]; H_+)$; this latter condition is implied by the hypotheses

$$\langle \mathbf{B}\mathbf{v}, \mathbf{v} \rangle \geqq 0 \quad \forall \mathbf{v} \in H_+,$$

$$(-1)^k \frac{d^k \beta}{dt^k}(t) \geqq 0 \quad \forall t \in [0, T], \qquad k = 0, 1, 2.$$

The existence and uniqueness proof for the initial-value problem (IV.229) in [120] follows, essentially, the pattern of the proofs used in London [103] and Barbu [7], [8] for initial-value problems associated with the abstract integral equation (IV.163) where $\mathbf{w}(t) \in \mathbf{g}(\mathbf{u}(t))$ with $\mathbf{g} = \partial\phi$ (a subdifferential); we simply indicate here that the basic idea in all of these closely related papers is that of replacing the equation under consideration by an equation that involves the so-called Yosida regularization of the multivalued map $\mathbf{g} = \partial\phi$, i.e., $\mathbf{g}_\lambda = \lambda^{-1}(\mathbf{I} - (\mathbf{I} + \lambda\mathbf{g})^{-1})$, $\lambda > 0$, and then letting the regularization parameter $\lambda \to 0^+$. Once the existence and uniqueness theorem for (IV.229) has been established it is not too difficult to actually make the identification between (IV.229) and the initial-history boundary value problem (IV.226)–(IV.228); we will simply note here that (IV.229) is equivalent to

(IV.229′)
$$\begin{aligned}&\mathbf{u}_t + a_0\mathbf{u} + \int_0^t a(t-s)\mathbf{u}(s)\,ds\\&\quad + \int_0^t \beta(t-s)\mathbf{B}\mathbf{u}(s)\,ds = \bar{\mathbf{f}}(t), \qquad 0 \leqq t \leqq T,\\&\mathbf{u}(0) = u_0,\end{aligned}$$

where $\bar{\mathbf{f}}(t) \in \partial\phi(\mathbf{u}(t))$, $0 \leqq t \leqq T$. If we use the definition of subdifferential, extended to the current abstract setting, we have, as a consequence of $\bar{\mathbf{f}}(t) \in \partial\phi(\mathbf{u}(t))$, that $\bar{\mathbf{f}}(t) \in H_-$ with

(IV.230)
$$\phi(\mathbf{v}) - \phi(\mathbf{u}) \geqq (\bar{\mathbf{f}}(t), \mathbf{v} - \mathbf{u}) \quad \forall \mathbf{v} \in \mathcal{D}(\phi)$$

where $(\cdot, \cdot)$ denotes the duality pairing on $H_- \times H_+$ and $\mathcal{D}(\phi) \subset H_+$ (so that $\mathbf{v} - \mathbf{u} \in H_+$). The appropriate proper, convex, lower semicontinuous ϕ, in terms of the application at hand, is given by

(IV.231)
$$\phi(\mathbf{v}) = \begin{cases} 0, & \mathbf{v} \in K,\\ +\infty, & \mathbf{v} \in H_+/K,\end{cases}$$

where $K \subset H_+$ is a closed convex subset; thus if $\mathbf{u}, \mathbf{v} \in K$, (IV.230) reduces to $(\bar{\mathbf{f}}(t), \mathbf{v} - \mathbf{u}) \leqq 0$, and (IV.229′) implies that solutions $\mathbf{u} \in K$ satisfy an integrodifferential inequality. In terms of the specific application to (IV.226)–(IV.228) one takes $H = (L^2(\Omega))$ and $H_+ = \{u \in H^1 | u = 0$, a.e. on $\Gamma_1\}$; for K we take

(IV.232)
$$K = \{u \in H_+ \,|\, |u(\mathbf{x})| + |\nabla_{\mathbf{x}} u(\mathbf{x})| \leqq \delta \text{ a.e. on } \Omega\},$$

where $\delta > 0$ is sufficiently small so that temperature fields $\theta \in K$ will satisfy the constitutive relations (IV.225) which result from linearization of the nonlinear constitutive relations in [64]. We now take $\mathbf{B}$ to be the bounded linear map which is defined by $\langle Bu, v\rangle_{L_2(\Omega)} = \int_\Omega (\partial u/\partial x_j)(\partial v/\partial x_j)\,d\mathbf{x}$, $\forall u, v \in H^1(\Omega)$, we then multiply (IV.226) by $v(\mathbf{x}, t) - \theta(\mathbf{x}, t)$, where $v \in L^2([0, T]; K)$, integrate over Ω using the boundary conditions (IV.227) and the initial condition (IV.228) and, finally, integrate over $[0, T]$ with respect to t. Via this process, we obtain from (IV.226)–(IV.228) an integrodifferential inequality of the same form as that which results by choosing ϕ according to (IV.231), letting both sides of (IV.229′), considered as elements of H_-, act on $v - u$, ($u, v \in K$), and then integrating over $[0, T]$ with respect to t; for complete details we refer the reader to the original paper of Naumann and, in particular to § 2.2 of [120].

Bibliography

[1] R. A. Adams, *Sobolev Spaces*, Academic Press, New York, 1975.

[2] S. Agmon, *Unicité et Convexité dans les Problémes Differentiels*, University of Montreal Press, Montreal, 1966.

[3] S. Agmon and L. Nirenberg, *Lower bounds and uniqueness theorems for solutions of differential equations in Hilbert space*, Comm. Pure Appl. Math., Vol. 20 (1967), pp. 207–229.

[4] J. M. Ball, *Finite time blow-up in nonlinear problems*, in Nonlinear Evolution Equations, M. G. Crandall, ed., Academic Press, New York, (1978), pp. 189–205.

[5] ———, *Remarks on blow-up and nonexistence theorems for nonlinear evolution equations*, Quart. J. Math., Oxford, Vol. 28 (1977), pp. 473–486.

[6] V. Barbu, *Nonlinear Semigroups and Differential Equations in Banach Spaces*, Noordhoff International Publishing, Groningen, 1976.

[7] V. Barbu, *Nonlinear Volterra equations in a Hilbert space*, SIAM J. Math. Anal., Vol. 6 (1975), pp. 728–741.

[8] ———, *On a nonlinear Volterra integral equation on a Hilbert space*, SIAM J. Math. Anal., Vol. 8 (1977), pp. 346–355.

[9] C. E. Beevers, *Hölder stability in linear anisotropic viscoelasticity*, in I.U.T.A.M. Symposium on the Mechanics of Viscoelastic Media and Bodies, Gothenberg, 1974.

[10] ———, *Uniqueness and stability in linear viscoelasticity*, ZAMP, Vol. 26 (1975), pp. 177–186.

[11] F. Bloom, *Exponential growth estimates for solutions of ill-posed integrodifferential initial-value problems in Hilbert space*, J. Math. Anal. Appl., in press.

[12] ———, *A lower bound for the norm of the solution of a nonlinear Volterra equation in one-dimensional viscoelasticity*, submitted for publication.

[13] ———, *Asymptotic bounds for solutions to a system of damped integrodifferential equations of electromagnetic theory*, J. Math. Anal. Appl., Vol. 73 (1980), pp. 524–542.

[14] ———, *Bounds for solutions to a class of damped integrodifferential equations associated with a theory of nonconducting material dielectrics*, SIAM J. Math. Anal., Vol. 11 (1980), pp. 265–291.

[15] ———, *Remarks on the asymptotic behavior of solutions to damped evolution equations in Hilbert space*, Proc. AMS, Vol. 75 (1979), pp. 25–31.

[16] ———, *Lower bounds for solutions to a class of nonlinear integrodifferential equations in Hilbert space*, Applicable Analysis, Vol. 10 (1980), pp. 295–307.

[17] ———, *Growth estimates for electric displacement fields and bounds for constitutive constants in the Maxwell-Hopkinson theory of dielectrics*, Int. J. Eng. Sci., Vol. 17 (1979), pp. 1–15.

[18] ———, *A note on growth estimates for the displacement vector in viscoelastic materials*, Quart. J. Mech. and Appl. Math., Vol. XXXI (1978), pp. 323–333.

[19] ———, *Concavity arguments and growth estimates for damped linear integrodifferential equations with applications to a class of holohedral isotropic dielectrics*, ZAMP, Vol. 29 (1978), pp. 644–663.

[20] ———, *Stability and growth estimates for electric fields in nonconducting material dielectrics*, J. Math. Anal. Appl., Vol. 67 (1979), pp. 296–322.

[21] ———, *Continuous data dependence for an abstract Volterra integrodifferential equation in Hilbert space with applications to viscoelasticity*, Annali della Scuola Normale (PISA), Vol. IV (1971), pp. 179–207.

[22] ———, *Growth estimates for solutions to initial-boundary value problems in viscoelasticity*, J. Math. Anal. Appl., Vol. 59 (1977), pp. 469–488.

[23] ———, *Continuous dependence on initial geometry for a class of abstract equations in Hilbert space*, J. Math. Anal. Appl., Vol. 58 (1977), pp. 293–297.

[24] ———, *Stability and growth estimates for Volterra integrodifferential equations in Hilbert space*, Bull. A.M.S., Vol. 82 (1976), pp. 603–606.

[25] ———, *Some stability theorems for an abstract equation in Hilbert space with applications to linear elastodynamics*, J. Math. Anal. Appl., Vol. 61 (1977), pp. 521–536.

[26] H. BREZIS, *Monotonicity methods in Hilbert spaces and some applications to nonlinear partial differential equations*, Contributions to Nonlinear Functional Analysis, E. H. Zarantonello, ed., Academic Press, New York (1971), pp. 101–156.

[27] ———, *Opérateurs Maximaux Monotones et Semigroupes de Contractions dans les Espaces de Hilbert*, North-Holland Publishing, Amsterdam, 1973.

[28] R. C. BROWNE, *Dynamic stability of one-dimensional nonlinearly viscoelastic bodies*, Arch. Rat. Mech. Anal., Vol. 68 (1978), pp. 257–282.

[29] L. BRUN, *Sur l'unicité en thermoélasticité dynamique et diverses expressions analogues á la formule de Clapeyron*, C. R. Acad. Sci. Paris, Vol. 261 (1965), pp. 2584–2587.

[30] ——— *Méthodes energétiques dans les systémes évolutifs linéaires, Premier partie: Séparation des énergies, Deuxiéme partie: Theorémes d'unicité*, J. Mechanique Vol. 8 (1969), pp. 125–166, 167–192.

[31] P. J. CHEN AND M. F. MCCARTHY, *The electrical responses of a dynamicalty loaded deformable dielectric with memory*, Arch. Rat. Mech. Anal., Vol. 62 (1976), pp. 353–366.

[32] P. J. CHEN, M. F. MCCARTHY AND T. R. O'LEARY, *One-dimensional shock and acceleration waves in deformable dielectric materials with memory*, Arch. Rat. Mech Anal., Vol. 62 (1976), pp. 189–207.

[33] B. D. COLEMAN, *On thermodynamics, strain impulses, and viscoelasticity*, Arch. Rat. Mech. Anal., Vol. 17 (1964), pp. 230–254.

[34] B. D. COLEMAN AND W. NOLL, *Foundations of linear viscoelasticity*, Rev. Mod. Phys., Vol. 33 (1961), pp. 239–249.

[35] M. G. CRANDALL, *An introduction to evolution governed by accretive operators*, in *Dynamical Systems*, Vol. I, Academic Press, 1976, pp. 131–165.

[36] M. G. CRANDALL AND J. A. NOHEL, *An abstract functional differential equation and a related nonlinear Volterra equation*, Israel Journal of Mathematics, Vol. 29 (1978), pp. 313–328.

[37] C. M. DAFERMOS, *Asymptotic stability in viscoelasticity*, Arch. Rat. Mech. Anal., Vol. 37 (1970), pp. 297–308.

[38] ———, *An abstract Volterra equation with applications to linear viscoelasticity*, J. Diff. Eqs., Vol. 7 (1970), pp. 554–569.

[39] ———, *Asymptotic behavior of solutions of evolution equations*, in *Nonlinear Evolution Equations*, M. G. Crandall, ed., Academic Press, New York, 1978, pp. 103–123.

[40] ———, *Contraction semigroups and the trend to equilibrium in continuum mechanics*, Proc. I.U.T.A.M./I.M.U. Conference on Applications of Functional Analysis to Mechanics, 1975.

[41] ———, *Wave equations with weak damping*, SIAM J. Appl. Math., Vol. 18 (1970), pp. 759–767.

[42] ———, *The mixed initial-boundary value problem for the equations of nonlinear one-dimensional viscoelasticity*, J. Diff. Eqs., Vol. 6 (1969), pp. 71–86.

[43] C. M. DAFERMOS AND J. NOHEL, *Energy methods for nonlinear hyperbolic volterra integrodifferential equations*, Comm. in PDE, Vol. 4 (1979), pp. 219–278.

[44] C. M. DAFERMOS AND M. SLEMROD, *Asymptotic behavior of nonlinear contraction semigroups*, J. Functional Anal., Vol. 13 (1973), pp. 97–106.

[45] P. DAVIS, *Hyperbolic integrodifferential equations arising in the electromagnetic theory of dielectrics*, J. Diff. Eqs., Vol. 18 (1975), pp. 170–178.

[46] W. A. DAY, *On the monotonicity of the relaxation functions of viscoelastic materials*, Proc. Camb. Phil. Soc., Vol. 67 (1970), pp. 503–508.

[47] ———, *Time-reversal and the symmetry of the relaxation function of a linear viscoelastic material*, Arch. Rat. Mech. Anal., Vol. 40 (1971), pp. 155–159.

[48] ———, *Restrictions on relaxation functions in linear viscoelasticity*, Quart. J. Mech. and Appl. Math., Vol. XXIV (1971), pp. 487–497.

[49] G. DUVAUT AND J. L. LIONS, *Inequalities in Mechanics and Physics*, Springer-Verlag, 1974.

[50] L. C. EVANS, *Application of nonlinear semigroup theory to certain partial differential equations*, in Nonlinear Evolutions Equations, M. G. Crandall, ed., Academic Press, 1978, pp. 163–188.

[51] R. E. EWING, *The approximation of certain parabolic equations backward in time by Sobolev equations*, SIAM J. Math. Anal., Vol. 6 (1975), pp. 283–294.

[52] A. FRIEDMAN, *Partial Differential Equations*, Holt, Rinehart, and Winston, New York, 1969.

[53] ———, *Remarks on nonlinear parabolic equations*, Proc. Symp. Appl. Math., Vol. 13 (1965), pp. 3–23.

[54] H. FUJITA, *On the blowing up of solutions to the Cauchy problem for* $u_t = \Delta u + u^{1+\alpha}$, J. Fac. Sci. Univ. Tokyo, Vol. 13 (1966), pp. 109–124.

[55] R. T. GLASSEY, *Blow up theorems for nonlinear wave equations*, Math. Z., Vol. 132 (1973), pp. 183–203.

[56] J. GOLDSTEIN, *Nonlinear semigroups and nonlinear partial differential equations*, Proc. Colóquio Brasileiro de Matemática Pocos de Caldas, 1975.

[57] ———, *Uniqueness for nonlinear Cauchy problems in Banach spaces*, Proc. AMS, Vol. 53 (1975), pp. 91–95.

[58] J. M. GREENBERG, R. C. MACCAMY, AND V. J. MIZEL, *On the existence, uniqueness, and stability of solutions of the equation* $\rho\chi_{tt} = E(\chi_x)\chi_{xx} + \lambda\chi_{xxt}$, J. Math. Mech., Vol. 17 (1968), pp. 707–728.

[59] M. E. GURTIN, *The linear theory of elasticity*, in Handbuch der Physik, Vol. VIa/2, pp. 1–295.

[60] M. E. GURTIN AND I. HERRERA, *On dissipation inequalities and linear viscoelasticity*, Quart. Appl. Math., Vol. 23 (1965), pp. 235–245.

[61] M. E. GURTIN AND R. C. MACCAMY, *Population dynamics with age dependence*, in Nonlinear Analysis and Mechanics III, R. J. Knops, ed., 1979, pp. 1–35.

[62] ———, *Some simple models for nonlinear age-dependent population dynamics*, Math. Biosciences (to appear).

[63] ———, *Nonlinear age-dependent population dynamics*, Arch. Rat. Mech. Anal., Vol. 3 (1974), pp. 281–300.

[64] M. E. GURTIN AND A. C. PIPKIN, *A general theory of heat conduction with finite wave speed*, Arch. Rat. Mech. Anal., Vol. 31 (1968), pp. 113–126.

[65] J. K. HALE, *Dynamical systems and stability*, J. Math. Anal. Appl., Vol. 26 (1969), pp. 39–59.

[66] K. B. HANNSGEN, *On a nonlinear volterra equation*, Michigan Math. J., Vol. 16 (1969), pp. 365–376.

[67] R. P. HERMANN AND R. R. NACHLINGER, *On uniqueness and wave propagation in nonlinear heat conductors with memory*, J. Math. Anal. Appl., Vol. 50 (1975), pp. 530–547.

[68] J. HOPKINSON, *The residual charge of the Leyden jar*, Phil. Trans. Roy. Soc. London, Vol. 167 (1877), pp. 599–626.

[69] F. JOHN, *Continuous dependence on data for solutions of partial differential equations with a prescribed bound*, Comm. Pure Appl. Math., Vol. 13 (1960), pp. 551–585.

[70] ———, *Formation of singularities in one-dimensional nonlinear wave propagation*, Comm. Pure Appl. Math., Vol. 27 (1974), pp. 377–405.

[71] S. KAPLAN, *On the growth of solutions of quasilinear parabolic equations*, Comm. Pure Appl. Math., Vol. 16 (1963), pp. 305–330.

[72] R. J. KNOPS, *Logarithmic convexity and other techniques applied to problems in continuum mechanics*, Symp. on Non-Well-Posed Problems and Logarithmic Convexity, Springer Lecture Notes, 316, Springer, New York, 1973, pp. 31–54.

[73] R. J. KNOPS, H. A. LEVINE, AND L. E. PAYNE, *Nonexistence, instability, and growth theorems for solutions to an abstract nonlinear equation with applications to elastodynamics*, Arch. Rat. Mech. Anal., Vol. 55 (1974), pp. 52–72.

[74] R. J. KNOPS AND L. E. PAYNE, *Uniqueness Theorems in Linear Elasticity*, Springer-Verlag, New York, pp. 1971.

[75] ———, *Uniqueness in classical elastodynamics*, Arch. Rat. Mech. Anal., Vol. 27 (1968), pp. 349–355.

[76] ———, *Stability in linear elasticity*, Int. J. Solids Struct., Vol. 4 (1968), pp. 1233–1242.

[77] ———, *Continuous data dependence for the equations of classical elastodynamics*, Proc. Camb. Phil. Soc., Vol. 66 (1969), pp. 481–491.

[78] ———, *On the uniqueness and continuous data dependence in dynamical problems of linear thermoelasticity*, Int. J. Solids Struct., Vol. 6 (1970), pp. 1173–1184.

[79] ———, *Growth estimates for solutions of evolutionary equations in Hilbert space with applications in elastodynamics*, Arch. Rat. Mech. Anal., Vol. 41 (1971), pp. 363–398.

[80] ———, *Hölder stability and logarithmic convexity*, in Proceedings of the I.U.T.A.M. Conference on Instability of Continuous Systems, Springer-Verlag, 1971, pp. 248–255.

[81] H. KÖNIG AND J. MEIXNER, *Lineare systeme und lineare transformationen*, Mat. Nachr., Vol. 19 (1958), pp. 256–323.

[82] V. LAKSHMIKANTHAM AND S. LEELA, *Differential and Integral Inequalities*, Vol. I, Academic Press, 1969.

[83] R. LATTES AND J. L. LIONS, *The Method of Quasireversibility, Applications to Partial Differential Equations*, American Elsevier, New York, 1969.

[84] M. M. LAVRENTIEV, *On the Cauchy problem for the Laplace equation*, Izvest. Akad., Nauk. SSSR, Ser. Math., Vol. 120 (1956), pp. 819–842.

[85] ———, *Some Improperly Posed Problems in Mathematical Physics*, Springer-Verlag, New York, 1967.

[86] P. D. LAX, *Development of singularities of solutions of nonlinear hyperbolic partial differential equations*, J. Math. Physics, Vol. 5 (1964), pp. 611–613.

[87] J. J. LEVIN, *The asymptotic behavior of the solution of a Volterra equation*, Proc. AMS, Vol. 14 (1963), pp. 534–541.

[88] J. J. LEVIN AND J. A. NOHEL, *Note on a nonlinear Volterra equation*, Proc. AMS, Vol. 14 (1963), pp. 924–929.

[89] H. A. LEVINE, *On a theorem of Knops and Payne in dynamical linear thermoelasticity*, Arch. Rat. Mech. Anal., Vol. 38 (1970), pp. 290–307.

[90] ———, *Uniqueness and growth of weak solutions to certain linear differential equations in Hilbert space*, J. Diff. Eqs., Vol. 17 (1975), pp. 73–81.

[91] ———, *Logarithmic convexity and the Cauchy problem for some abstract second order differential inequalities*, J. Diff. Eqs., Vol. 8 (1970), pp. 34–55.

[92] ———, *Logarithmic convexity, first order differential inequalities, and some applications*, Trans. Amer. Math. Soc., Vol. 152 (1970), pp. 299–319.

[93] ———, *Some uniqueness and growth theorems in the Cauchy problem for* $Pu_{tt}+Mu_t+Nu=0$, Math. Z., Vol. 126 (1972), pp. 345–360.

[94] ———, *Instability and nonexistence of global solutions to nonlinear wave equations of the form* $Pu_{tt}=-Au+\mathcal{F}(u)$, Trans. Amer. Math. Soc., Vol. 1950 (1974), pp. 1–21.

[95] ———, *Some nonexistence and instability theorems for solutions of formally parabolic equations of the form* $Pu_t=-Au+\mathcal{F}(u)$, Arch. Rat. Mech. Anal., Vol. 5 (1973), pp. 371–38.

[96] ———, *Some additional remarks on the nonexistence of global solutions to nonlinear wave equations*, SIAM J. Math. Anal., Vol. 5 (1974), pp. 138–146.

[97] ———, *On the nonexistence of global solutions to a nonlinear Euler-Poisson-Darboux equation*, J. Math. Anal. Appl., Vol. 48 (1970), pp. 646–651.

[98] ———, *Growth of solutions of generalized nonlinear Euler-Poisson-Darboux equations*, Arch. Rat. Mech. Anal., Vol. 61 (1976), pp. 77–89.

[99] H. A. LEVINE AND L. E. PAYNE, *Nonexistence of global weak solutions for classes of nonlinear wave and parabolic equations*, J. Math. Anal. and Appl., Vol. 55 (1976), pp. 329–334.

[100] ———, *On the nonexistence of entire solutions to nonlinear second order elliptic equations*, SIAM J. Math. Anal., Vol. 7 (1976), pp. 337–343.

[101] ———, *Nonexistence theorems for the heat equation with nonlinear boundary conditions and for the porous medium equation backward in time*, J. Diff. Eqs., Vol. 16 (1974), pp. 319–334.

[102] ———, *Some nonexistence theorems for initial boundary value problems with nonlinear boundary constraints*, Proc. AMS, Vol. 46 (1974), pp. 277–284.

[103] S. O. LONDON, *On an integral equation in a Hilbert space*, SIAM J. Math. Anal., Vol. 8 (1977), pp. 950–970.

[104] R. C. MACCAMY, *A model for one-dimensional, nonlinear viscoelasticity*, Quart. Appl. Math., Vol. 35 (1977), pp. 21–33.

[105] ———, *An integrodifferential equation with applications in heat flow*, Quart. Appl. Math., Vol. 35 (1977), pp. 1–19.

[106] ———, *Existence, uniqueness, and stability of* $u_{tt}=\partial/\partial x\ (\sigma(u_x)+\lambda(u_x)u_{xt})$, Ind. Univ. Math. J., Vol. 20 (1970), pp. 231–238.

[107] ———, *Exponential stability for a class of functional differential equations*, Arch. Rat. Mech. Anal., Vol. 40 (1971), pp. 120–138.

[108] ———, *Nonlinear Volterra equations on a Hilbert space*, J. Diff. Eqs., Vol 16 (1974), pp. 373–383.

[109] ———, *Stability theorems for a class of functional differential equations*, SIAM J. Appl. Math., Vol. 30 (1976), pp. 557–576.

[110] R. C. MACCAMY AND J. S. W. WONG, *Stability theorems for some functional equations*, Trans. Amer. Math. Soc., Vol. 164 (1972), pp. 1–37.

[111] A. MATSUMARA, *Global existence and asymptotics of the solutions of the second order quasilinear hyperbolic equations with the first order dissipation* (to appear).

[112] J. C. MAXWELL, *A Treatise on Electricity and Magnetism* (1873) Reprinted by Dover Press, New York.

[113] R. K. MILLER, *Volterra integral equations in a Banach space*, Funkcialaj Ekvaciaj, Vol. 18 (1975), pp. 163–193.

[114] ———, *Nonlinear Volterra Integral Equations*, W. A. Benjamin, New York, 1971.

[115] R. K. MILLER AND R. L. WHEELER, *Asymptotic behavior for a linear Volterra equation in Hilbert space*, J. Diff. Eqs., Vol. 23 (1977), pp. 270–284.

[116] K. MILLER, *Stabilized quasireversibility and other nearly-best-possible methods for non-well-posed problems*, Symp. on Non-well-posed Problems and Logarithmic Convexity, Springer Lecture Notes 316, Springer, New York, 1973, pp. 161–176.

[117] A. MURRAY, *Uniqueness and continuous dependence for the equations of elastodynamics with a strain energy function*, Arch. Rat. Mech. Anal., Vol. 47 (1972), pp. 195–204.

[118] A. MURRAY AND M. H. PROTTER, *The asymptotic behavior of solutions of second order systems of partial differential equations*, J. Diff. Eqs., Vol. 13 (1973), pp. 57–80.

[119] R. R. NACHLINGER AND J. W. NUNZIATO, *Stability of uniform temperature fields in linear heat conductors with memory*, Int. J. Eng. Sci., Vol. 14 (1976), pp. 693–701.

[120] J. NAUMANN, *On a class of first-order evolution inequalities arising in heat conduction with memory*, SIAM J. Math. Anal., Vol. 10 (1979), pp. 1144–1160.

[121] T. NISHIDA, *Quasilinear wave equations with the dissipation*, in Nonlinear Hyperbolic Equations and Related Topics, Chapter II, Publications de la Faculté d'Orsay, France, 1978.

[122] J. A. NOHEL, *A forced quasilinear wave equation with dissipation*, MRC Technical Summary Report No. 1799, also, Proc. of EQUADIFF. IV, Lecture Notes in Mathematics, 703, Springer-Verlag, 1977, pp. 318–327.

[123] J. A. NOHEL AND D. F. SHEA, *Frequency domain methods for Volterra equations*, Adv. in Math., Vol. 22 (1976), pp. 278–304.

[124] L. E. PAYNE, *On some nonwell-posed problems for partial differential equations*, Numerical Solution of Nonlinear Differential Equations, M.R.C. Conference, University of Wisconsin, John Wiley Pub., New York (1966), pp. 239–263.

[125] ———, *Some general remarks on improperly posed problems for partial differential equations*, Symp. on Non-well-posed Problems and Logarithmic Convexity, Springer Lecture Notes 316, Springer, New York, 1973, pp. 1–30.

[126] ———, *Improperly Posed Problems in Partial Differential Equations*, CBMS Regional Conference Series in Applied Mathematics 22, Society for Industrial and Applied Mathematics, Philadelphia, 1975.

[127] A. PAZY, *Semigroups of Linear Operators and Applications to Partial Differential Equations*, Lecture Notes 10, University of Maryland, 1974.

[128] ———, *Semigroups of nonlinear contractions in Hilbert space*, in Nonlinear Analysis and Mechanics, Vol. III, Pitman Pub., 1979, pp. 36–134.

[129] V. M. POPOV, *Hyperstability of Control Systems*, Springer, Berlin, 1973.

[130] ———, *Dichotomy and stability by frequency-domain methods*, Proc. IEEE, Vol. 62 (1974), pp. 548–562.

[131] C. PUCCI, *Sui problemi di Cauchy non ben posti*, Rend. Accad. Naz. Lincei, Vol. 18 (1955), pp. 473–477.

[132] M. H. PROTTER, *Properties of solutions of parabolic equations and inequalities*, Canad. J. Math., Vol. 13 (1961), pp. 331–345.

[133] D. L. RUSSELL, *Decay rates for weakly damped systems in Hilbert space obtained with control-theoretic methods*, J. Diff. Eqs., Vol. 19 (1975), pp. 344–370.

[134] R. E. SHOWALTER, *The final value problem for evolution equations*, J. Math. Anal. Appl., Vol. 47 (1974), pp. 563–572.

[135] ———, *Quasireversibility of first and second order parabolic equations*, in Improperly Posed Boundary Value Problems, A. Carasso and A. Stone, ed., Pitman Pub., Canton, 1975.

[136] L. S. SHU AND E. T. ONAT, *On anisotropic linear viscoelastic solids*, Proc. Fourth Symp. on Naval Structural Mechanics. Reprinted in Mechanics and Chemistry of Solid Propellants, Pergamon, Oxford and New York, 1966.

[137] M. SLEMROD, *A hereditary partial differential equation with applications in the theory of simple fluids*, Arch. Rat. Mech. Anal., Vol. 62 (1976), pp. 303–321.

[138] ———, *An energy stability method for simple fluids*, Arch. Rat. Mech. Anal., Vol. 68 (1978), pp. 1–18.
[139] ———, *Instability of steady shearing flows in a nonlinear viscoelastic fluid*, Arch. Rat. Mech. Anal., Vol. 68 (1978), pp. 211–225.
[140] ———, *Stability of steady shearing flows in a nonlinear viscoelastic fluid*, Proc. Roy. Soc. Edinburgh, to appear.
[141] ———, *Damped conservation laws in continuum mechanics*, in Nonlinear Analysis and Mechanics, Vol. III, Pitman Pub., 1978, pp. 135–173.
[142] A. N. TIKHONOV AND V. Y. ARSENIN (F. John, translation editor), *Solution of Ill-Posed Problems*, John Wiley, New York, 1977.
[143] R. A. TOUPIN AND R. S. RIVLIN, *Linear functional electromagnetic constitutive relations and plane waves in a hemihedral isotropic material*, Arch. Rat. Mech. Anal., Vol. 6 (1960), pp. 188–197.
[144] M. TSUTSUMI, *Existence and nonexistence of global solutions for nonlinear parabolic equations*, Pub. RIMS Kyoto U., Vol. 8 (1972/73), pp. 221–229.
[145] V. VOLTERRA, *Sur les equations integrodifférentielles et leurs applications*, Acta. Math., Vol. 35 (1912), pp. 295–356.
[146] ———, *Theory of Functionals*, (1928), Reprinted by Dover Press, New York.
[147] S. Zaidman, *Bounded solutions of some abstract differential equations*, proc. AMS, Vol. 23 (1969), pp. 340–342.
[148] O. STAFFENS, *On a nonlinear hyperbolic Volterra equation*, SIAM J. Math. Anal., Vol. 11 (1980), pp. 793–812.
[149] C. M. DAFERMOS AND J. A. NOHEL, *A Nonlinear Hyperbolic Volterra Equation in Viscoelasticity*, MRC Technical Summary Report 2095, University of Wisconsin, Madison, June, 1980.

Index

A